Lecture Notes in Economics and Mathematical Systems

Managing Editors: M. Beckmann and H. P. Künzi

Operations Research

132

Ursula H. Funke

Mathematical Models in Marketing

A Collection of Abstracts

With a Preface by M. Beckmann

Springer-Verlag Berlin Heidelberg GmbH 1976

Author

Prof. Dr. Ursula H. Funke
Department of Economics
Brown University
Providence R.I. 02912/USA

Library of Congress Cataloging in Publication Data

Funke, Ursula, 1939-
    Mathematical models in marketing.

    (Lecture notes in economics and mathematical systems :
Operations research ; 132)
    Bibliography: p.
    Includes indexes.
    1. Marketing--Mathematical models. I. Title.
II. Series: Lecture notes in economics and mathematical
systems ; 132.
HF5415.125.F86   658.8'001'51   76-44509

AMS Subject Classifications (1970): 90 A 10, 90 A 15, 90 B 99

ISBN 978-3-540-07869-2   ISBN 978-3-642-51565-1 (eBook)
DOI 10.1007/978-3-642-51565-1

PREFACE

Among the most active fields in Operations Research in recent years Marketing has been outstanding. One index of Operations Research activity in marketing is the number of books and articles on marketing subjects containing mathematical models.

This collection of abstracts includes from the

1940's	1	model,
1950's	12	models,
1960's	80	models,
1970's (so far)	58	models.

Here and in the following we have excluded mere statistical models of the regression or analysis of variance types. A difficulty facing the reader is not just the abundance of articles and the reappearance of basically the same models in various forms, but their dispersion over a great number of journals both in the marketing and in the Operations Research fields. (22 have been covered here).

As one remedy to this situation there have appeared several books of readings in which key articles have been reprinted, sometimes with editorial comment. Some of these are:

F.M. Bass, et. al., (eds.), Mathematical Models and Methods in Marketing, Homewood, Illinois: R.D. Irwin, Inc., 1961.

R.E. Frank, A.A. Kuehn, and W.F. Massy, Quantitative Techniques in Marketing Analysis, Text and Readings, Homewood, Illinois: R.D. Irwin, Inc., 1962.

B. Montgomery and G.L. Urban, Applications of Management Science in Marketing, Englewood-Cliffs: Prentice-Hall, Inc., 1970.

R.L. Day and T.E. Ness, Marketing Models: Behavioral Science Applications, Scranton-Toronto-London: International Textbook Company, 1971.

L. Day and J. Parsons, <u>Marketing Models</u>: <u>Quantitative</u>
<u>Applications</u>, Scranton-Toronto-London: International Textbook
Company, 1971.

The aims of the present monograph are more ambitious. It
is an attempt to collect all significant articles and books
that contain mathematical models on marketing subjects. It
seems to us that the greatest usefulness is achieved not by
reprinting but by abstracting from these articles the types of
mathematical model used. This has been done, wherever possible,
in the authors' own notation and words (but without their
assistance). The format used and the classification that has
emerged are explained in the introduction.

It is impossible to summarize the state of the art as it
emerges from these abstracts in a few paragraphs, and no such
attempt will be made here. The field is in a state of flux,
and no common framework is apparent in terms of which all
theoretical ideas can be readily organized. This would signal
the need for efforts to develop such a general theoretical
framework on the basis of a unified theory of behavior. As
far as sellers behavior goes, the well-known economic model
of profit maximizing behavior and its variants of the sales
maximization and satisficing types would seem to offer promising
starting points.

Things are more tricky at the consumer's or buyer's end.
Here it would seem that the economic theory of choice as
formulated in utility theory would have to be brought in. But
at present, utility theory is not formulated in a way that allows
for brand switching, or responses to advertising, packaging or
other forms of marketing effort.

Ursula Funke and I have made an attempt at reformulating
utility models for purposes of marketing theory in the paper,
"Product Attraction, Marketing Effort and Sales: Towards a

Utility Model of Market Behavior," [#41] to which we refer the interested reader.

It remains for me to record the pleasure I had in collaborating with Ursula Funke on this project. Both the article, which summarizes our theoretical results, and this collection of abstracts, which was preliminary to our theoretical efforts, have been sponsored by Grant No. Be 272124 from the "Deutsche Forschungsgemeinschaft." We should like to express our sincere gratitude for this support.

Providence, April 1976

Martin J. Beckmann

INTRODUCTION

Mathematical models can be classified in a number of ways, e.g., static and dynamic; deterministic and stochastic; linear and nonlinear; individual and aggregate; descriptive, predictive, and normative; according to the mathematical technique applied or according to the problem area in which they are used.

In marketing, the level of sophistication of the mathematical models varies considerably, so that a number of models will be meaningful to a marketing specialist without an extensive mathematical background. To make it easier for the nontechnical user we have chosen to classify the models included in this collection according to the major marketing problem areas in which they are applied.

Since the emphasis lies on mathematical models, we shall not as a rule present statistical models, flow chart models, computer models, or the empirical testing aspects of these theories. We have also excluded competitive bidding, inventory and transportation models since these areas do not form the core of the marketing field.

The mathematical formulation of human behavior and particularly consumer behavior is only in its infancy. In marketing the chief development of mathematical models began after World War II with a major thrust in the early 1960's. So far, it seems to have been mainly an Anglo-Saxon affair and we have, therefore, concentrated on the American and British literature. (See, however, entries #41, 88, 94, 113, 118.)

The models included in this collection have been selected on the basis of being either meaningful in terms of actual application or historically important or worthwhile for their own sake. Within the nine subject categories they are arranged chronologically except in cases where one article is an extension of another. References at the end of the abstract refer to closely related models.

Since the model categories frequently overlap and some models treat several subjects simultaneously (example: "Advertising and Promotion Effects on Consumer Response to New Products"), the classification of some models is somewhat arbitrary. The reader is referred to the subject index.

Part 1 contains models of consumer purchasing behavior, the center of marketing. The first subclass consists of brand choice models in general; new product models are listed in Part 4.1.

Brand choice models are as a rule stochastic models. The development of stochastic models of buying behavior began in the late 1950's. There have been four major model-building approaches: zero-order models, Markov models, linear learning models, and probabilistic diffusion models. Each presumes a different consumer behavior process. Whereas the zero-order model assumes that past brand choices do not affect future brand choices, the two-state Markov model postulates that only the most recent purchase affects the current brand choice decision. The learning model presumes that brand choice is dependent upon the complete history of past purchases. The probability diffusion model, finally, is a zero-order model in which the probability of choosing a particular brand may change between purchases; it allows nonstationarity in the probability of brand choice, but, in contrast to the linear learning and Markov models, assumes that this nonstationarity is not due to purchase event feedback.

In the late 1960's marketing theorists began to look to the behavioral sciences and formulated buyer attitude models. These deal with the relationship between consumers' attitude toward a product and brand choice behavior.

The category Other Consumer Behavior Models contains product attribute models, models where the demand is affected by dealer or store location, consumer utility models, a model of demand with variable consumer preferences, and a model of consumer behavior with and without knowledge of the price distribution.

Part 2 considers Advertising and Sales Promotion models where the term sales promotion covers a wide variety of sales stimulating devices, such as temporary price reductions, premiums, coupons, and sampling. The technique of programming dominates the models in Media Selection. The abundance of models in Advertising Expenditure reflects the importance of advertising in marketing research and in most firms' budgets. Some theoretical results justify the common empirical use of a percentage-of-sales-rule in advertising budget decisions (see, for instance, model #56). A large number of models in the category Sales Response to Advertising treats a central question in advertising: the measurement problem, as well as the duration and stability of advertising's effects. The category Other Models of Advertising and Sales Promotion comprises models of promotional competition and the diffusion of knowledge.

Part 3 considers the subject of pricing whose importance to the firm is not mirrored by the number of models represented here. There are two reasons for this. First, the mathematical models of general price theory are widely published and assumed known. Second, many advertising, new product, and marketing mix models contain the pricing variable.

Four out of five new products fail. With market research focussing on this problem it is not surprising that the category Product Models consists mostly of new product models. They range from the simple Fourt-Woodlock model to Massy's STEAM model.

Part 5 is directed at the problem area of sales forecasting. Sales forecasting or market share expressions are found in models of all categories. Besides the five models presented here, other forecasting models are to be found especially in the New Product category (e.g., #96, 108, and 109).

The Facility Location Models of Part 6 treat the location of a warehouse or retail outlet as a decision variable of the firm.

In most economies personal selling represents a larger marketing expenditure than advertising. Model building started early in this field, and the Sales Force Model category contains the earliest paper in this collection (1943). Most articles deal with salesmen scheduling and routing, i.e., with the problem of allocating sales effort among customers, geographical areas, and time, whereas the Montgomery-Silk-Zaragoza model investigates the problem of allocating selling effort across a firm's product line.

Part 8, Marketing Mix Models, comprises models of market response to more than one marketing variable. Models which consider several variables but with the emphasis on one variable, for example advertising, are listed under that variable.

Part 9, finally, contains two models of market simulation of which MATE represents business games, and one model of trade area boundaries.

This collection does not claim to be exhaustive, rather it is intended to be representative. Its main purpose is to provide a survey of the state of the art which may be useful to the marketing researcher, to the general Operations Research practitioner, and to the teacher of management science courses.

I am grateful to the Economics Department of Brown University for providing a stimulating environment and the facilities to undertake this research. My gratitude extends to Nancy Kimelman for helping with the indices and to the department's capable secretary, Marion Anthony.

I am deeply indebted to Martin J. Beckmann, colleague, "Meister," and friend. I thank Wolfgang for his understanding and patience and dedicate this volume to my parents, Hertha and Erwin Rehbinder.

Providence, April 1976

Ursula H. Funke

CONTENTS

List of Models

Subject: Brand Choice

Title: Brand Preferences and Simple Markov Processes

Author: Richard B. Maffei

Source: Operations Research, Vol. 8, No. 2, March-April 1960, 210-18

Summary: Brand preference information combined with a simple two-dimensional Markov process is used to study characteristics of market dynamics. Advertising alters temporarily the brand preference structure of the consuming public. Discussions of time relations, period-to-period changes in market shares, gains and losses resulting from promotional activity and rapidity of convergence to new steady-state values are considered. Sensitivity characteristics of the relations are commented upon.

Model: Let

p_{ij} = the probability that a given person, or group of persons, will purchase Brand j after having bought Brand i in the previous period.

$N_i(t+T)$ = the total number of units of Brand i purchased at time $t+T$.

N_i^* = the number of units of product i that will be bought period by period in the steady state.

N = the total number of units of product sold in the market period by period.

$$N = N_1(t) + N_2(t) = N_1(t+q) + N_2(t+q).$$

Time Relations

Transient conditions. Suppose at time $T = 0$ purchases of Brands 1 and 2 were $N_1(t)$ and $N_2(t)$ respectively. It follows that

Reference: #5.

$$N_1(t+1) = p_{11}N_1(t) + p_{21}N_2(t),$$

$$N_2(t+1) = p_{12}N_1(t) + p_{22}N_2(t). \tag{1}$$

$$N_1(t+1) = (p_{11} - p_{21})N_1(t) + p_{21}N. \tag{2}$$

If we let $s = p_{11} - p_{21}$ and $N' = p_{21}N,$

$$N_1(t+T) = s^T N_1(t) + N' \sum_{r=0}^{T-1} s^r. \quad (T = 1,2,\ldots) \tag{3}$$

<u>Steady-state conditions.</u>

(a) $s = 1.$

$$N_1(t+T) = N_1(t) \quad \text{and} \quad N_1(\infty) = N_1(t). \tag{4}$$

(b) $0 < s < 1.$

$$N_1(\infty) = [p_{21}/(p_{12} + p_{21})]N = K_1 N = N_1^*. \tag{5}$$

(c) $s = 0.$

$$N_1(\infty) = N' = p_{21}N = p_{11}N. \tag{6}$$

(d) $-1 < s < 0, s = -u.$

$$N_1(\infty) = K_1 N = N_1^*. \tag{7}$$

(e) $s = -1.$

$$N_1(t+T) = N_1(t), \quad\quad\quad\quad\quad\quad \text{(T even)}$$

$$N_1(t+T) = -N_1(t) + N' = -N_1(t) + N = N_2(t). \quad \text{(T odd)} \tag{8}$$

<u>Absolute Change Relations</u>

From (3) we obtain

$$D_1(t+T) = s^T[p_{12}N_1(t) - p_{21}N_2(t)]. \tag{9}$$

Cumulated Gain (or Loss) Relations

Starting with the definition

$$G_1(t+q) = N_1(t+q) - N_1(t),$$

sum over T+1 periods to obtain

$$\sum_{q=0}^{q=T} G_1(t+q) = \sum_{q=0}^{q=T} N_1(t+q) - (T+1)N_1(t). \qquad (10)$$

Equation (10) can then be written as

$$\sum_{q=0}^{q=T} G_1(t+q)$$
$$= \{[N_1(t) - N_1^*]/(1-s)\}[1 - s^{T+1} - (T+1)(1-s)]. \qquad (11)$$

Convergence Times

Method 1. For specific values of s we have found that $N_i(\infty) = K_i N = N_i^*$, for $i = 1,2$. Defining the discrepancy in terms of some criterion level, c, we can specify that

$$N_i(t+T) \leqq c\, N_i(\infty). \qquad (12)$$

(a) $0 < s < 1$.

$$N_i(t+T) = s^T N_i(t) + N_i^*(1-s^T)$$
$$= s^T[N_i(t) - N_i^*] + N_i^*. \qquad (13)$$

$$s^T \leqq N_i^*(c-1)/[N_i(t) - N_i^*]. \qquad (14)$$

(b) $-1 < s < 0$.

$$s^T \leqq N_i^*(c-1)/[N_i(t) - N_i^*] \quad \text{for T even only.} \qquad (15)$$

Method 2. The second way to deal with convergence is less ambiguous, but mathematically more sophisticated. As before we define $N_i(t+q) = s^q[N_i(t) - N_i^*] + N_i^*$. Now by defining the difference $\Delta N_i(t+q) = N_i(t+q) - N_i^*$, we obtain

$$\Delta N_i(t+q) = s^q[N_i(t) - N_i^*] = s^q \, \Delta N_i(t). \qquad (16)$$

Summing over all q,

$$\sum_{q=0}^{q=T} \Delta N_i(t+q) = \Delta N_i(t) \sum_{q=0}^{q=T} s^q \qquad (17)$$

which reduces to

$$\sum_{q=0}^{q=T} \Delta N_i(t+q) = [\Delta N_i(t)/(1-s)](1- s^{T+1}).$$

From (11)

$$\sum_{q=0}^{q=T} G_i(t+q) = \sum_{q=0}^{q=T} \Delta N_i(t+q) - (T+1)\Delta N_i(t). \qquad (18)$$

Imposing the conditions of Method 2,

$$\sum_{q=0}^{q=T} \Delta N_i(t+q) \leq c \, [\Delta N_i(t)/(1-s). \quad (0 < c < 1) \qquad (19)$$

It follows that

$$[\Delta N_i(t)/(1-s)](1- s^{T+1}) \leq c \, \Delta N_i(t)/(1-s),$$

or finally that $\qquad\qquad s^{T+1} \leq 1- c. \qquad (20)$

This simple result, which holds for $0 < s < 1$ and $-1 < s < 0$, specifies that the time necessary for a certain level of convergence to be attained is independent of initial conditions and depends only upon the values of s and c.

Subject: Brand Choice

Title: Customer Behavior as a Markov Process

Authors: Jerome D. Herniter and John F. Magee

Source: Operations Research, Vol. 9, No. 1, January-February 1961,
          105-22

Summary: First-order Markov processes are applied to the study of
          customer populations. In particular, theoretical
          development of ergodic matrices.

Model: Let

$$\overline{N}(n) \, P = \overline{N}(n+1). \tag{1}$$

where

N = total number of customers

n = time period

$\overline{N}(n)$ = vector of distribution of the population among
     the states i at time period n

P = transition matrix $\{p_{ij}\}$

$$\sum_{j=1}^{s} p_{ij} = 1.$$

Define the transform of P^n and $\overline{N}(n)$ as

$$P^n \leftrightarrow \mathcal{P}(z) = \sum_{n=0}^{n=\infty} P^n z^n, \tag{2}$$

$$\overline{N}(n) \leftrightarrow \overline{\pi}(z) = \sum_{n=0}^{n=\infty} \overline{N}(n) \, z^n. \tag{3}$$

Transforming (1),

$$\sum_{n=0}^{n=\infty} \overline{N}(n) \, z^n P = (1/z) \, \{ \sum_{n=0}^{n=\infty} \overline{N}(n) \, z^n - \overline{N}(0) \}$$

or $\quad \overline{\pi}(z) = \overline{N}(0) \, [I - zP]^{-1}, \tag{4}$

$$\mathcal{P}(z) = [I - zP]^{-1}. \tag{5}$$

For the two-state system let the P-matrix be

$$P = \left\| \begin{matrix} 1-\alpha & \alpha \\ \beta & 1-\beta \end{matrix} \right\|$$

Expanding by means of partial fractions,

$$P^n = \frac{1}{\alpha+\beta} \left\{ \left|\begin{matrix} \beta & \alpha \\ \beta & \alpha \end{matrix}\right| + (1-\alpha-\beta)^n \left|\begin{matrix} \alpha & -\alpha \\ -\beta & \beta \end{matrix}\right| \right\}. \quad (7)$$

If the initial vector is $\bar{N}(0) = \left|\left| N_1(0) \quad N - N_1(0) \right|\right|$,

$$N(n) = [N\beta/(\alpha+\beta)][1 - (1-\alpha-\beta)^n] + N_1(0)(1-\alpha-\beta)^n. \quad (8)$$

The effect of the initial vector decays geometrically and the steady-state vector is approached geometrically. If $\alpha + \beta > 1$ the approach will be oscillatory. The general, completely ergodic stochastic matrix can be expressed as one steady-state matrix plus a transient component that vanishes for large n.

The vector of profits is

$$\left|\left|\begin{matrix} p_1 - c \\ p_2 - c \end{matrix}\right|\right| = \left|\left|\begin{matrix} v_1 \\ v_2 \end{matrix}\right|\right| = \bar{V}$$

where the expected gross profit per customer is p_1 from active customers, p_2 from nonactive customers $(p_2 < p_1)$; c is the cost per customer per period of a continuing promotion.

The expected total profit in period n is

$$R(n) = \bar{N}(n)\,\bar{V}. \quad (9)$$

Using (8) we find that

$$R(n) = [N/(\alpha+\beta)](\beta v_1 + \alpha v_2)$$
$$- [\beta N/(\alpha+\beta) - N_1(0)](v_1 - v_2)(1-\alpha-\beta)^n. \quad (10)$$

The first component is the steady-state profit

$$g = [N/(\alpha+\beta)](\beta v_1 + \alpha v_2). \quad (11)$$

The transient component decays geometrically.

In general, P^n for an n-dimensional completely ergodic matrix can be represented as

$$P^n = S + T(n).$$ (12)

Since the expected profit in period n is

$$R(n) = \overline{N}(n)\overline{V} = \overline{N}(0)P^n\overline{V} = \overline{N}(0)S\overline{V} + \overline{N}(0)T(n)\overline{V},$$

and $\overline{N}(0)S$ is the steady-state vector, $\overline{N}*$,

$$R(n) = \overline{N}*\overline{V} + \overline{N}(0)T(n)\overline{V}.$$ (13)

Thus the profit in period n is composed of a steady component, which is independent of time and the initial vector, plus a transient component, which vanishes for large n.

If the company expects a rate of return equal to r per period on its investments, the present worth of all future profits arising from the continuing promotion is

$$W = \sum_{n=0}^{n=\infty} [1/(1+r)]^n R(n).$$ (14)

In the two-state case

$$W = \frac{N}{\alpha+\beta}(\beta v_1 + \alpha v_2)(\frac{1+r}{r}) - [\frac{\beta N}{\alpha+\beta} - N_1(0)](v_1 - v_2)(\frac{1+r}{r+\alpha+\beta}).$$ (15)

For the selection of optimum strategies, consider first the case of <u>directed promotion</u>. The alternatives for each state may be: (a) calls by salesmen, (b) direct mail, (c) a combination of (a) and (b), and (d) no promotion.

The basic recursion relation for the ith state is

$$w_i^*(n) = \text{Max}_\ell \{v_i(\ell) + [1/(1+r)] \sum_j P_{ij}(\ell)w_j^*(n+1)\}.$$ (16)

That is, we select that policy ℓ in period n which will maximize the sum of the immediate profits, $v_i(\ell)$, and the discounted future profits,

$$[1/(1+r)] \sum_j p_{ij}(\ell) \ w_j^*(n+1).$$

If the profit resulting from alternative ℓ arises after one period, the recursion relation is

$$w_i^*(n) = \text{Max}_\ell \ \{[1/(1+r)] \sum_j p_{ij}(\ell) + w_j^*(n+1)]\}. \quad (17)$$

Determination of the optimum policy

(a) by the 'flooding technique':

$$g_i(n) = g_i(n-1). \tag{19}$$

If the transition matrices are ergodic, then, when (19) holds, $q_i(n) = q$ for all i. If discounting is used, $g = 0$;

(b) by the policy iteration technique:

The basic iteration equations, assuming the matrices are ergodic, are

$$g + w_i = q_i + \sum_j p_{ij} w_j, \quad (i = 1, \ldots, N) \tag{20}$$

where g is the gain and q_i is the immediate profit in the ith state.   The iteration method is as follows:

1. Use q_i and p_{ij} for a given policy in equation (20) and solve for the relative values g and w_i by setting $w_N = 0$. This will yield N simultaneous equations with N unknowns.

2. For each state i, find the alternative ℓ', that $q_i(\ell) + \sum_j p_{ij}(\ell) w_j$, using the previously determined values, v_i. Then ℓ' is the alternative selected; $q_i(\ell')$ replaces q_i, $p_{ij}(\ell')$ replaces p_{ij} in (20), and step 1 is repeated. The iteration is completed when two successive iterations yield the same set of alternatives.

When discounting is to be used, (20) becomes

$$w_i = q_i + [1/(1+r)] \sum_{j=1}^{j=N} P_{ij} w_j. \quad (i = 1, \ldots, N) \quad (21)$$

In broadcast promotion, all states receive the same treatment and the distribution of the population among the states must be considered.

The operational characteristics of the mathematical model suggested by Vidale and Wolfe can be derived from the two-state Markov model. The Vidale-Wolfe model [#72] assumes that the transition rate α (λ in their notation) from customer to noncustomer is fixed, independent of the promotional effort, while the intensity of promotion affects only the transition rate, β, from noncustomer to customer status. The transient and steady-state behavior of the Vidale-Wolfe model is directly analogous to the behavior of the Markov model under the following specific assumptions about the nature of the transition matrices:

No Advertising   Advertising at rate A

$$P_1 = \left\| \begin{matrix} 1-\lambda & \lambda \\ 0 & 1 \end{matrix} \right\| \qquad\qquad P_2 = \left\| \begin{matrix} 1-\lambda & \lambda \\ r^A/M & 1 - r^A/M \end{matrix} \right\|$$

Reference: #72.

Subject: Brand Choice

Title: Brand Choice as a Probability Process

Author: Ronald E. Frank

Source: Journal of Business, Vol. 35, No. 1, January 1962,
372-89

Summary: A probability model is used for analyzing consumer brand
choice. The model is applied to data on coffee.

Model: The following questions are investigated:

I.   Given that a customer has made a run of 1, 2, 3,...
n purchases of a certain brand, what is the relative
frequency of buying the same brand at next purchase?

II. Given that a customer left a certain brand 1,2,3,... n
purchases ago, what is the relative frequency of buying
it at the next purchase?

The procedure used is:

I. 1. Partition every family's purchase record into the
runs of which it is composed.

2. The first and last runs are omitted from the
calculations because their exact length cannot be
determined from the available data.

3. The remaining runs are then classified by brand
and by number of purchases to form a frequency
distribution of the number of runs that are 1,2,3,... n
purchases long for each brand.

4. Each of these distributions is then converted into a
cumulative distribution of the number of runs that
are more than n purchases long.

5. The estimated probability of staying with a given
brand (P_s) on the n+1 purchase, after having purchased
the brand for n consecutive times is:

$$P_{s,n+1} = \frac{\text{Number of runs more than n+1 purchases long}}{\text{Number of runs more than n purchases long}}$$

II.L.Divide each family's purchase record into X and non-X(0)
runs of which it is composed.

2. The first and last runs, if they are non-X runs,
are omitted from the calculations.

3. The remaining non-X runs are then classified by
number of purchases to form a frequency distribution
of the number of purchases 1,2,3,... n that it
takes to return to Brand X after leaving it.

4. Each of these distributions is converted into a
cumulative distribution of the number of runs more
than n purchases long.

5. The probability of returning to Brand X (P_r) on the n+1 purchase after having left the brand for n consecutive purchases is:

$$P_{r,n+1} = \frac{\text{Number of runs exactly } n \text{ purchases in length}}{\text{Number of runs more than } n-1 \text{ purchases in length}}$$

Empirical result: For the majority of families purchasing a given coffee brand one could just as well assume that their probability of purchasing the brand remained constant.

Subject: Brand Choice

Title: The Dynamics of Brand Loyalty: A Markovian Approach

Authors: Frank Harary and Benjamin Lipstein

Source: Operations Research, Vol. 10, No. 1, January-February, 1962, 19-40

Summary: Finite stationary Markov chains and graph theory applied to brand switching and brand loyalty.

Model: A finite Markov chain consists of a collection of n events or states E_1, E_2, ..., E_n that satisfies the following conditions:

1. There is a given distribution of initial probabilities $(a_1, a_2, ..., a_n)$ where a_k is the probability that the first event is E_k.

2. In addition, there is a given matrix of transition probabilities:

$$P = \left\| \begin{matrix} p_{11}p_{12} & \cdots & p_{1n} \\ p_{21}p_{22} & \cdots & p_{2n} \\ p_{n1}p_{n2} & \cdots & p_{nn} \end{matrix} \right\|,$$

where the number p_{ij} is the conditional probability that if the present event is E_i, then the next event is E_j.

3. In order for the numbers a_i and p_{ij} to have probabilistic validity, it is also necessary that:

$$0 \leqq a_i \leqq 1 \quad \text{and} \quad \sum_{i=1}^{i=n} a_i = 1,$$

$$0 \leqq p_{ij} \leqq 1 \quad \text{and} \quad \sum_{j=1}^{j=n} p_{ij} = 1.$$

for any i = 1 to n.

BRANDS AS CHAINS

A brand chain is a Markov chain in which the states are brands and the transition probabilities tell the likelihood of consumers moving from one brand to another. A numerical example is given.

(a) The Steady State

The absolute probabilities are given by

$$\alpha_1 = \alpha_0 P,$$
$$\alpha_2 = \alpha_1 P = \alpha_0 P^2,$$

and in general $\alpha_{t+1} = \alpha_t P = \alpha_0 P^{t+1}$.

In brand switching, the absolute probabilities are brand shares derived from the consumer panel. As the power n of P becomes large, the rows of the matrix P^n approach a constant vector called the steady state.
The steady-state predictions of brand shares can be useful for evaluating advertising and promotion activity.

(b) Average Absorbing Time before Trying the Brand

This involves designating a specific brand in question as an absorbing state and computing the number of time periods on the average required for it to capture the entire market. Starting with the transition probability matrix P, the row and column of the brand in question are ruled out, leaving a residual matrix designated as Q. The average absorbing time is given by the fundamental matrix, namely the inverse of $(I-Q)$.

These average absorbing times are useful guides of the average number of purchase periods required for consumers to have tried the absorbing brand.

(c) New Product Introduction

These four measures (1) brand shares, (2) new triers, (3) repeat-buying rates, and (4) hard-core buyers provide a dynamic description of how a new brand evolves in test markets.

The stability index for brand i is the ratio in the brand movement matrix of the ith column sum to the ith row sum.

The stability index gives a view of how a new product first disturbs the market and then becomes a part of the market. An effectively introduced new product disrupts old loyalties in the market and the proportion of hard-core buyers. After a specified amount of time, which will vary by product category, the market tends to return to a modified equilibrium as measured by the position of hard-core buyers. At that point the new product has become part of the market.

Subject: Brand Choice

Title: A Mathematical Model for Marketing

Authors: Peter A. Longton and Bernard J. Warner

Source: Metra, Vol. 1, No. 3, 1962, 297-310

Summary: A model of brand choice behavior is developed. Then the probability distribution of gene frequency is transferred into that of market share and the notions of systematic and stochastic drift are introduced into marketing.

Model: A MODEL FOR PURCHASING SEQUENCES

Let there be a set of k alternative brands with each of which is associated a probability of purchase, p_j, $(j = 1,2, \ldots k)$ where $0 \leq p_j \leq 1$ and

$$\sum_{j=1}^{k} p_j = 1.$$

Consider processes of order one.

Let p_{ij} be the conditional probability of purchasing the jth brand on any given occasion after having bought brand i on the immediately preceding occasion. Since $i = 1, \ldots k$, $j = 1, \ldots k$, there are k^2 transition probabilities forming a consumers' brand-shifting matrix, P:

$$P = \begin{pmatrix} p_{11} & p_{12} & \cdots & p_{1k} \\ p_{21} & p_{22} & \cdots & p_{2k} \\ \cdot & & & \\ \cdot & & & \\ \cdot & & & \\ p_{k1} & p_{k2} & \cdots & p_{kk} \end{pmatrix}$$

Now if $s_i(t)$ is the proportion of purchasers of brand i on the tth occasion, then the proportion of purchasers who buy i on the tth occasion and j on the $(t+1)^{th}$ occasion will be, on average, $s_i(t)p_{ij}$. Since the purchasers of brand j on the $(t+1)^{th}$ occasion may have purchased any of the k brands the tth occasion, we find:

$$s_j(t+1) = s_1(t)p_{1j} + s_2(t)p_{2j} + \ldots + s_k(t)p_{kj}$$
$$= \sum_i s_i(t)p_{ij}.$$

We may rewrite this as:

$$s_{t+1} = s_t\,P$$

where s_t is a row vector.

Then:

$$s_{t+1} = s_t P = s_{t-1}P^2 = \ldots = s_1 P^t$$

Except in rather special cases, powers of brand-shifting matrices tend to a limiting matrix with identical rows, as the power becomes large. This limiting matrix is of exactly the same form as the brand-shifting matrix for a zero-order process. Thus, the brand shares tend to limits given by a row of the limiting matrix, irrespective of the initial brand-shares in the market.

A general formula for these limiting brand-shares is complex. Let us set up a model for the purchasing habits of individuals by defining coefficients of detentivity, d_1, d_2, ..., d_k, and purchasing pressures, w_1, w_2, ..., w_k. Detentivity can be interpreted as brand loyalty; the purchasing pressure to buy any particular brand is exerted by a spectrum of factors ranging from the commodity itself to advertising. It will be assumed that of those buying brand i on the tth occasion, a proportion d_i buy the same brand on the (t+1)th occasion without being subject to the purchasing pressures. A proportion are not detained by brand i, and of these a proportion w_j buy j on the (t+1)th occasion. Thus

$$0 \le d_i \le 1, \quad 0 \le w_j \le 1, \quad \text{and} \quad \sum_{j=1}^{k} w_j = 1$$

and the elements of the brand-shifting matrix are

$$
p_{ij} = \begin{cases} d_i + (1-d_i)w_j, & i = j \\ \\ (1 - d_i)w_j, & i \neq j \end{cases}
$$

If the limiting value of $s_j(t)$ as $t \to \infty$ is s_j and the limiting vector is s given by:

$$s = (s_1, s_2, \ldots s_k)$$

then: $\quad s = sP$

i.e., $\quad s(I-P) = 0$

where I is the unit matrix. This is a set of k homogeneous equations in the k unknowns $s_1, s_2, \ldots s_k$, and the solution is found to be:

$$s_j \; \alpha \; \frac{(1-d_1)(1-d_2) \ldots (1-d_k)w_j}{(1-d_j)}$$

Since $\sum_j s_j = 1$, it follows that:

$$s_j = \frac{w_j/(1-d_j)}{w_1/(1-d_1) + w_2/(1-d_2) + \ldots + w_k/(1-d_k)}$$

Rewrite the first brand formula as:

$$1/s_1 = 1 + [(1-d_1)/w_1] \; [w_2/(1-d_2) + w_3/(1-d_3) + \ldots + w_k/(1-d_k)].$$

Thus, s_1, the market share at equilibrium, is a function of the detentivities $d_1, d_2, \ldots d_k$, and the purchasing pressures $w_1, w_2, \ldots w_k$. A special case of this model arises when the detentivities are equal. For the case of two brands, see MAFFEI [#1].

AN EQUATION FOR RELATIVE GROWTH

Consider the relative growth, g, for sales of brand 1 over one cycle of the process previously described. Let the number of purchasers of brand j on the tth occasion be $n_j(t)$. Then the number, of these purchasers who will buy brand 1 on the $(t+1)$th occasion is:

$$n_1(t)d_1 + w_1 \sum_j n_j(t)(1-d_j)$$

However, new purchasers may be entering the market; suppose that the total market is $N(t)$ and that it has a relative growth G. Then the number of new purchasers on the $(t+1)$th occasion will be NG, and it will be assumed that they are subjected to the same purchasing pressures as those already in the market but not detailed by their previous purchase. Hence:

$$g_1 = \frac{n_1(t+1) - n_1(t)}{n_1(t)}$$

$$= \frac{n_1 d_1 + w_1 \sum_j n_j(1-d_j) + NGw_1 - n_1}{n_1}$$

where:

$$n_j = n_j(t), \quad N = N(t)$$

Hence:

$$g_1 = w_1(1+G)N/n_1 - (w_1/n_1) \sum_j n_j d_j + (1-d_1).$$

PROBABILITY DISTRIBUTION OF MARKET SHARE

The model is essentially the same as that used to describe the distribution of gene frequencies in population genetics.

Let $\Phi(p,t)$ be the probability distribution of p at time t. We approximate $\Phi(p,t)$ by a histogram with class-width Δp and a probability of $\Phi(p,t)\,\Delta p$ in the class with boundaries $p \pm \frac{1}{2}\,\Delta p$.

Let $m(p)\Delta t$ be the probability that p moves to the next higher class, centered on $p + \Delta p$, in time Δt, due to systematic drift.

Let $v(p)\Delta t$ be the probability that p moves outside its class in time Δt due to stochastic drift, with equal chances of moving to each of the classes centered on $p + \frac{1}{2}\,\Delta p$ and $p - \frac{1}{2}\,\Delta p$.

Then:

$$
\begin{aligned}
\Phi(p,t+\Delta t)\Delta p = \;& \Phi(p,t)\Delta p \\
& - m(p)\Delta t\;\Phi(p,t)\Delta p \\
& - v(p)\Delta t\;\Phi(p,t)\Delta p \\
& + m(p-\Delta p)\Delta t\;\Phi(p-\Delta p,t)\Delta p \\
& + \tfrac{1}{2}\,v(p+\Delta p)\Delta t\;\Phi(p+\Delta p,t)\Delta p \\
& + \tfrac{1}{2}\,v(p-\Delta p)\Delta t\;\Phi(p-\Delta p,t)\Delta p.
\end{aligned}
$$

Let $\sigma^2(p)\Delta t$ be the variance of the change in p due to stochastic drift in time Δt, i.e.,

$$
\sigma^2(p) = (\Delta p)^2\,v(p)
$$

and $\mu(p)\Delta t$ be the mean change in p due to systematic drift in time Δt, i.e.,

$$
\mu(p) = \Delta p.m(p)
$$

Then:

$$
\frac{\Phi(p,t+\Delta t) - \Phi(p,t)}{\Delta t} = \frac{-(\mu(p)\,\Phi(p,t)\; - \;\mu(p-\Delta p)\;\Phi(p-\Delta p)}{\Delta p}
$$

$$
+\;\frac{\tfrac{1}{2}(\sigma^2(p+\Delta p)\,\Phi(p+\Delta p) - 2\sigma^2(p)\,\Phi(p) + \sigma^2(p-\Delta p)\,\Phi(p-\Delta p)}{\Delta^2 p},
$$

and taking the limit as $\Delta t \rightarrow 0$, $\Delta p \rightarrow 0$

$$\frac{\partial \phi(p,t)}{\partial t} = \frac{-\partial}{\partial p} (\mu(p)\phi(p,t)) + \frac{\partial^2}{2\partial p^2} (\sigma^2(p)\phi(p,t))$$

This differential equation is equivalent to the Fokker-Plank equation in Physics. In marketing the systematic pressure might represent the effect of an advertising campaign and the random drift might represent variation due to other effects acting on a group of finite size.

Reference: #1.

Subject: Brand Choice

Title: The Demand for Branded Goods as Estimated from Consumer
           Panel Data

Author: Lester G. Telser

Source: The Review of Economics and Statistics, Vol. 44, No. 3,
           August 1962, 300-24

Summary: Estimation of a demand schedule facing a firm. Application
           of a theory of stochastic processes to consumer panel
           data for nationally advertised brands of frequently purchased
           consumer goods (orange juice, coffee, margarine).

Model: Let

$$a_{ij} = f_{ij}(P_{1t}, P_{2t}, \ldots, P_{it}, \ldots, P_{jt}, \ldots, P_{nt}) \quad (1)$$

where

a_{ij} = probability that purchases of brand i during
period $t-1$ are transferred to brand j during
period t

P_{it} = price of brand i during period t.

Adding a null brand,

$$a_{io} = f_{io}(P_{jt}) \qquad j = 0,1, \ldots, n \qquad (2)$$

$$\sum_{j=0}^{n} a_{ij} = 1 \qquad \text{for all } i = 0,1, \ldots, n. \qquad (3)$$

Let

$z_{i,t}$ = unconditional probability that brand i is
purchased at time t.

Then

$$z_{i,t} = \sum_{j=0}^{n} z_{j,t-1}\, a_{jt}. \qquad (4)$$

For given a_{ij}'s, the unconditional probabilities of
purchase approach equilibrium or limiting values, z_i^*, that
satisfy (4). Hence

$$z_i^* = \sum_{j} z_j^*\, a_{ji} \qquad i,j = 0,1, \ldots, n. \qquad (5)$$

A natural measure of $z_{i,t}$ is the market share of brand i,

$m_{i,t}$, so that by definition

$$z_{i,t} = m_{i,t} \quad \text{for} \quad i = 1, \ldots, n. \tag{6}$$

If

$$\sum_{i=1}^{n} z_{i,t} = 1 \tag{7}$$

then

$$\sum_{j=1}^{n} a_{ij} = 1 \quad \text{for} \quad i = 1, \ldots, n. \tag{8}$$

Substituting market shares for the unconditional purchase probabilities (provided the movement of purchases to and from the outside does not alter the market shares) we may write

$$m_{it} = \sum_{j=1}^{n} a_{ji} m_{j,t-1} \quad i,j = 1, \ldots, n. \tag{9}$$

The market share of brand i during period $t-1$ is

$$m_{i,t-1} = \frac{\sum_{j=1}^{n} S_{ij} + S_{io}}{\sum_{i,j=1}^{n} S_{ij} + S_{.o}} = \frac{S_{i.} + S_{io}}{S_{..} + S_{.o}} \tag{10}$$

$S_{..}$ is total sales of the product retained in both periods t and $t-1$. (The dot indicates summation over the subscript it replaces.)

If

$$\frac{S_{i.}}{S_{..}} = \frac{S_{i.} + S_{io}}{S_{..} + S_{.o}} = m_{i,t-1} \tag{11}$$

then

$$\frac{S_{i.}}{S_{..}} = \frac{S_{io}}{S_{.o}} = m_{i,t-1}. \tag{12}$$

Hence, the market share of brand i is the same regardless of whether transfers to the outside are included in the calculation. Moreover, all brands have the same chance of transitions to the outside since

$$a_{io} = \frac{S_{io}}{S_{i.} + S_{io}} = \frac{S_{.o}}{S_{..} + S_{.o}} = \lambda. \tag{13}$$

Therefore, the amount of sales each brand loses to the outside in period t-1 is a constant proportion λ of its total sales in that period.

Let a'_{ji} be the probability of transition from i to j conditional on the event that the product class does not disappear. Then

$$a'_{ij} = \frac{S_{ij}}{S_{i.}}, \tag{14}$$

$$a_{ij} = (1 - a_{io})a'_{ij} = (1 - \lambda)a'_{ij} \tag{15}$$

and

$$m_{i,t} = \frac{S_{.i} + S_{oi}}{S_{..} + S_{o.}} = \frac{S_{.i}}{S_{..}} = \frac{S_{oi}}{S_{o.}}. \tag{16}$$

Thus, each brand gains sales S_{oi} that are a constant proportion of its total sales in period t.

Modifying (9) to include a function of relevant outside variables, $D_{i,t}$,

$$m_{it} = \sum_{j} m_{j,t-1} a_{ji} + D_{i,t}. \tag{17}$$

The effects of price changes on market shares resemble price effects on absolute quantities. (9) can be written more fully as

$$m_{it} = \sum_{j=1}^{n} m_{j,t-1} \, f_{ji}(P_{k,t})_j \quad i,k = 1, \ldots, n. \quad (18)$$

Assuming prices are varied independently,

$$\frac{\partial m_{it}}{\partial P_{it}} = \sum_{j=1}^{n} m_{j,t-1} \, \frac{\partial f_{ji}(P_{k,t})}{\partial P_{i,t}} \quad (19)$$

and

$$\sum_{j} \frac{\partial m_{jt}}{\partial P_{it}} > 0, \quad j \neq i. \quad (20)$$

To determine whether a brand's market share varies directly with a noncorresponding price we take

$$\frac{\partial m_{it}}{\partial P_{\ell,t}} = \sum_{j} m_{j,t-1} \, \frac{\partial f_{ji}(P_{kt})}{\partial P_{\ell,t}} \quad (21)$$

and

$$\sum_{\substack{i \\ i \neq \ell}} \frac{\partial f_{ji}}{\partial P_{\ell}} = - \frac{\partial f_{j\ell}}{\partial P_{\ell}} > 0. \quad (22)$$

Some of the terms on the left-hand side of (22) could be negative. Thus, the signs of the partial derivatives of the right-hand side of (21) could be negative and the share of brand i could vary inversely with the price of brand ℓ. We are assured by (20) that this could not be true of all brands since a direct relation between the share of one brand and the price of another must dominate.

The long-term response of all market shares to a change in the price of one brand is calculated by comparative statics, i.e., by comparing the equilibrium market shares corresponding to the new prices with the initial equilibrium values.

Subject: Brand Choice

Title: Consumer Brand Choice--A Learning Process?

Author: Alfred A. Kuehn

Source: R.E. Frank, et. al. (eds.), Quantitative Techniques in
           Marketing Analysis, Homewood, Illinois: Richard D. Irwin, Inc.,
           1961, 390-403

Summary: A stochastic learning model describes consumer brand shifting.
           Brand shares are predicted from the sequence, rhythm, and
           frequency of consumers' past purchases. Data on frozen orange
           juice concentrates is used as empirical work.

Model: To illustrate how this brand shifting model describes

           changes in the consumer's probability of purchasing any

           given brand as a result of his purchases of that brand

           (e.g., Brand A) and competing brands (e.g., Brand X), let

           us examine the effect of the four-purchase sequence XAAX

           on a consumer with initial probability $P_{A,1}$ (Figure 1).

           The model is defined in terms of four parameters,

           namely, the intercepts and slopes of the two lines referred

           to in Figure 1 as the Purchase Operator and the Rejection

           Operator. If the brand in question is purchased by the

           consumer on a given buying occasion, the consumer's

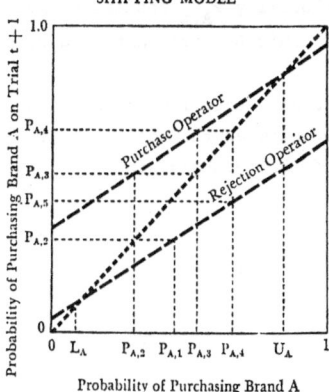

FIGURE 1
STOCHASTIC (PROBABILISTIC) BRAND
SHIFTING MODEL

Probability of Purchasing Brand A
on Trial t

probability of again buying the same brand the next time that type product is purchased is read from the Purchaser Operator. If the brand is rejected by the consumer on a given buying occasion, the consumer's probability of buying that brand when he next buys that type of product is read from the Rejection Operator. Thus in Figure 1 our hypothetical consumer begins on trial 1 with the probability $P_{A,1}$ of buying Brand A. The consumer chooses some other brand (X) on trial 1, however, and thus his probability of buying Brand A on trial 2 ($P_{A,2}$) is obtained from the Rejection Operator, resulting in a slight reduction in the probability of purchasing A on the next trial. On trial 2, however, the consumer does purchase Brand A and thus increases the likelihood of his again buying the brand on the next occasion (trial 3) to $P_{A,3}$. Continuing in this fashion, the consumer again buys A on trial 3, thereby increasing his probability of purchasing Brand A on trial 4 to $P_{A,4}$. He again rejects A on trial 4, however, decreasing his probability of buying A on trial 5 to $P_{A,5}$.

Two characteristics of the model should be noted: (1) The probability $P_{A,t}$ approaches but never exceeds the upper limit U_A with repeated purchasing of the brand, and (2) the probability $P_{A,t}$ approaches but never drops below the lower limit L_A with continued rejection of the brand.

The Purchase and Rejection Operators are functions of the time elapsed between the consumer's t^{th} and $t+1^{st}$ purchases and of the merchandising activities of competitors.

References: #18 and 23.

Subject: Brand Choice

Title: Stochastic Process Models of Consumer Behavior

Author: Ronald A. Howard

Source: Journal of Advertising Research, Vol. 3, No. 3, September 1963, 35-42

Summary: Three problem areas of the Markov chain analysis are being commented on: the problem of describing inter-purchase times, the problem of aggregation, and the problem of estimation of transition matrices using prior knowledge.

Model: <u>The Aggregation Problem and the Vector Markov Process</u>

      Consider how a whole population of units behaves when it is subjected to random forces. Let us call this kind of process the vector Markov process.

      Let $c_i(n)$ be the number of customers of brand i at time n. We assume that each customer in an N-brand market makes transitions from one brand to another at successive instants in time according to some transition probability matrix P. The problem is to determine the statistics of the number of customers of each brand i at time n if we have some initial number of customers of each brand i at time 0, $c_i(0)$.

      The number of customers of brand j at time n will be given by the binomial probability distribution with probability of success $\phi_{kj}(n)$ which is easily derivable from the transition matrix for the Markov process, P and number of trials c_k. The total number of customers of brand j at time n will have a distribution that is the convolution of all these binomial distributions:

$$p\ \{c_j(n) = m\} = \overset{N}{\underset{i=1}{*}}\ p_B[m|c_i, \phi_{ij}(n)].$$

Here $*$ is a symbol for manifold convolution and $p_B[m|c_i, \phi_{ij}(n)]$ is the binomial probability of m successes in c_i trials with probability of success $\phi_{ij}(n)$.

From a more general point of view, we would like to know

$$p \{c_1(n) = n_1, c_2(n) = n_2, \ldots, c_N(n) = n_N\}$$

Let $\bar{c}_j(n)$ be the expected number of customers in brand j at time n. Then

$$\bar{c}_j(n) = \sum_{i=1}^{N} c_i \phi_{ij}(n)$$

In row vector-matrix notation:

$$\bar{\underline{c}}(n) = \underline{c}(0) \Phi(n) = \underline{c}(0) P^n$$

Since the variance of a binomial distribution is equal to the number of trials times the probability of success times one minus the probability of success we obtain

$$\overset{v}{c}_j(n) = \sum_{i=1}^{N} c_i \phi_{ij}(n)[1 - \phi_{ij}(n)]$$

If we define a quantity $v_{ij}(n)$ by

$$v_{ij}(n) = \phi_{ij}(n)[1 - \phi_{ij}(n)]$$

then we can write $\overset{v}{c}_j(n)$ in the form

$$\overset{v}{c}_j(n) = \sum_{i=1}^{N} c_i v_{ij}(n)$$

Finally, by using a variance matrix $V(n)$ with elements $v_{ij}(n)$ we can write this equation in matrix form as

$$\overset{v}{\underline{c}}(n) = \underline{c}(0) \, V(n).$$

Random Interpurchase Times--The Semi-Markov Process

The consumer purchase pattern is far from regular. However, there is no need to assume that the time between purchases is a constant and thereby create the necessity for fictitious "No Purchase" brands: all we must do is expand our concept of a Markov process.

The regular Markov process is defined by transition probabilities. This transition is assumed to require one time period, such as a week or a month. Let the random variable τ_{ij} be the time required for a transition from i to j. It is selected from a density function $h_{ij}(\cdot)$ that depends on the transition to be made.

A semi-Markov process is described by a transition probability matrix P and by a holding time matrix H with elements $h_{ij}(\cdot)$. The element $\phi_{ij}(t)$ of its interval transition probability matrix $\phi(t)$ is the probability that a system that entered state i at time zero will be in state j at time t. The interval transition probabilities satisfy the equation

$$\phi_{ij}(t) = \delta_{ij} \left[1 - \sum_{k=1}^{N} p_{ik} \int_{o}^{t} d\tau\, h_{ik}(\tau) \right]$$

$$+ \sum_{k=1}^{N} p_{ik} \int_{o}^{t} d\tau\, h_{ik}(\tau) \phi_{kj}(t-\tau)\, t \geq 0,\ 1 \leq i,j \leq N$$

$$\delta_{ij} = \begin{cases} 1 & i = j \\ 0 & i \neq j \end{cases}$$

Prior knowledge about the transition probability matrix can be expressed in terms of continuous distributions over the probabilities in this matrix. Combining the prior knowledge with experimental results yields a new set of distributions with generally smaller variances.

Subject: Brand Choice

Title: A Latent Markov Process Within the Individual

Author: James S. Coleman

Source: Models of Change and Response Uncertainty, Englewood Cliffs,
           New Jersey: Prentice-Hall, Inc., 1964, Chapter Two, Sec-
           tion 1, 16-19

Summary: Having extracted the variation due to response uncertainty
           the process and measurement of change are examined.

Model: Let the unit which undergoes change be a response element
           of which each individual has a large set; m_1, m_2, ..., m_s
           response elements are associated with responses 1, 2, ..., s,
           respectively. If $Pr\{R_i\}$ is the probability of giving
           response i, then

$$Pr\{R_i\} = \frac{m_i}{\sum\limits_{j=1}^{s} m_j}. \tag{1}$$

Each element has a transition rate q_{ij} from each state i
to each other state j. If the probability of an
element's being in state i at time t is v_{it}, then for
a set of s possible states of the element, each element's
behavior is governed by

$$\frac{dv_{1t}}{dt} = q_{11}v_{1t} + \cdots + q_{i1}v_{it} + \cdots + q_{i1}v_{st}$$

$$\vdots$$

$$\frac{dv_{it}}{dt} = q_{1i}v_{1t} + \cdots + q_{ii}v_{it} + \cdots + q_{si}v_{st} \tag{2}$$

$$\vdots$$

$$\frac{dv_{st}}{dt} = q_{1s}v_{1t} + \cdots + q_{is}v_{it} + \cdots + q_{ss}v_{st}.$$

Here the q_{ij}'s are transition rates for the elements; and q_{ii} is defined as $-\sum_{j=1}^{s} q_{ij}$. The solution of (2) is

$$
\begin{aligned}
v_{1t} &= v_{10}r_{11t} + v_{20}r_{21t} + \cdots + v_{so}r_{s1t} \\
&\quad \cdot \\
&\quad \cdot \\
&\quad \cdot \\
v_{st} &= v_{10}r_{1st} + v_{20}r_{2st} + \cdots + v_{so}r_{sst}
\end{aligned}
\tag{3}
$$

where r_{ijt} is a function of the q_{ij}'s and t.
As time continues, each v_j becomes less dependent upon its original value and more dependent on values of other v_i's. The function that relates r_{ijt} to the q_{ij}'s is an infinite series, the first four terms of which are

$$
r_{ijt} = \delta_{ij} + tq_{ij} + \frac{t^2}{2}\sum_{k=1}^{s} q_{ik}q_{kj} + \frac{t^3}{3!}\sum_{k=1}^{s}\sum_{h=1}^{s} q_{ik}q_{kh}q_{hj} + \cdots
\tag{4}
$$

where δ_{ij} is the Kroneker delta, 0 if $j \neq 1$, 1 if $j = i$. This result may alternatively be expressed in matrix notation

$$
V'_t = V'_o \, e^{Qt}
\tag{5}
$$

where e^{Qt} is a matrix that is the sum of the infinite series

$$
e^{Qt} = I + Qt + \frac{Q^2 t^2}{2!} + \frac{Q^3 t^3}{3!} + \cdots
\tag{6}
$$

Since

$$
v_{it} \simeq Pr\{R_i\} = \frac{m_{it}}{\sum_{j=1}^{s} m_{jt}}
\tag{7}
$$

v_{it} is the probability that this person will give response i at time t.

References: #11 and 18.

Subject: Brand Choice

Title: The Neglected Use of Data

Author: A.S.C. Ehrenberg

Source: Journal of Advertising Research, Vol. 7, No. 2, June
           1967, 2-7

Summary: The problem of assessing the likely success or failure
           of a product is discussed using two practical cases.

Model: <u>Intentions-to-Buy and Brand-Usage</u>

The percentage I of respondents expressing an
intention-to-buy any brand is found to be approximately
proportional to the square root of the percentage U
claiming current usage of that brand, i.e.,

$$I = K \sqrt{U},$$

where K is a constant which is the same for the
different brands in any one product-field. For
intentions-values I ranging from nearly 0 to almost
100 per cent, the average fit of this relationship
between U and I is within a mean deviation
$|I - K\sqrt{U}|$ of about three percentage points.

Subject: Brand Choice

Title: A Stochastic Response Model with Application to Brand
          Choice

Author: David B. Montgomery

Source: Management Science, Vol. 15, No. 7, March 1969,
          323-37

Summary: Development of a heterogeneous, nonstationary zero response
          model, an extension of a class of stochastic response models
          presented by Coleman [#9]. The data base for empirical work
          is dentifrice purchase records.

Model:

Assumptions:

Each respondent's response probability is generated by
the same basic stochastic process over a sequence of
responses.

At any given response occasion t (integer), different
respondents may have different response probabilities.

At any given response occasion t, there are two mutually
exclusive and collectively exhaustive response alternatives,
A and B.

Each respondent possesses N (possibly an infinite number of)
hypothetical response elements.

At any given response occasion t, each of an individual's
response elements is uniquely associated with either
response A or B.

If at response occasion t an individual has i of his
elements associated with response A, his probability of
making response A at response occasion t will be

$$P \{A \text{ at } t \mid i \text{ elements associated with A at } t\} = \frac{i}{N}.$$

For an individual respondent, the actual response he makes
at t-1 does not affect the probability that he will make
response A at t.

Individual respondents respond independently of one
another.

If at time t an individual is in state i (i = 0, 1 ...,
N-1), the probability of the transition $i \rightarrow i + 1$ in the
interval (t, t + Δt) is $\lambda_i \Delta t + 0(\Delta t)$.

If at any time t an individual is in state i (i = 1, 2,
..., N), the probability of the transition $i \rightarrow i - 1$ in
the interval (t, t + Δt) is $\mu_i \Delta t + 0(\Delta t)$.

The probability of a transition to other than a neighboring state is $0(\Delta t)$.

The process is stationary.

Then

$$\frac{dp_o(t)}{dt} = -\lambda_o p_o(t) + \mu_1 p_1(t) \qquad \text{for} \quad i = 0$$

$$\frac{dp_i(t)}{dt} = -(\lambda_i + \mu_i)p_i(t) + \lambda_{i-1}p_{i-1}(t) \tag{1}$$

$$+ \mu_{i+1}p_{i+1}(t) \qquad \text{for} \quad 0 < i < N$$

$$\frac{dp_N(t)}{dt} = -\mu_N p_N(t) + \lambda_{N-1}p_{N-1}(t) \qquad \text{for} \quad i = N.$$

Further assume that

each element associated with response B has transition intensity α toward becoming associated with response A.

each element associated with response A has transition intensity β toward becoming associated with response B.

the transition intensity of each element is increased by an amount γ for each element associated with the opposite response.

Then

$$\lambda_i = (\alpha + i\gamma)(N-i) \tag{2}$$

and

$$\mu_i = [\beta + (N-i)\gamma]i. \tag{3}$$

Substituting (2) and (3) into (1):

$$\frac{dp_o(t)}{dt} = -N\alpha p_o(t) + [\beta + (N-1)\gamma]p_1(t) \quad \text{for} \quad i = 0$$

$$\frac{dp_i(t)}{dt} = -\{(N-i)(\alpha + i\gamma)$$

$$+ i[\beta + (N-i)\gamma]\}p_i(t) \tag{4}$$

$$+ (N-i+1)[\alpha + (i-1)\gamma]p_{i-1}(t)$$

$$+ (i+1)[\beta + (N-i-1)\gamma]p_{i+1}(t) \quad \text{for} \quad 0 < i < N$$

$$\frac{dp_N(t)}{dt} = -N\beta p_N(t) + [\alpha + (N-1)\gamma]p_{N-1}(t) \quad \text{for} \quad i = N.$$

From (4) the steady-state distribution of p_i is:

$$p_i = \binom{N}{i} \frac{\Gamma[(\alpha/\gamma) + i]\Gamma[N + (\beta/\gamma) - i]\Gamma[(\alpha + \beta)/\gamma]}{\Gamma[\alpha/\gamma]\Gamma[N + (\alpha + \beta)/\gamma]\Gamma[\beta/\gamma]}$$

$$i = 0, 1, \ldots, N. \quad (5)$$

Using (5) the steady-state distribution of response probability X is given by

$$f(X) = \frac{\Gamma[(\alpha + \beta)/\gamma]X^{(\alpha/\gamma)-1}[1-X]^{(\beta/\gamma)-1}}{\Gamma[\alpha/\gamma]\Gamma[\beta/\gamma]} \quad (6)$$

which is the beta distribution having mean, $\alpha/(\alpha+\beta)$, and variance $\alpha\beta\gamma/(\alpha + \beta)^2(\alpha + \beta + \gamma)$.

The mean value function is

$$E[X(t)] = E[X(0)] \exp [- (\alpha + \beta)t]$$

$$+ \frac{\alpha}{(\alpha+\beta)} \{1 - \exp [- (\alpha + \beta)t]\} \quad (7)$$

where $E[X(0)]$ is the initial condition.

Suppose for some individual we know $X(0)$. Then $E[X(0)] = X(0)$ and

$$E[X(t)|X(0)] = X(0) \exp [- (\alpha + \beta)t]$$

$$+ \frac{\alpha}{(\alpha+\beta)} \{1 - \exp [- (\alpha + \beta)t]\}. \quad (8)$$

For the diffusion limit of the process the infinitesimal mean displacement $a(X)$ and the infinitesimal variance $2b(X)$ are defined as:

$$\lim_{h \to 0} \frac{E[X(t+h) - X(t)|X(t)]}{h} = a(X) \quad (9)$$

$$\lim_{h \to 0} \frac{Var[X(t+h) - X(t)|X(t)]}{h} = 2b(X). \quad (10)$$

The infinitesimal mean and variance of the model are given by

$$a(X) = \alpha - (\alpha + \beta)X(t) \quad (11)$$

and

$$2b(X) = 2\gamma[X(t) - X^2(t)], \quad (12)$$

respectively.

Let $f(t,X)$ denote the probability density function of $X(t)$. Kolmogorov has shown that $f(t,X)$ must satisfy the Fokker-Plank diffusion equation

$$\frac{\partial f(t,X)}{\partial t} = \frac{\partial^2 \{b(X) f(t,X)\}}{\partial X^2} - \frac{\partial \{a(X) f(t,X)\}}{\partial X} \qquad (13)$$

where $a(X)$ and $b(X)$ are the quantities defined in (9) and (10).

Using (11) and (12) it is shown that (6) satisfies (13).

Let

  M = number of individual respondents in a sample

  $f[X(t)]$ = the probability density function of the population of individuals with respect to their response probability, $X(t)$, at t.

Lemma 1. If $P(t)$ denotes the expected proportion of the M respondents making response A at t, then

$$P(t) = E[X(t)] = \int_0^1 E[X(t)|X(0)] f[X(0)] dX(0) \qquad (14)$$

and the observed proportion $Q(t)$ is an unbiased estimate of $P(t) = E[X(t)]$.

Lemma 2. If $P(0,t)$ denotes the expected proportion of the M respondents making response A at 0 and A again at t, then

$$P(0,t) = E[X(0,t)] = \int_0^1 E[X(t)|X(0)] X(0) f[X(0)] dX(0) \qquad (15)$$

and the observed proportion $Q(0,t)$ is an unbiased estimate of $P(0,t) = E[X(0,t)]$.

References: #9, 18, 19, 20.

Subject: Brand Choice

Title: A Mathematical Model of Consumer Behavior

Author: Benjamin Lipstein

Source: Journal of Marketing Research, Vol. 2, August 1965, 259-65

Summary: The model of consumer behavior relates advertising effort to attitude changes and consumer purchases, using a nonstationary Markov process. An analytical method of handling nonstationary stochastic matrices is developed. The behavior of brand shares under different market conditions is considered along with numerical examples of how the model operates.

Model:

Let

$$a_{ij} = [r_{ij}d_j(p_j + e_{ij})],$$

where

a_{ij} = probability that consumer having bought i in the last period with buy j in the next period

r_{ij} = equivalent element in reaction matrix

d_j = availability in distribution of brand j

p_j = degree to which price changes in j will cause consumers to buy j

e_{ij} = error component with expectation of zero

$$\rho_o A = \alpha_o$$

where

ρ = row vector representing brand share preference at a moment in time

A = switching matrix

α = market brand share vector.

The product of the brand share preference vector ρ and the attitude, predisposition or preference matrix R gives rise to brand share preference in the next period.

$$\rho_o R = \rho_1.$$

When R_o, the preference matrix in period zero, is different from R_1, the preference matrix in period 1, we introduce a causative matrix C such that:

$$R_o C_o = R_1.$$

$$R_1 C_1 = R_2.$$

The value of C can be derived from the above relationship:

$$C = R_o^{-1} R_1$$

C is a measure of the effectiveness of the marketing innovation which caused R_o to change to R_1. Most typically this would be due to new advertising effort, though it could also come about through product improvement or other marketing innovations.

The efficiency of advertising effort will vary with the quality of the copy, media plans, and total expenditures. While the effect of a campaign can be observed through the elements of the C matrix, it is possible to relate advertising effort to the causative matrix by means of

$$c_j = K^j b$$

where c_j = jth column of the C matrix, $(b_1 b_2 \ldots b_n)$ = vector of advertising expenditures and K^1, K^2 etc., = matrices of regression coefficients relating the advertising expenditures to the columns of the C matrix. The jth superscript of K^j relates to the jth column of C. Hence, the K^j matrix is a measure of how advertising expenditures bring about these rates of change in attitude for the jth brand.

$$\alpha_t = \rho_o \, R^t A.$$

If R does not change over time, any change in share will come from the A matrix, which in turn can be due to distribution or price.

$$\alpha_t = \rho_o \, (R_1) \, (R_1 C_1) \, (R_1 C_1 C_2) \, (R_1 C_1 C_2 C_3)$$
$$\dots (R_1 C_1 C_2 C_3 \dots C_{t-1})A.$$

Brand share in the current period t is a function of the cumulative effects of attitude change modified by market conditions in the current period. Boundary conditions for the system are:

1. If R is constant and A is known, then α_t (brand share) can be estimated. Furthermore, if α_{t-1} is different from α_t, the change must be due to A.

2. If R changes such that C is a constant with eigenvalues less than 1, α_t will be dependent on the limiting vectors of C and A. The system tends toward equilibrium with brand shares approaching the steady state.

3. If R changes such that C is a constant with eigenvalues greater than 1, the system is not likely to persist. Changes in R are likely to force the system back toward equilibrium. The consequence of C with roots greater than 1 is that one or more brands will absorb the entire market.

4. If R changes such that C changes but the eigenvalues of C are less than 1, the system can continue indefinitely, fluctuating around an equilibrium.

40

Subject: Brand Choice

Title: New Models of Consumer Loyalty Behavior: Aids to Setting and Evaluating Marketing Plans

Author: Donald G. Morrison

Source: Marketing and Economic Development, Peter D. Bennett (ed.), Proceedings of the 1965 Fall Conference of the American Marketing Association, Chicago, 1965, 323-27

Summary: The compound Markov models help obtain quantitative measures of the influence of recent purchase decisions on the current brand choice, plus a measure of the heterogeneity in the population. The models provide a means for comparing the behavior of consumers toward different products, as well as a method for contrasting various market segments within a product category. In addition, they permit some insight into brand loyalty. The model is applied to coffee purchases.

Model: Brand Loyal Model

The brand loyal population of consumers is defined by:

1. each individual is a first order 0-1 process with transition matrix

$$\begin{array}{c c} & \begin{array}{c c} 1 & \quad\; 0 \end{array} \\ \begin{array}{c} 1 \\ \\ 0 \end{array} & \left[\begin{array}{c c} p & 1-p \\ \\ kp & 1-kp \end{array}\right] \end{array} , \quad \text{where}$$

2. p is distributed beta (α,β) among the families in the population, and

3. k is a constant, the same for all families, $0 < k < 1$.

High loyalty is directed towards a particular brand. Therefore, this model is classified as the brand loyal model.

Last Purchase Loyal Model

The last purchase loyal population of consumers is defined by:

1. each individual family is a 0-1 first-order process with transition matrix

$$
\begin{array}{cc}
1 & 0
\end{array}
$$

$$
\begin{array}{c}
1 \\
\\
\\
0
\end{array}
\begin{bmatrix}
p & 1-p \\
\\
\\
1-kp & kp
\end{bmatrix}
, \quad \text{where}
$$

2. p is distributed beta (α,β) among the individuals in the population, and

3. k is a constant, the <u>same</u> for <u>all</u> individuals, $0 < k < 1$.

The last purchase loyal population acts in the opposite manner from the brand loyal population of consumers. An individual with a high p is more loyal to the brand he last purchased (be it brand 0 or brand 1) than a person with a lower p. Here, if high loyalty exists, <u>that particular consumer demonstrates loyalty toward the brand he happened to have last purchased</u>. Accordingly, this model is called the Last Purchase Loyal model.

$b(p)$ is assumed to be a beta distribution, i.e.,

$$
\begin{aligned}
b(p) &= \frac{\Gamma(\alpha+\beta)}{\Gamma(\alpha)\Gamma(\beta)}\, p^{\alpha-1}(1-p)^{\beta-1} = \text{beta } (\alpha,b), \quad 0 < p < 1, \\
&= 0 \qquad\qquad\qquad\qquad\quad \text{otherwise.}
\end{aligned}
$$

For this distribution the mean and variance are

$$
E(p) = \frac{\alpha}{\alpha+\beta},
$$

$$
\text{Var}(p) = \frac{\alpha\beta}{(\alpha+\beta+1)(\alpha+\beta)^2}.
$$

Much more general densities are possible, and techniques for working with arbitrary densities $b(p)$ have been developed. The Bernoulli model is a special case of the brand loyal model: When $k = 1.0$ the probability of a purchase of brand 1 on the next trial is equal to p regardless of the last purchase decision.

Subject: Brand Choice

Title: Dynamic Inference

Author: Ronald A. Howard

Source: Operations Research, Vol. 13, No. 5, September-October
          1965, 712-33

Summary: We consider a model for dynamic uncertain processes.
          The underlying statistical parameters of a stochastic
          process that produces observable outputs are themselves
          allowed to change at times generated by another stochastic
          process. We would like to make probability assignments
          to future outputs of the process, given only the past
          outputs. We develop the inferential relations for the case
          where the changes of parameters are governed by a
          renewal process, and where the process that generates
          observables depends only on its present parameters. We
          illustrate these results using an example with a
          Bernoulli observable distribution, a beta parameter
          distribution, and a geometric distribution for the time
          between parameter changes. Possible applications of the
          general class of dynamic inferences models range
          from marketing to antisubmarine warfare.

Model: Let

ρ_k = time between the kth and $k+1^{th}$ change
      preceding $i = 0$

c_k = time index at which the kth change before
      $i = 0$ occurs.

Then

$$\rho_k = c_{k+1} - c_k \qquad (1)$$

or $\{\rho|\varepsilon\} = \{\rho|\varepsilon\}$ where $\{\rho|\varepsilon\}$ = change interval distribution

At every point i the process has some parameter vector
$\mu(i)$. When a change point k occurs, the value of $\mu(c_k)$
is selected from a prior $\{\mu|\varepsilon\}$.

$$\{\mu(c_k)|\varepsilon\} = \{\mu|\varepsilon\}. \qquad (2)$$

This value for μ is used until the next change point
occurs,

$$\mu(c_k) = \mu(c_k-1) = \mu(c_k-2) = \ldots = \mu(c_{k-1}+1). \qquad (3)$$

Assume all we can observe is a variable $x(i)$ at each time
point i. The value of $x(i)$ observed at the time point i
is selected independently for every i from a given

probability distribution $\{x|\mu\varepsilon\}$ with parameters $\mu(i)$,

$$\{x(i)|\mu(i)\varepsilon\} = \{x|\mu\varepsilon\}. \tag{4}$$

We shall let $x(a,b)$, $b \geq a$, be the vector of observable variables from time a to time b into the past,

$$x(a,b) = [x(a), x(a+1), \ldots, x(b)].$$

If in any expression we have $x(a,b)$, $a > b$, we shall for convenience merely consider the vector $x(a,b)$ to be absent from the expression. We shall assume that we are able to observe the x's from $i=1$ to some point $i=m$.

Our problem is this. Having established distributions $\{\rho|\varepsilon\}$, $\{\mu|\varepsilon\}$, and $\{x|\mu\varepsilon\}$ and having specified a method for selecting the time point $i = 0$, what can we infer from the observable vector $x(1,m)$ about the observable variable $x(0)$ that will be generated at time zero? What can we say about when the changes occurred?

By basic probability theory we can write $\{u|S\}$ in the form

$$\{u|S\} = S_v \{u|vS\}\{v|S\}, \tag{5}$$

Using Bayes' theorem,

$$\{u|vS\} = \{v|uS\}\{u|S\}/S_u \{v|uS\}\{u|S\}. \tag{6}$$

Our goal is to find $\{x(0)|x\varepsilon\}$ in terms of the basic distribution of the process.

First, we expand in c, the vector of change times,

$$\{x(0)|x\varepsilon\} = S_c \{x(0)|cx\varepsilon\}\{c|x\varepsilon\}. \tag{7}$$

$$\{x(0)|cx\varepsilon\} = \{x(0)|c_1 x(1,c_1)\varepsilon\}. \tag{8}$$

$$\{x(0)|c_1 x(1,c_1)\varepsilon\} = S_{\mu(0)}\{x(0)/\mu(0)c_1 x(1,c_1)\varepsilon\}x\{\mu(0)|c_1 x(1,c_1)\varepsilon\}.$$

The probability assignment on $x(0)$ given the parameter
vector is found from one of our original distributions,

$$\{x(0)\,|\,\mu(0)c_1 x(1,c_1)\varepsilon\} = \{x(0)\,|\,\mu(0)\varepsilon\} = \{x = x(0)\,|\,\mu = \mu(0)\varepsilon\}. \quad (10)$$

By Bayes' theorem,

$$\{\mu(0)\,|\,c_1 x(1,c_1)\varepsilon\} = \{x(1,c_1)\,|\,c_1 \mu(0)\varepsilon\}\{\mu(0)\,|\,c_1 \varepsilon\}/S_{\mu(0)}$$
$$\text{Numerator.} \quad (11)$$

We write the first term in the numerator in the form

$$\{x(1,c_1)\,|\,c_1 \mu(0)\varepsilon\} = \{x(1)\,|\,\mu(1) = \mu(0)\varepsilon\}\{x(2)\,|\,\mu(2) = \mu(0)\varepsilon\}$$
$$x\{x(c_1)\,|\,\mu(c_1) = \mu(0)\varepsilon\}. \quad (12)$$

Writing

$$\{\mu(0)c_1\varepsilon\} = \{\mu\,|\,\varepsilon\} \quad (13)$$

and using Bayes' theorem,

$$\{c\,|\,x\varepsilon\} = \{x\,|\,c\varepsilon\}\{c\,|\,\varepsilon\}/S_c \quad \text{Numerator.} \quad (14)$$

Because the x's are independent after every change, we
can write

$$\{x\,|\,c\varepsilon\} = \{x(1,c_1)\,|\,c\varepsilon\}\{x(c_1+1,c_2)\,|\,c\varepsilon\}\ldots\{x(c_\ell+1,m)\,|\,c\varepsilon\}, \quad (15)$$

where c_ℓ is the time of the last (earliest) change in the
vector x. We expand a typical factor
$\{x(c_k+1,c_{k+1})\,|\,c\varepsilon\}$, $1 \leqq k \leqq \ell-1$, in $\mu(c_{k+1})$,

$$\{x(c_k+1,c_{k+1})\,|\,c\varepsilon\} = S_{\mu(c_{k+1})} \{x(c_k+1,c_{k+1})\,|\,\mu(c_{k+1})c\varepsilon\}$$
$$\{\mu(c_{k+1})\,|\,c\varepsilon\}. \quad (k = 1,2,3,\ldots,\ell-1). \quad (16)$$

The quantity $\{x(c_k+1,c_{k+1})\,|\,\mu(c_{k+1})c\varepsilon\}$ is like that in (12),

$$\{\mathbf{x}(c_k+1, c_{k+1}) \,|\, \mu(c_{k+1}) \mathbf{c}\varepsilon\} = \{\mathbf{x}(c_k+1) \,|\, \mu(c_k+1)$$

$$= \mu(c_{k+1})\varepsilon\} \mathbf{x}\{x(c_k+2) \,|\, \mu(c_k+2) = \mu(c_{k+1})\varepsilon\} \qquad (17)$$

$$\mathbf{x}\{x(c_{k+1}) \,|\, \mu(c_{k+1})\varepsilon\}. \quad (1 \leq k \leq \ell-1)$$

For the first factor of (15) we use (12) integrated with respect to $\{\mu(0)\,|\,\varepsilon\}$. For the last factor $\{\mathbf{x}(c_\ell+1, m)\,|\,\mathbf{c}\varepsilon\}$ we compute only those quantities on the right in (17) that correspond to the vector $\mathbf{x}(c_\ell+1, m)$.

We use the prior distribution,

$$\{\mu(c_{k+1}) \,|\, \mathbf{c}\varepsilon\} = \{\mu\,|\,\varepsilon\}. \tag{18}$$

We write

$$\{\mathbf{c}\,|\,\varepsilon\} = \{c_1, c_2 \ldots c_\ell\,|\,\varepsilon\} = \{c_1\,|\,\varepsilon\}\{\rho = c_2 - c_1\,|\,\varepsilon\}\{\rho = c_3 - c_2\,|\,\varepsilon\}$$

$$\ldots \{\rho = c_\ell - c_{\ell-1}\,|\,\varepsilon\}\{\rho > m - c_\ell\,|\,\varepsilon\}. \tag{19}$$

Suppose that the process that located $i = 0$ was independent of the renewal process that generates changes. Then the probability distribution of the change interval ρ_0 in which $i = 0$ lies is going to be different from $\{\rho\,|\,\varepsilon\}$ because a longer change interval is more likely to be selected. Let R represent the selection of a particular change interval or cycle by this random selection process. Then by Bayes' theorem

$$\{\rho_0 | \text{R}\varepsilon\} = \{\text{R}\,|\,\rho_0\varepsilon\}\{\rho_0\,|\,\varepsilon\}/\sum\nolimits_{\rho_0} \quad \text{Numerator.} \tag{20}$$

Since the probability of selecting a particular cycle given that its length is ρ_0 is proportional to ρ_0,

$$\{\text{R}\,|\,\rho_0\varepsilon\} = k_{\rho_0}, \tag{21}$$

where k is a proportionality constant. Then equation (20) becomes

$$\{\rho_o | R\epsilon\} = k_{\rho_o} \{\rho_o | \epsilon\}/k \sum_{\rho_o} \{\rho_o | \rho_o | \epsilon\} = \rho_o \{\rho_o | \epsilon\} / \langle \rho_o | \epsilon \rangle. \quad (22)$$

The symbol $\langle x | S \rangle$ represents the expectation of the random variable x on the state of information S.

Since the point $i = 0$ is equally likely to be placed anywhere in the change interval of length ρ_o by the selection process R, we have

$$\{c_1 | \rho_o \ R\epsilon\} = 1/\rho_o. \quad (0 \leqq c_1 \leqq \rho_o - 1) \quad (23)$$

Then we can write

$$\{c_1 | R\epsilon\} = \int_{\rho_o = 1}^{\rho_o = \infty} \{c_1 | \rho_o \ R\epsilon\}\{\rho_o | R\epsilon\} \quad (24)$$

$$= \{\rho \geqq c_1 + 1 | \epsilon\}/\langle \rho | \epsilon \rangle. \quad (c_1 = 0,1,2, \ldots).$$

Thus the probability that the last change took place at time c_1 given that $i = 0$ was picked by a process independent of the change process is equal to the probability that the interchange time is greater than or equal to $c_1 + 1$ divided by the mean of the interchange time probability mass function. Thus for this one particular method of selecting $i = 0$ we have a simple way to calculate the $\{c_1 | \epsilon\}$ needed in (19).

In order to obtain the probability distribution to assign to $x(0)$ in the light of the vector $\mathbf{x}$ in (7) we use Figure 1 showing what equations must be substituted into what equations to produce (7). Since the origins of this tree, equations (10), (12), (13), (17), (18), and (24) depend only on the assumed distributions ($\{\rho | \epsilon\}$, $\{\mu | \epsilon\}$, and $\{x | \mu\epsilon\}$) the problem is solved. The natural sequence of the solution is to start with the tips of the tree and gradually work back to obtain the final result.

Reference: #20.

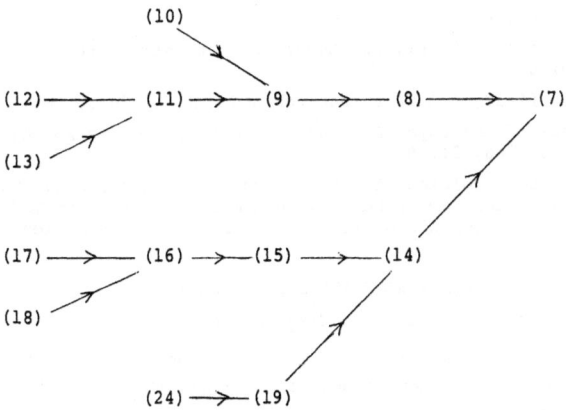

Figure 1
Road Map for Successive Equation Substitutions

Subject: Brand Choice

Title: Progress on a Simplified Model of Stationary Purchasing Behaviour

Authors: C. Chatfield, A.S.C. Ehrenberg, and G.J. Goodhardt

Source: Journal of the Royal Statistical Society, Series A, Vol. 129, Part 3, 1966, 317-60

Summary: Two closely related stochastic models are described: The two-parameter NBD or negative binomial distribution model, and the simpler LSD or logarithmic series distribution model.

Model: THE NEGATIVE BINOMIAL DISTRIBUTION (NBD)

Purchasing in Two or More Time-Periods

To analyze stationary consumer purchasing behaviour over successive (equal) periods of time, the following two-parameter model of a compound Poisson form can be postulated:

1. Purchases of a given consumer in successive time-periods should follow a Poisson distribution.

2. The average rates of purchasing of different consumers in the long-run should differ, their distribution being a Γ-distribution with k degrees of freedom.

The distribution of the amounts bought by different consumers in any one time-period should follow a negative binomial distribution with a mean m and an "exponent" k. The probability generating function of the NBD is $(1+m/k - mu/k)^{-k}$, and the proportion of informants who buy 0, 1, 2, 3, ..., etc. units is equal to the successive coefficients of powers of the dummy variable u in the expansion of this expression. The proportion buying r units is given by:

$$P_r = (1 + \frac{m}{k})^{-k} \quad \frac{\Gamma(k+r)}{\Gamma(r+1)\Gamma(k)} \quad (\frac{m}{m+k})^r.$$

The multivariate NBD can be represented by the probability generating function of the distribution of people buying r_i units in the ith period out of t periods of lengths T_i which is

$$\{1 + a \sum_{i=1}^{t} T_i(1-u_i)\}^{-k},$$

where $a = m/k$ and m is the average amount bought in some time-period of "unit" length and u_i are dummy variables; k should be constant for any one stationary brand.

The values of b_R (= proportion of consumers who buy the brand in both periods) and m_R (= average amount bought by repeat buyers per time period) can be expressed by formulae which only involve m and k, the latter being calculated from $b = 1 - (1 + m/k)^{-k}$.

Purchasing in One Time Period

For larger values of the standard deviation (δ), the theoretical value proved to be generally higher than the observed one(s). These discrepancies were extremely regular and systematic. There was always a variance discrepancy of the form $\sigma - s \doteq m$ and an apparently good fit of the NBD is obtained simply in those cases in which the mean m is very small compared with s or σ. Only the phenomenon of "shelving" (a rather sudden drop in the observed frequency of purchase at certain values) seems capable of accounting for the variance discrepancy.

THE LOGARITHMIC SERIES DISTRIBUTION (LSD)
Purchasing Behavior in a Single Time Period

As the parameters k and m of an NBD are made to tend to zero in such a way that their ratio $a = m/k$ tends towards a nonzero limit, the nonzero part of the NBD tends towards the so-called logarithmic series distribution or LSD. This means that the nonzero part of an observed NBD with a low enough k can be approximated by a LSD.

It follows that consumer purchasing data which can be successfully described by the two-parameter NBD can also be described by the one-parameter LSD, together with one other parameter, the proportion of buyers (or of nonbuyers). In the NBD model, the number of zeros (1-b) and the mean m

operate in an interrelated sort of way and lead to relatively complex formulae involving both statistics (e.g., $b = 1 - (1+a)^{-k}$, where $a = m/k$). The LSD provides the possibility of describing consumer purchasing data by two <u>independent</u> parameters, namely one parameter of the LSD (e.g., its mean w) and the now quite separate proportion of nonbuyers $p_0 = 1 - b$. This gives scope for considerable simplification and development in the handling of consumer purchasing data.

The LSD is a particular statistical distribution of probabilities pr for positive numbers r. Thus

$$pr = \frac{-q^r}{r \ln(1-q)}$$

for $r \geq 1$, where q is a convenient way of expressing the single parameter of the distribution. (Logarithms are to base e.)

For $r \geq 1$, $\sum pr = 1$, since $\sum \frac{q^r}{r} = -\ln(1-q)$.

The probabilities p_r are therefore essentially the terms in the expansion of the logarithmic expression $\ln(1-q)$.

In terms of the parameter q, the mean w of the LSD is given by

$$w = \frac{-q}{(1-q) \ln(1-q)}.$$

In consumer purchasing data, the mean w is the average rate of buying per buyer. Thus $w = m/b$ where m is the mean rate of buying per informant and b the proportion of informants who buy.

The purchasing data where the LSD does not fit is the same data for which the NBD itself did not give a good fit, i.e., where there was the "variance discrepancy." For such data the theoretical variance σ^2 of the fitted LSD distribution tends again to be larger than the observed variance s^2 of the truncated nonzero distribution of purchases.

The variance σ^2 of the LSD is given by

$$\sigma^2 = \frac{-q\{1+q/\ln(1-q)\}}{(1-q)^2 \ln(1-q)}.$$

There is again a variance discrepancy of the form $\sigma^2 > s^2$. It appears, however, to behave differently from that for the NBD.

Purchasing in Two or More Time Periods

Since $a_T = T_a$,

$$q_T = \frac{T_q}{1+(T-1)q}, \quad w_T = \frac{Ta}{\ln(1+Ta)}, \quad \text{and} \quad \frac{w_T}{w} = \frac{T \ln(1+a)}{\ln(1+Ta)}.$$

For the range of w mainly found in consumer purchasing data such as $1\cdot5 < w < 20$, we have

$$\frac{w_T-1}{w-1} \doteqdot T^{0\cdot82}.$$

Since the proportion of buyers is given by $b_T = m_T/w_T$, we can write

$$b_T = \frac{T_m}{w_T} = \frac{T_w}{w_T} b \quad \text{and} \quad \frac{b_T}{b} = \frac{\ln(1+Ta)}{\ln(1+a)}.$$

Using the approximate relationship $\frac{w_T-1}{w-1} = T^{0\cdot82}$, we have

$$\frac{b_T}{b} \doteqdot \frac{T_w}{1+(w-1)T^{0\cdot82}}.$$

In the special case of double-length periods the proportion b_R of repeat-buyers is

$$b_R = b \left\{2 - \frac{\ln(1+2a)}{\ln(1+a)}\right\} = b \left\{1 + \frac{\ln(1+q)}{\ln(1-q)}\right\}.$$

The proportion of all the purchases made in a time-period which is accounted for by repeat-buyers is given by

$$\frac{m_R}{m} = q.$$

The average amount w_R bought per repeat-buyer is

$$w_R = \frac{-q^2}{(1-q) \ln(1-q^2)},$$

and the average amounts w_L or w_N bought per lost or new buyer are given by

$$w_L = w_N = \frac{q}{\ln(1+q)}$$

$$\frac{w_R}{w} \doteq 1 \cdot 2$$

for the range $1 \cdot 5 < w < 20$. Under stationary conditions and for values of w which are greater than about 2,

$$w_L = w_N \doteq 1 \cdot 4.$$

The underlying stochastic LSD model is the following:

1. There exists a proportion of "never-buyers."

2. Purchases of any one buyer in successive time-periods follow a Poisson distribution with a certain long-run average, μ say.

3. The long-run average rates of purchasing μ of different "buyers" in the market follow a <u>truncated</u> Γ-distribution, i.e., that the frequency of any particular value μ is given by

    $$(ce^{-\mu/a}/\mu)d\mu, \quad \text{for} \quad \delta \leq \mu \leq \infty.$$

    Here δ is some very small number, c is a constant chosen so that

    $$\int_{\delta}^{\infty} (ce^{-\mu/a}/\mu)d\mu = 1,$$

    and a is a parameter of the distribution.

Subject: Brand Choice

Title: The Beta-Binomial Model for Consumer Purchasing Behaviour

Authors: C. Chatfield and G.J. Goodhardt

Source: Applied Statistics, Vol. 19, No. 3, 1970, 240-48

Summary: A model for consumer purchasing behaviour is proposed
          which is based on the beta-binomial distribution.
          The model is tested for several nondurable consumer
          products, and the results give insight into discrepancies
          from a previously used model based on the negative
          binomial distribution (NBD).

Model: The NBD/LSD model of consumer purchasing has in the past
          proved to be of considerable practical application in
          marketing, particularly in the analysis of repeat purchasing
          behaviour. However, in some instances, particularly when
          heavily purchased products are involved, the NBD model fails
          to describe the detailed purchasing behaviour in one period
          (the variance discrepancy).

          The beta binomial distribution (BBD) model was
          developed in an attempt to deal with this variance
          discrepancy.

          We consider first the implications of the Poisson part
          of the NBD model on the number of weeks in which a particular
          consumer makes at least one purchase. Suppose this
          consumer's personal Poisson distribution is such that she
          makes an average of λ purchases per week. Then the
          probability that she makes at least one purchase in any
          one week is $1-e^{-\lambda} = p$ (say) and this is independent of
          purchases in any other week. Thus the number of weeks
          out of n in which this consumer makes at least one purchase
          will be binomially distributed with parameters n and p.
          However, the value of p will vary from consumer to consumer.

          The distribution of $1-e^{-\lambda}$ when λ follows a gamma
          distribution proved mathematically unrewarding and so we
          tried the beta distribution:

          $$f_B(p) = \frac{p^{\alpha-1}(1-p)^{\beta-1}}{B(\alpha,\beta)}, \qquad 0 \le p \le 1,$$

where α, β are positive constants and

$$B(\alpha, \beta) = \int_{0}^{1} x^{\alpha-1}(1-x)^{\beta-1} dx.$$

Just as the Poisson distribution can be obtained as a limit of the binomial distribution, the gamma distribution can be obtained from the beta distribution by putting $p(\beta-1) = z$ and letting $\beta \rightarrow \infty$, $p \rightarrow 0$ in such a way that $z \rightarrow$ constant.

Thus the NBD can be regarded as an unbounded limiting form of the distribution obtained by compounding a binomial distribution with a beta distribution.

This suggested the following model for consumer purchasing behaviour:

1. The probability, p, that a given consumer will buy at least one unit of the brand in a particular week is constant and independent of previous purchases. Thus, in a time-period of n weeks, the number of weeks in which the consumer buys at least one unit will follow a binomial distribution with parameters n and p.

2. The probability, p, varies from consumer to consumer, and has a beta distribution in the whole population.

An immediate consequence of these two assumptions is that the distribution of weeks will be described by a beta binomial distribution. The proportion of the population who buy on exactly r out of n weeks is given by

$$P(r) = \int_{0}^{1} f_B(p) \binom{n}{r} p^r (1-p)^{n-r} dp$$

$$= \frac{\binom{n}{r} B(\alpha+r, n+\beta-r)}{B(\alpha, \beta)}, \quad r = 0, 1, \ldots.$$

Like the NBD the BBD can be derived in several ways
and is called the negative hypergeometric or Polya
distribution in addition to the beta binomial or compound
binomial distribution.

The mean and variance of the BBD are given by

$$\mu = \frac{n\alpha}{\alpha+\beta}$$

and

$$\sigma^2 = \frac{n\alpha\beta(n+\alpha+\beta)}{(\alpha+\beta)^2(1+\alpha+\beta)}.$$

The distributions of weeks can be divided into two
general categories. First, those for which $P(n)$ is
"small," so that the BBD decreases monotonically as r
goes from 0 to n. Secondly, those for which $P(n)$ is
relatively large, either because the distribution is
U-shaped with $P(n) > P(n-1)$, or because the distribution
exhibits a plateau or shelf with the probabilities
decreasing very slowly for large values of r.

The value of the parameter β provides an easy way
of distinguishing between the two types of distribution.
Distributions of the first type give a value of β bigger
than about 2, whereas those of the second type give a
value less than about 2. However, no clear-cut classification
is possible. When the distribution exhibits a plateau
effect the value of β usually lies between 1 and 2, but
when the distribution is U-shaped, we have $P(n) > P(n-1)$,
which is true when

$$\beta < 1 - \frac{(1-\alpha)}{n}$$

$$< 1 \quad \text{when} \quad \alpha < 1.$$

It has been found that for the second type of distribution,
the corresponding distribution of purchases gives a variance
discrepancy when fitted by a NBD. But for the first type of

distribution, the NBD gives a good fit to the corresponding distribution of purchases.

For lightly bought brands, the distribution of weeks will have a "small" value of $P(n)$ and both the NBD and BBD models will apply. But for heavily bought products, the distribution of weeks will have a "large" value of $P(n)$ and the corresponding distribution of purchases will exhibit the shelving effect. Because of this, the NBD variance discrepancy occurs.

THE MULTIVARIATE BBD

Suppose there are t time-periods of length n_i weeks ($i = 1, \ldots, t$). Then the proportion of consumers buying in exactly r_i weeks in the ith period is given by

$$P(r_1, \ldots, r_i, \ldots, r_t) = \frac{B\{\alpha + \sum_i r_i, \beta + \sum_i (n_i - r_i)\}}{B(\alpha, \beta) \Pi_i \{(n_i + 1) B(r_i + 1, n_i - r_i + 1)\}},$$

$$i = 1, \ldots, r; \quad r_i = 0, 1, \ldots, n_i.$$

If the time-periods are divided into the first s and the remaining $t-s$, then the conditional distribution of $(r_{s+1}, \ldots, r_t)$ given values of $(r_1, \ldots, r_s)$ is itself a multivariate BBD with parameters

$$\alpha' = \alpha + \sum_{i=1}^{s} r_i, \quad \beta' = \beta + \sum_{i=1}^{s} (n_i - r_i).$$

PREDICTIONS FROM THE BBD MODEL

If a BBD has been fitted to the distribution of weeks obtained from the sampled period, then the BBD parameters, α, β are such that

$$1 - b = B(\alpha, n + \beta)/B(\alpha, \beta).$$

Under stationary conditions,

$$1 - b_T = B(\alpha, T + \beta)/B(\alpha, \beta).$$

In two consecutive equal time-periods of n weeks
the population can be divided into four subgroups. A
"repeat" buyer buys in both periods, a "lost" or "lapsed"
buyer in period I but not in period II, a "new" buyer
in period II but not in period I, and a nonbuyer in
neither. The subscripts L, N and R refer to lost, new
and repeat buyers.

Under stationary conditions

$$b_N = b_L = b_2 - b = b - b_R.$$

Denoting by m_N^* the mean number of weeks in period II in
which a purchase was made by a new buyer averaged over
the whole population we obtain

$$m_N^* = \int_o^1 f_B(p)(1-p)^n \, np \, dp$$

$$= nB(\alpha+1, \, n+\beta)/B(\alpha,\beta)$$

$$= m_L^*.$$

By subtraction we have

$$m_R^* = \mu - m_N^*.$$

The repeat buying formulae can be obtained directly from
the multivariate BBD with $t = 2$ and $n_1 = n_2 = n$.

Subject: Brand Choice

Title: A Stochastic Interpretation of the Heavy Half

Author: Donald G. Morrison

Source: Journal of Marketing Research, Vol. 5, May 1968, 194-98

Summary: Purchase concentration is sufficiently skewed to
           consider the heavy users as a distinct market segment.
           Models based on equal average consumption and a model
           allowing consumers to have different average consumption
           rates are developed.

Model:

Heavy-Half Proportions Under Identical Average Consumption

X_i = ith person's consumption

$E[X]$ = average consumption for all people

$E[XHH]$ = average consumption given that the person is in
           the heavy half

$E[XLH]$ = average consumption given that the person is in
           the light half

ξ = proportion of total consumption accounted for by
           the heavy half

General Formula for Heavy-Half Consumption

$$\xi = \frac{E[X|HH]}{2E[X]} \tag{1}$$

Uniform Distribution

$$f(x) = \frac{1}{b} \quad \text{for} \quad 0 \leq x \leq b$$

$$E[X|HH] = \frac{3}{4} b \tag{2}$$

$$\xi_{U(0,b)} = .750$$

Exponential Distribution

$$f(x) = \frac{1}{\lambda} \exp\left(- (1/\lambda)x\right) \quad \text{for} \quad x \geq 0$$

$$\xi_{exp} = \frac{1.692\lambda}{2\lambda}$$

$$= .846 \tag{3}$$

Normal Distribution

$$f(x) = \frac{1}{2\pi\sigma} \exp\left(-\frac{1}{2} \frac{(x-\mu)^2}{\sigma^2}\right) \quad -\infty < x < \infty$$

$$E[X|HH] = \mu + \sqrt{\frac{2}{\Pi}} \sigma$$

$$\xi_{nor} = .500 + .399 \frac{\sigma}{\mu} \tag{4}$$

Equation (4), the expression for ξ_{nor}, should be a reasonable approximation for most distributions that have a known mean and variance.

Heavy-Half Proportions Allowing for Different Average Consumption

True heavy-half proportion:

$$\xi_{real} = .500 + .4 \frac{\sigma_\mu}{\mu} \tag{5}$$

Apparent heavy-half proportion:

$$\xi_{apparent} = .500 + .4 \frac{\sqrt{\sigma_\mu^2 \quad \sigma_\varepsilon^2}}{\mu} \tag{6}$$

Since $\sigma_\varepsilon^2 > 0$, the apparent heavy half will always be greater than the true heavy half.

Effect of time on the random component variance:

$$\xi_{apparent} = .500 + .399 \sqrt{\frac{t^2\sigma_\mu + t\sigma_\varepsilon^2}{t\mu}} \tag{7}$$

Subject: Brand Choice

Title: A Dual-Effects Model of Brand Choice

Author: J. Morgan Jones

Source: Journal of Marketing Research, Vol. 7, No. 4, November 1970, 458-64

Summary: A model which accounts for both purchase event feedback and change due to an external effect. Its two most closely related antecedents are the linear learning model [#7] and the stationary probability diffusion model [#11].

Model: THE LINEAR LEARNING MODEL

Kuehn [#7] proposed a model of consumer brand choice behavior in which the probability of purchase at purchase occasion n was a linear function of the probability at occasion n-1 and the outcome of the purchase (brand selected) at n-1. In symbols:

$$p_n = \begin{cases} \alpha + \beta + \lambda p_{n-1} & \text{if Brand A is purchased at n-1} \\ \\ \alpha + \lambda p_{n-1} & \text{if any other brand is purchased at n-1.} \end{cases} \quad (1)$$

THE PROBABILITY DIFFUSION MODEL

Montgomery [#11] developed a stationary probability diffusion model from a hypothetical construct similar to Coleman's Latent Markov Model [#9]. Assume that each consumer has a number, N, of hypothetical elements, and each element is associated with only one of two possible responses, A and B, and the elements are randomly changing allegiance, the mechanism for change being a stationary birth-death process with:

$$\lambda_i = (N - i)[\alpha + \gamma^i]$$
$$\mu_i = i[\beta + \gamma(N - i)] \quad (2)$$

where α, β, and γ are constants.

If Response A represents a purchase of Brand A and Response B a purchase of any brand other than A, a binary

choice model results, if the probability of Response A can
be described. This probability is:

$$p(t) = i/N,$$ (3)

the proportion of elements associated with Response A at
some time t. If $N \to \infty$, the model changes to a probability diffusion model.

As an alternative to the study of $p(t)$, Montgomery
developed the expected probability of purchase at time t:

$$E[p(t)] = m(t).$$ (4)

For a case with many consumers, each with the same initial
probability of purchase, their average probability of
purchase at some time t would be approximately $m(t)$.
Thus, $m(t)$ provides a good measure of how consumers are
reacting, on the average, and Montgomery showed that:

$$m(t) = p_j(t_o)e^{-(\alpha+\beta)(t-t_o)}
+ \frac{\alpha}{\alpha+\beta} [1 - e^{-(\alpha+\beta)(t-t_o)}]$$ (5)

where:

    t_o is the time at which the model was first applied
       (initial time)

$p_j(t_o)$ is the initial probability of purchase for
       Individual J.

This model proposes the change independent of purchase
events.

A PROBABILITY DIFFUSION MODEL WITH PURCHASE EVENT FEEDBACK

The new model to be discussed will be a generalization
of Montgomery's evolutionary model, and will be designed to
account for changes due to both purchase event feedback
and an external effect.

Let the propensities for change in the birth-death process be altered following purchases. For example,

$$\lambda_i = (N - i)[\alpha_n + \gamma^i]$$
$$\mu_i = i[\beta_n + \gamma(N - i)], \tag{6}$$

where n is the index denoting the nth purchase.

There are many mechanisms by which α_n and β_n might change. One of the simplest is to propose that α_n changes only following purchases of Brand A and that β_n changes only following purchases of Brand B. In particular, let:

$$\alpha_n = \begin{cases} \alpha_{n-1} + \lambda & \text{if Brand A was purchased at } n \\ \alpha_{n-1} & \text{if Brand B was purchased at } n \end{cases} \tag{7}$$

and

$$\beta_n = \begin{cases} \beta_{n-1} & \text{if Brand A was purchased at } n \\ \beta_{n-1} + \phi & \text{if Brand B was purchased at } n. \end{cases} \tag{8}$$

Thus, for any given value of i, the propensity for one more elements to be associated with Response A changes following each purchase of Brand A. Furthermore, from the nature of the change, each purchase will effect the same amount of change in the propensity for future transfer of elements.

With the exception of these changes, the development of the dual-effects model parallels that of the Montgomery model. In particular, let $N \to \infty$ remove the jerky behavior from $p(t)$, and then study the expected value of $p(t)$, again denoted $m(t)$, because $p(t)$ evolves randomly. For this model it can be shown that $m(t)$ will be an exponential function in the times between purchases, $t_{n-1} \leq t < t_n$.

The dual-effects model allows for evolutionary behavior in the absence of purchases, while effecting a change in the expected probability of purchases following purchase events.

COMPARISON OF THE THREE MODELS

    (a) For an individual consumer.

        The dual-effects model can:

        1. Evolve through time as a result of a change in the market environment at t_o;

        2. Account for purchase event feedback by changing the (mean) probability of purchase for $t > t_n$.

    (b) For a population of consumers.

        The population-wide mean value function for the linear learning model is

$$M(t_{n+1}) = \phi^{n+1} M(t_o) + (1 - \phi^{n+1})s \qquad (10)$$

where

$\phi = \beta + \lambda$; $s =$ long-run expected brand share

$$= \frac{\alpha}{1 - (\beta + \lambda)}$$

and $M(t) =$ mean probability of purchase for the entire consumer population.

For the probability diffusion model:

$$M(t) = E[m(t)] = M(t_o)e^{-(\alpha+\beta)(t-t_o)}$$
$$+ \frac{\alpha}{\alpha+\beta} [1 - e^{-(\alpha+\beta)(t-t_o)}] . \qquad (11)$$

Both functions (11 and 12) are monotonic. However, suppose that there is a market with two brands, one of which is preferred in terms of purchase event feedback. If there is a change in market environment which enhances the appeal of the less-preferred brand, two opposing forces will occur in the market.

Whereas a model whose overall mean value function was monotonic would be a poor choice, the dual-effects model can represent this situation very well. Although it is very difficult to describe the model's population-wide mean value function, since it is a weighted sum of exponential functions, the model's ability to treat two (perhaps opposing) influences can be readily seen. References: #7, 9, 11.

Subject: Brand Choice

Title: A Stochastic Model for Adaptive Behavior in a Dynamic Situation

Author: J. Morgan Jones

Source: Management Science, Vol. 17, No. 7, March 1971, 484-97

Summary: A stochastic model of adaptive behavior in a dynamic situation is developed which describes the probability of choosing a certain brand as a function of time and the purchase history since some (arbitrary) time origin. The model is developed from a hypothetical construct of behavior, and, in the limit, becomes a nonstationary probability diffusion process.

Model: Suppose there are a number, N, of elements associated with each household. Each household has two possible responses, A and B. At any time, each of the N elements is associated with one of the two responses. Let $p(t)$ be the probability of response A occurring at time t, and let $p(t) = i/N$.

The nonstationary probability mechanism is based upon the following axioms:

1. The probability of transition of the random variable from state i to state $i+1$ in a small time interval h is $\int_{t}^{t+h} \lambda_i(x)\,dx + o(h)$.

2. The probability of transition of the random variable from state i to state $i-1$ in a small time interval h is $\int_{t}^{t+h} \mu_i(x)\,dx + o(h)$.

3. The probability of transition of the random variable to states other than $i-1$, i, or $i+1$ is $o(h)$.

4. The transition intensities, $\lambda_i(t)$ and $\mu_i(t)$, are nonstationary. Furthermore, for all $h > 0$

$$\lim_{h \to 0} (\int_{t}^{t+h} \lambda_i(x)\,dx)/h = \lambda_i(t); \quad \lim_{h \to 0} (\int_{t}^{t+h} \mu_i(x)\,dx)/h = \mu_i(t).$$

The two desired effects can be achieved by appropriately choosing $\lambda_i(t)$ and $\mu_i(t)$. These in turn can be determined by specifying the behavior of the elements associated with the responses A and B. If an element is associated with response A, we assume the propensity for any single element to change association is the sum of two attractions. First, the element has a nonstationary attraction to change, $\alpha(t)$. Second, the elements already associated with response A provide a stationary attraction which is proportional to their number, γi. Since there are N-i elements associated with response B, and each element has a propensity to change of $\alpha(t) + \gamma i$, we have

$$\lambda_i(t) = (N - i)[\alpha(t) + \gamma i].$$

By similar reasoning, assume the self-propensity of an element to change its association from response A to response B is the nonstationary quantity $\beta(t)$. Also assume the attractive force has the same constant of proportionality, γ. We then have

$$\mu_i(t) = i[\beta(t) + (N-i)\gamma].$$

Finally, we assume that $\alpha(t)$ and $\beta(t)$ are step functions, and constant between purchase occasions. Since the derivation of the probability distribution of $p(t)$ over the population is quite difficult, we derive the mean value function, $m(t)$, making use of the identity $E[Y] = E[E[Y|X]]$, where the inner expectation is over Y and the outer expectation over X. In particular,

$$m(t+h) - m(t) = E[E[p(t+h) - p(t)|p(t)]].$$
$$= E[(1/N) \int_t^{t+h} \{\lambda_i(x) - \mu_i(x)\}dx + o(h)],$$

with the expectation being taken over all values of $p(t)$. If h is sufficiently small, $t_k \leqq t + h < t_{k+1}$,

$$m(t+h) - m(t) = E[(1/N)\{\lambda_i(t) - \mu_i(t)\}h + o(h)].$$

Dividing both sides by h and taking the limit as $h \to 0$, we have

$$dm(t)/dt = E[(\lambda_i(t) - \mu_i(t))/N] \quad \text{for} \quad t_k \leqq t < t_{k+1}.$$

A NONSTATIONARY PROBABILITY DIFFUSION MODEL OF CONSUMER BEHAVIOR

If we let $N \to \infty$, the model changes from a nonstationary birth-death process to a nonstationary probability diffusion process.

$$\lim_{N\to\infty} E[(\lambda_i(t) - \mu_i(t))/N] = \alpha(t) - m(t)[\alpha(t) + \beta(t)].$$

Thus

$$dm(t)/dt = \alpha(t) - m(t)[\alpha(t) + \beta(t)].$$

The solution, given the initial condition $m(t) = m(t_k)$ at $t = t_k$ is

$$m(t) = m(t_k) \exp (-[\alpha(t) + \beta(t)](t - t_k))$$

$$+ \alpha(t)[1 - \exp(-[\alpha(t) + \beta(t)](t - t_k))]/[\alpha(t) + \beta(t)]$$

for $t_k \leqq t < t_{k+1}$.

$m(t)$ can be expressed as a function of $m(t_0)$ at any time since Theorem 1 is proved.

Theorem 1. $m(t)$ is continuous from the left at $t = t_k$ for $k = 1,2,3, \ldots$.

Furthermore, it has been shown that $m(t)$ is an exponential function for $t \, \varepsilon \, [t_k, \, t_{k+1}]$. Thus, we see that in terms of $m(t)$ the model has the desired properties. Namely:

1. The mean probability of purchase changes in the absence of purchases; and

2. The mean probability of purchase increases (or
    decreases) following purchase events.

For a derivation of the <u>overall</u> mean value function let
$m(t|B)$ denote the mean value function of Category B of
the population, and $m(t|A)$ the mean value function of
Category A. Then

$$m(t) = m(t|B) \, \Pr[H_1 = B] + m(t|A) \, \Pr[H_1 = A],$$

where $m(t)$ is the mean value function for the population
as a whole, and H_1 is a variable denoting the purchase
history after one purchase. Letting $r = \Pr[H_1 = B]$, we
have

$$m(t) = m(t|B)r + m(t|A)(1-r).$$

Theorem 2. If the probability of purchase of Brand A is
independent of the interpurchase time for all consumers,
the expected value of the market share at time t is
the mean value function of $F(p(t))$, the distribution of
the purchase probabilities at time t. The theorem
is proved.

For a study of how purchase event feedback mechanisms
affect the probability of purchase of any consumer we again
observe $m(t)$ since $p(t)$ is a random variable.
Let

$$\left. \begin{aligned} a_k &= \alpha(t) + \beta(t) \\[2mm] b_k &= \alpha(t)/[\alpha(t) + \beta(t)] \end{aligned} \right\} \quad \text{for} \quad t_k \leqq t < t_{k+1}.$$

Then

$$m(t) = m(t_k)e^{-a_k(t-t_k)} + b_k[1 - e^{-a_k(t-t_k)}].$$

If purchase event feedback does not occur on the kth purchase,

$$\text{pr}(\tau) = m(t_k)e^{-a_{k-1}\tau} + b_{k-1}[1 - e^{-a_{k-1}\tau}]$$

where

$$\tau = t - t_k \quad \text{and}$$

$\text{pr}(\tau) = m(t)$ before the most recent purchase.

If feedback does occur, we have

$$m(\tau) = m(t_k)e^{-a_k\tau} + b_k[1 - e^{-a_k\tau}].$$

Theorem 3. If either

(a) $a_k > a_{k-1}$, $b_k > b_{k-1}$ and $m(t_k) \leqq b_k$, or

(b) $a_k < a_{k-1}$, $b_k > b_{k-1}$ and $m(t_k) \geqq b_{k-1}$

then $m(\tau) > \text{pr}(\tau)$ for all $\tau > 0$. Similarly, if either

(c) $a_k > a_{k-1}$, $b_k < b_{k-1}$ and $m(t_k) \geqq b_k$, or

(d) $a_k < a_{k-1}$, $b_k < b_{k-1}$ and $m(t_k) \leqq b_{k-1}$,

then $m(\tau) < \text{pr}(\tau)$ for all $\tau > 0$.

In the absence of external influence the current model is very similar to the linear learning model. It differs only in that the probability of purchase is discontinuous at times of purchase in the linear learning model, while it is a continuous exponential approximation to this discontinuity in the current model. The Montgomery model [#11] is a special case of the current model.

Reference: #11.

Subject: Brand Choice

Title: The New-Trier Stochastic Model of Brand Choice

Author: David A. Aaker

Source: Management Science, Vol. 17, No. 5, April 1971,
           B-435-450

Summary: The new-trier model, a description and empirical evaluation
           of a new nonstationary, heterogeneous, stochastic model of
           brand choice, models the purchasing process following the
           first purchase of an unfamiliar (to the purchaser) brand.

Model: The new-trier model is a brand-choice stochastic model of
           new-trier purchasing behavior.   The purchases are
           described by a binary 0-1 number indicating that the new
           trier repurchased the new brand (1) or that he purchased
           another brand (0).   The model focuses on the probability
           of repurchasing the new brand on subsequent purchase
           occasions.

           A trial period immediately follows the first purchase
           of the brand, culminating in a brand decision.   The
           trial period includes c-1 purchases beyond the first
           purchase, where c is an integer random variable assumed
           to be distributed geometrically:

$$\{c\} = \tau(1-\tau)^{c-1} \qquad c = 1,2,3,\ \ldots \qquad (1)$$

           where the braces denote a probability distribution. During
           the trial period the new trier is assumed to follow a
           Bernoulli model with a parameter p_b.   At the end of the
           trial period, an abrupt change in purchase probability
           is permitted to occur. After the trial period, the buyer
           is assumed to follow another Bernoulli model with
           parameter p_a.   The model distributes p_b and p_a over
           the population, using two independent beta distributions,
           the parameters of which are BR and BN (before decision)
           and AR and AN (after decision).   The model, as it is now
           developed, is similar to Howard's dynamic inference
           model [#14].

Introducing the decay of p_a, the model includes a rejection probability--a probability that $p_a = 0$. The p_a beta distribution is employed only if rejection has not occurred--that is, if $p_a > 0$. Further, the cumulative rejection probability is permitted to increase geometrically (with parameter γ) in time from an initial value, ϕ, to a final value, α. Let n index discrete time, and $p(n)$ be the new-trier's probability of purchasing the brand at time n. Time is defined here to be an integer value corresponding to purchase occasions. The origin ($n = 0$) is the first purchase of the new brand. Note that $p(n)$ is p_b when $n < c$ and p_a when $n \geq c$. If $\{p(n)\}$ = the probability distribution of $p(n)$, then:

$$\{p(n)\} = \{\Gamma(BN)[p(n)]BR^{-1}[1 - p(n)]BN - BR^{+1}\}/\{\Gamma(BR)\Gamma(BN - BR)\}$$

$$0 < p(n) < 1; \quad n = 1,2, \ldots, c-1$$

$$= [1 - \alpha + (\alpha + \phi)\gamma^{n-c}] \cdot \qquad (2)$$

$$\{\Gamma(AN)[p(n)]AR^{-1}[1 - p(n)]AN - AR^{+1}\}/[\Gamma(AR)\Gamma(AN - AR)]$$

$$0 < p(n) < 1; \quad n = c, c+1, \ldots$$

$$= \phi + (\alpha - \phi)(1 - \gamma^{n-c}) \qquad p(n) = 0; \quad n = c, c+1, \ldots.$$

The model is compared to Montgomery's probability diffusion model [#11] and tested empirically. Two special cases of the model are also considered: (a) decay is eliminated due to successive trials, and (b) the trial consisted only of the use experience following the initial purchase.

References: #11, 14.

Subject: Brand Choice

Title: A Composite Heterogeneous Model for Brand Choice Behavior

Author: J. Morgan Jones

Source: Management Science, Vol. 19, No. 5, January 1973, 499-509

Summary: This model of consumer brand choice behavior is a composite of the Bernoulli, Markov, and Linear Learning Models. The properties of the model are explored, and a parameter estimation technique is developed. A special case of the model is developed for situations in which insufficient data are available to estimate the parameters of the complete model.

Model: Previously proposed brand choice models have always assumed that each consumer in the population obeys the same mechanism of behavior. We propose a composite model which allows each consumer to obey one of three mechanisms: Bernoulli, Markov, and Linear Learning. In addition, we are extending the heterogeneity of the model to include $p(0)$ and the parameters of the models.

The Heterogeneous Bernoulli Model

The model restricts the consumer to a binary choice between Brand A, and Brand B, which is a collection of all brands other than A. It assumes that the probability of purchase for Brand A at time t, $p(t)$, is constant over time and independent of past purchase events.

Let the cumulative probability distribution of p be:

$$F(p) = w_1 \quad \text{if } p = 0,$$
$$= w_1 + (1 - w_1)G(p) \quad \text{if } 0 < p \leqq 1, \tag{1}$$

where

$$0 \leqq w_1 \leqq 1 \tag{2}$$

and

$G(p)$ is a cumulative Beta distribution,

$$= \frac{\Gamma(a_1 + b_1)}{\Gamma(a_1)\Gamma(b_1)} \int_0^p x^{a_1 - 1} (1 - x)^{b_1 - 1} dx. \tag{3}$$

The mean probability of purchase, $E(p)$, is

$$E[p] = \int_0^1 p\,dF(p) = (1 - w_1)a_1/(a_1 + b_1).$$

The Heterogeneous Markov Model

A binary choice Markov model has the initial unconditional probability of purchase for Brand A, p, and a matrix of conditional probabilities,

$$
\begin{array}{cc}
& A \qquad\quad B \\
\begin{array}{c} A \\ \\ B \end{array} &
\left|
\begin{array}{cc}
\phi_{AA} & \phi_{AB} \\
\\
\phi_{BA} & \phi_{BB}
\end{array}
\right| = \phi,
\end{array}
$$

where ϕ_{ij} is the probability of purchasing Brand j at $t+1$, given Brand i was purchased at t. We will assume stationary transition probabilities.

To incorporate heterogeneity, we assume that p is distributed as Beta with a mass point at zero, as given in equations (1), (2), and (3) above, but with parameters w_2, a_2, and b_2. Further, we reparameterize the transition matrix,

$$
\phi = \left|
\begin{array}{cc}
\gamma + \delta & 1 - \gamma - \delta \\
\\
\gamma & 1 - \gamma
\end{array}
\right|,
$$

and assume that γ and δ are jointly distributed according to a Dirichlet distribution:

$$
\begin{aligned}
f(\gamma,\delta) &= \frac{\Gamma(c+d+e)}{\Gamma(c)\Gamma(d)\Gamma(e)}\ \gamma^{c-1}\delta^{d-1}(1-\gamma-\delta)^{e-1} \\
&\qquad \text{if} \quad \gamma \geq 0,\ \delta \geq 0,\ \gamma + \delta \leq 1, \qquad\qquad (4) \\
&= 0 \quad \text{otherwise},
\end{aligned}
$$

which is independent of the value of p.

$$E[\pi] = c/(c+e).$$

The Heterogeneous Linear Learning Model

This binary choice model has the initial probability of purchase, p, and the laws for changing $p(t)$ following a purchase:

$$p(t+1) = \alpha + \beta + \lambda p(t) \quad \text{if Brand A was purchased at time } t,$$

$$= \alpha + \lambda p(t) \quad \text{if Brand B was purchased at time } t.$$

To account for heterogeneity, we again assume p is distributed as Beta with a mass point at zero, as in equations (1), (2), and (3), with parameters w_3, a_3 and b_3. In addition, and independent of the value of p, we assume that α, β and λ are Dirichlet distributed with

$$f(\alpha,\beta,\lambda) = \frac{\Gamma(j+k+m+n)}{\Gamma(j)\,\Gamma(k)\,\Gamma(m)\,\Gamma(n)} \; \alpha^{j-1}\beta^{k-1}\lambda^{m-1}(1-\alpha-\beta-\lambda)^{n-1}$$

$$(5)$$

$$\text{if} \quad \alpha \geq 0, \; \beta \geq 0, \; \lambda \geq 0, \text{ and } \; \alpha + \beta + \lambda \leq 1,$$

$$= 0 \quad \text{otherwise.}$$

The long-run mean probability of purchase is

$$\mu = \alpha/(1 - (\beta + \lambda))$$

and the expected mean probability of purchase is

$$E[\mu] = j/(j + n).$$

A Composite, Heterogeneous Model of the Population

The models which have been developed above can be combined into a single model for the population by assuming that a proportion z_1 of the population exhibits heterogeneous Bernoulli behavior, a proportion z_2 exhibits heterogeneous Markov behavior, and the rest of the population $1 - z_1 - z_2$ exhibits heterogeneous linear learning behavior. Thus, we have developed a composite, heterogeneous model for the population. In this model, two families need not exhibit the same behavioral mechanism, and even if they do they may have no values (such as p) in common.

PARAMETER ESTIMATION TECHNIQUE

For the purchase string AB we have

$P(AB|p$ and Bernoulli Model$) = p(1-p)$,

$P(AB|p,$ and a Markov Model with γ and $\delta)$

$$= p(1 - \gamma - \delta),$$

and

$P(AB|p,$ and a Linear Learning Model with α, β and $\lambda)$

$$= p(1 - \alpha - \beta - \lambda p).$$

Then we obtain

$$P(AB|\text{Bernoulli mech.}) = (1 - w_1) [\frac{a_1}{a_1 + b_1} - \frac{a_1(a_1 + 1)}{(a_1 + b_1)(a_1 + b_1 + 1)}],$$

$$P(AB|\text{Markov mech.}) = (1 - w_2) \cdot \frac{a_2}{a_2 + b_2} \cdot \frac{e}{c + d + e},$$

and

$$P(AB|\text{Linear Learning mech.}) = (1 - w_3) \{\frac{a_3}{a_3 + b_3} - \frac{a_3}{a_3 + b_3} \cdot \frac{j}{j+k+m+n}$$

$$- \frac{a_3}{a_3 + b_3} \cdot \frac{k}{j+k+m+n} - \frac{a_3(a_3 + 1)}{(a_3 + b_3)(a_3 + b_3 + 1)} \cdot \frac{m}{j+k+m+n}\}.$$

Finally,

$$P(AB) = z_1 P(AB|\text{Bernoulli mechanism}) + z_2 P(AB|\text{Markov machanism})$$

$$+ (1 - z_1 - z_2) P(AB|\text{Linear Learning mechanism}).$$

Analogous techniques can be used to develop theoretical purchase probabilities for longer purchase strings; these can be inserted in a goodness-of-fit expression.
Estimations of the long-run overall mean probability of purchase are performed.

Subject: Brand Choice

Title: An Entropy Model of Brand Purchase Behavior

Author: Jerome D. Herniter

Source: Journal of Marketing Research, Vol. 10, November 1973, 361-75

Summary: A probabilistic model of consumer purchase behavior for frequently purchased, low cost items. The concept of maximum entropy is used to specify the model. The only empirical data required are market shares; all other brand selection statistics, such as repeat and switch rates, are derived quantities.

Model:

There are two interpretations of entropy. The first is that maximum entropy is an inherent characteristic of a probabilistic system at equilibrium. The second interpretation states that by maximizing entropy one obtains the only unbiased encoding of our information concerning the system. Both interpretations lead to the same mathematics and consequently to the same results.

Entropy of a System

The entropy of a system consisting of n possible states is:

$$S = -k \sum_{i=1}^{n} p_i \log p_i$$

where p_i is the probability the system is in the state i. Consider a two-brand market (brands 1 and 2) and a customer whose preference for brand 1 is α. When purchase is made, there are two possible states: a purchase of brand 1 or a purchase of brand 2. Let $b = i$ be the event that brand i is purchased.

The entropy of the system is:

$$S = -[p(b = 1) \ln p(b = 1) + p(b = 2) \ln p(b = 2)].$$

Assuming the probability of purchasing a brand is numerically equal to the preference for the brand.

$$S = -[\alpha \ln \alpha + (1 - \alpha) \ln (1 - \alpha)].$$

The customer is selected at random from a population. There is a distribution of α over the population, $f(\alpha)$. The state of this system is defined by the customer's parameter value, α, and the brand purchased; there are an infinite number of states. The entropy of the system is:

$$S = -\int_0^1 f(\alpha) \ln f(\alpha) \, d\alpha - \int_0^1 f(\alpha) \, [\alpha \ln \alpha$$

$$+ (1 - \alpha) \ln (1 - \alpha)] d\alpha.$$

The problem now becomes one of determining the distribution $f(\alpha)$ which yields the maximum entropy of the system.

For the two-brand case:

$$f(\alpha) = \alpha^{-\alpha} (1 - \alpha)^{-(1-\alpha)} e^{-\lambda}$$

$\lambda = .5165$. The entropy of this system may be shown to be $S = \lambda$ for this example.

Brand Selection

Assume a marketing model that deals exclusively with brand selection under "stable" market conditions. (The effects of advertising, price, and purchase timing are not discussed here.) The market used is a compound multinomial model with heterogeneous populations, assumed to be known and unchanging. The customers have preferences for any combination of N brands.

In a two-brand market there are three categories: (1) customers who have a preference only for brand 1, (2) customers who have a preference only for brand (2), and (3) customers who have preferences for both brands.

Let i = index of the category

C_i = event a customer is in category i

g_i = probability a customer is in category i,
$$p(C_i) = g_i$$

$b(n)$ = a random variable whose value is 1 if the customer purchases brand 1 on the nth purchase occasion and 2 if she purchases brand 2

α = parameter which defines the customer's preference for brand 1 if she is in category 3

$f(\alpha)$ = distribution of α, $0 \leq \alpha \leq 1$

m_1 = market share of brand 1

m_2 = market share of brand 2.

We wish to obtain the maximum entropy subject to the following constraints:

$$\sum_{i=1}^{3} g_i = 1 \tag{1}$$

$$\int_0^1 f(\alpha)d\alpha = 1. \tag{2}$$

The market share constraints are:

$$p[b(n) = 1] = g_1 + g_3 \int_0^1 \alpha f(\alpha)d\alpha = m_1 \quad \text{and} \tag{3}$$

$$p[b(n) = 2] = g_2 + g_3 \int_0^1 (1 - \alpha) f(\alpha)d\alpha = m_2. \tag{4}$$

Results:

$$f(\alpha) = \alpha^{-\alpha}(1 - \alpha)^{-(1-\alpha)} e^{-(\mu+\xi_2)} e^{-\alpha(\xi_1-\xi_2)}. \tag{5}$$

$$g_3 = e^{-\eta+\mu} \tag{6}$$

$$g_1/g_2 = e^{-(\xi_1 - \xi_2)} \tag{7}$$

$$g_2/g_3 = e^{-(\mu + \xi_2)}. \tag{8}$$

From (7) and (8) it follows that:

$$f(\alpha) = \frac{1}{g_3} \alpha^{-\alpha} (1 - \alpha)^{-(1-\alpha)} g_2^{(1-\alpha)} g_1^{\alpha}. \tag{9}$$

When $\alpha = 0$, $f(\alpha) = g_2/g_3$. Similarly, when $\alpha = 1$, $f(1) = g_1/g_3$. The normalization constraints become:

$$\left(\frac{g_2}{g_3}\right) \int_0^1 \alpha^{-\alpha} (1 - \alpha)^{-(1-\alpha)} \left(\frac{g_1}{g_2}\right) d\alpha = 1, \tag{10}$$

$$\sum_{i=1}^{3} g_i = 1, \tag{11}$$

and for the brand 1 market share constraint:

$$g_1 + g_3(\alpha) = m_1. \tag{12}$$

The maximum entropy for two brands is:

$$S = -m_1 \ln g_1 - m_2 \ln g_2. \tag{13}$$

For an n brand market:

$$S = -\sum_{i=1}^{n} m_i \ln g_i \tag{14}$$

Joint Probabilities and Switching

The joint probability of purchasing brand 1 twice is:

$$p[b(n) = 1, b(n-1) = 1] \tag{15}$$

$$= g_1 + g_3(\alpha^2).$$

The repeat purchase probability for brand 2 is:

$$p[b(n) = 2, b(n-1) = 2] = g_2 + g_3[(1-\alpha)^2]. \quad (16)$$

Probability of switching from brand 1 to brand 2:

$$p[b(n) = 2, b(n-1) = 1] = g_3[\alpha(1-\alpha)]. \quad (17)$$

First-order transition matrix:

$$p[b(n) = 1 \mid b(n-1) = 1] \quad (18)$$

$$= [g_1 + g_3(\alpha^2)]/[g_1 + g_3(\alpha)]$$

$$p[b(n) = 2 \mid b(n-1) = 1] \quad (19)$$

$$= g_3[\alpha(1-\alpha)]/[g_1 + g_3(\alpha)]$$

$$p[b(n) = 1 \mid b(n-1) = 2] \quad (20)$$

$$= g_3[\alpha(1-\alpha)]/[g_2 + g_3(1-\alpha)]$$

$$p[b(n) = 2 \mid b(n-1) = 2] \quad (21)$$

$$= [g_2 + g_3((1-\alpha)^2)]/[g_2 + g_3(1-\alpha)].$$

Store Shopping Model

We are concerned only with store visits in which a purchase is made. The model is obtained by maximizing the entropy of the store shopping system subject to the market share of purchases constraints and to the normalization constraints.

Brand-Store Model

The expected market shares of brands are the same for all stores. The entropy of the system is:

$$S = - \sum_{j=1}^{K} d_j \ln q_j - \sum_{i=1}^{N} m_i \ln q_i \qquad (22)$$

where

q_j = probability a customer is in store category j

d_j = share of purchases made in store j

m_{ji} = brand i's share of purchases in store j.

The transition matrices, joint probabilities, and total switching for brands will be identical with those obtained when store selection is ignored.

If brand selection is completely dependent on store selection, the entropy of the system is:

$$S = - \sum_{j=1}^{K} d_j \ln q_j - \sum_{j=1}^{K} d_j \sum_{i=1}^{N} m_{ji} \ln g_{ji}. \qquad (23)$$

Subject: Brand Choice

Title: A Modified Linear Learning Model of Buyer Behavior

Author: Gary L. Lilien

Source: Management Science, Vol. 20, No. 7, March 1974,
1027-36

Summary: A stochastic model of individual buyer behavior is
developed from a set of postulates about the buying
process. The postulates are shown to imply a
linear learning model modified by a term to explain
response to pricing stimuli. Thus, a customer's
purchasing probability is modelled as a combination of
the effect of his past purchasing behavior plus the
effect of price-variation in the market. Methods are
developed to calculate short- and long-term
probabilistic properties of the process. A method
for parameter estimation is included.

Model:

Kuehn's Simple Linear Learning Model (SLL) [7] assumes that
a customer's past-purchase tendency to buy a brand is a
<u>linear function</u> of his prepurchase tendency. With
respect to a particular brand, j, he can either buy j
at time t, with associated probability $P_j(t)$ or a brand
other than j with probability $1 - P_j(t)$. $P_j(t+1)$ is
then related to $P_j(t)$ as follows:

$$P_j(t+1) = \alpha + \beta + \lambda P_j(t) \quad \text{if} \quad j \text{ is purchased at } t,$$
$$= \alpha + \lambda P_j(t) \quad \text{if} \quad j \text{ is not purchased at } t. \tag{1}$$

Now assume the following:

1. The market breaks down naturally into two sets of
brands, "Premium" and "Standard."

2. Both brands can be characterized by a single price
at a given time.

3. If a particular customer is unaffected by price
variation, then his purchasing behavior can be
explained by the SLL.

4. Changes in price are perceived by a customer immediately,
and price levels only affect customers during the
particular period when they are in effect.

5. Customers act rationally, and they all perceive Premium
as "superior" to Standard.

6. The effect of price variation can be explained totally
by Premium-Standard price difference; actual Premium
(or Standard) price level has no effect.

7. There exists across families some (prior) joint
    distribution of initial probability of Premium
    purchase, P_o, and of price receptivity, C (roughly,
    the fraction of a purchaser's decision determined
    by price), which we will call $F_{C,P_o}(c,p)$.

    Families will be assumed to sample a particular c*
    and p_o^* from this distribution.

8. The parameters of the model, once chosen for a
    particular family, are constant over time.

Let

X_t = 1 = purchase of Premium

X_t = 0 = purchase of Standard

δ_t = Premium price minus Standard price at t

P_t = $Pr\{X_t = 1\}$

C = price consciousness of customer--roughly
    the fraction of his behavior determined
    by price

$\phi(\delta)$ = value of the price response function when
    price difference = δ.

Then our Modified Linear Learning Model (MLL) has the
following form:

$$P_{t+1} = (1-C)(\alpha + \beta X_t + \lambda P_t) + C\phi(\delta_{t+1}). \qquad (2)$$

For a given value of C, there exist limits independent
of δ in which P_t will be contained after finite time
with probability one. Call these limits $L(\infty)$ and $U(-\infty)$
respectively.

$$L(\infty) = [(1-C)\alpha]/[1 - (1-C)\lambda],$$
$$U(-\infty) = [(1-C)(\alpha + \beta) + C]/[1 - (1-C)\lambda]. \qquad (3)$$

If P_t is not conditioned on past events, then it can be
defined as a random variable. If P_o and C are known
for a particular family and the price history $\{\delta_t\}$
has also been recorded, then successive purchases result
merely in successive applications of (2). Then if we
know T (the length of time the process has been in
operation) and the parameters of the model, the value of
P_t, $t \leqq T$, can be calculated directly as described in (2).

We might, however, want to estimate P_t from knowledge of the parameters only, i.e., unconditioned on actual history. To this end, EP_t will develop. First, we calculate $E[P_{t+1}|P_t = p_t]$ which we will denote as $E[P_{t+1}|p_t]$.

Since $P_{t+1} = Pr(X_{t+1} = 1)$,

$$E[P_{t+1}|p_t, \delta_{t+1}] = (1-C)(\alpha + (\beta + \lambda)p_t) + C\phi(\delta_{t+1}). \quad (4)$$

Repeatedly using the law of conditional probabilities, we obtain:

$$E(P_{t+1}|p_0, \Delta_{t+1}) = \sum_{i=0}^{t} [(1-C)(\beta + \lambda)]^i (\alpha(1-C) + C\phi(\delta_{i+1}))$$
$$(5)$$
$$+ [(1-C)(\beta + \lambda)]^{t+1} p_0,$$

where Δ_{t+1} = vector of price differences $(\delta_1, \delta_2, \ldots, \delta_{t+1})$.

If prices were to remain stable, we get

$$E(P_{t+1}|p_0, \delta) = [(1-C)(\beta + \lambda)]^{t+1} p_0 + [\alpha(1-C)$$
$$+ C\phi(\delta)] \sum_{i=0}^{t} [(1-C)(\beta + \lambda)]^i. \quad (6)$$

The limit in (6) as $t \to \infty$ has meaning:

$$\lim_{t \to \infty} E(P_{t+1}|p_0, \delta) = [(1-C)\alpha + C\phi(\delta)]/[1- (1-C)(\beta + \lambda)]$$
$$(7)$$
$$= Q_\infty(\delta, C).$$

(7) now represents the expected long-term market share for Premium purchased by an individual with price receptivity C, independent of p_0.

Reference: #7.

Subject: Buyer Attitude

Title: Analog Experiments with a Model of Consumer Attitude Change

Authors: James M. Carman and Francesco M. Nicosia

Source: L.G. Smith (ed.), Reflections on Progress in Marketing, Proceedings, Chicago: American Marketing Association, 1965, 246-57

Summary: A model of consumer buying behavior that contains four variables and six parameters is described by a system of four linear differential equations.

Model: A Model of Consumer Attitude Change

The model can be described by a system of four linear differential equations:

$$\dot{B} = \frac{dB(t)}{dt} = b[M(t) - \beta B(t)] \tag{1}$$

$$M(t) = mA(t) \tag{2}$$

$$\dot{A} = \frac{dA(t)}{dt} = a[B(t) - \alpha A(t)] + cC(t) \tag{3}$$

$$C(t) = \overline{C} \tag{4}$$

where: $B(t)$ = behavior, i.e., buying at time (t).

$M(t)$ = the consumer's motivation toward a particular brand which results from some variety of complex interactions of various stimuli, not advertising alone, at time (t).

$A(t)$ = the consumer's attitude toward the brand that contributes to $M(t)$. The transformation of an attitude, A, into motivation, M, is postulated to be instantaneous. Changes in B feed back into A through (3).

$C(t)$ = some specific stimulus under study, such as advertising. C is the only exogenous variable defined by this particular model.

$b, \beta, m, a, \alpha, c$ are the "personality" parameters of the equations of the model. These parameters plus $\overline{C}$ are assumed to be constant with respect to time. $B(t)$, $M(t)$, $A(t)$, C are all defined as ≥ 0.

Assume that

$$0 < \frac{1}{\beta} \leq 1 \qquad \text{or} \qquad \beta \geq 1 \qquad\qquad (5)$$

$$0 < \frac{1}{\alpha} < 1 \qquad \text{or} \qquad \alpha > 1$$

$$0 < m < 1$$

$$a, b > 0$$

$$c \gtrless 0.$$

The conditions of equilibrium at time t are

$$\frac{dB(t)}{dt} = 0 \qquad\qquad (6.1)$$

$$\frac{dA(t)}{dt} = 0. \qquad\qquad (6.2)$$

Under the restrictions on the coefficients' values, the stability conditions are

$$b\beta + a\alpha > 0 \qquad\qquad (6.3)$$

$$ab(\alpha\beta - m) > 0 \qquad \text{or} \qquad \beta\alpha > m. \qquad\qquad (6.4)$$

For experiments with the analog computer the model was simplified by substituting (2) in (1) and (4) in (3):

$$\dot{B} = bmA(t) - b\beta B(t) = a_1 A(t) - b_1 B(t) \qquad\qquad (7.1)$$

$$\dot{A} = aB(t) - a\alpha A(t) + c\bar{C} = a_2 B(t) - b_2 A(t) + c_2 \bar{C}. \qquad (7.2)$$

The equations (7) were then simulated with the circuitry of the computer. The solution techniques and experimental runs are described.

Subject: Buyer Attitude

Title: A Behavior Theory Approach to the Relations between Beliefs about an Object and the Attitude Toward the Object

Author: Martin Fishbein

Source: M. Fishbein (ed.), Readings in Attitude Theory and Measurement, New York: John Wiley & Sons, Inc., 1967, 389-400

Summary: Behavior theory models of the relationships between the beliefs about an object and the attitude toward the object.

Model: An individual's attitude toward any object is a function of (1) the strength of his beliefs about the object (i.e., those beliefs in his response hierarchy) and (2) the evaluative aspect of those beliefs (i.e., the evaluation of the associated responses). Algebraically, this is:

$$A_o = \sum_1^N B_i a_i$$

where

A_o = the attitude toward object o,

B_i = the strength of belief i about o, that is, the probability or improbability that o is associated with some other concept x_i,

a_i = the evaluative aspect of B_i, that is, the evaluation of x_i,

N = the number of beliefs about o, that is, the number of responses in the individual's habit-family hierarchy.

This is similar to predictions made by others. For example, M.J. Rosenberg [Cognitive Structure and Attitudinal Affect, Journal of Abnormal and Social Psychology, 53 (1956), 367-72] developed a theory whose central equation may be expressed as follows:

$$A_o = \sum_{i=1}^N I_i V_i$$

where A_o = the attitude toward the object, I_i = the belief or probability that the object will lead to or block the attainment of a given valued state "i." V_i = the "value importance" or the amount of affect expected from valued state "i," and N = the number of beliefs.

The central equation of R.B. Zajonc's theory [Structure of the Cognitive Field. Unpublished doctoral dissertation, University of Michigan, 1954] can be defined as:

$$A_O = Va(o) = \frac{\sum_{i=1}^{N} Va(a) \; Prom(a)}{(N^2 + N)/2}$$

where A_O = attitude toward the object. Va(o) = the valence of the object. Va(a) = the valence of the characteristic "a," Prom(a) = the prominence of characteristic "a," and N = the number of characteristics.

References: #30 and 32.

Subject: Buyer Attitude

Title: A Two-Dimensional Concept of Brand Loyalty

Author: George S. Day

Source: Journal of Advertising Research, Vol. 9, No. 3, 1969, 29-35

Summary: Development of a two-dimensional brand loyalty concept compared to a purely behavioral definition.

Model: The equation satisfies the requirements that the brand loyalty score be brand specific and positively related to the degree of favorableness toward the brand:

$$L_i = \frac{P[B_i]}{kA_i^n} = f(X_a, X_b, \ldots X_j).$$

where:

L_i = the brand loyalty score for the ith buyer of brand m,

$P[B_i]$ = proportion of total purchases of the product that buyers devoted to brand m over the period of the study,

A_i = the attitude toward brand m at the beginning of the study (scaled so that a low value represents a favorable attitude),

$X_a, X_b, \ldots X_j$ = descriptive variables to be fitted to L_i by least squares,

k, n = constants whose values are varied by trial and error to maximize the fit between L_i and $X_a, X_b, \ldots X_j$.

Subject: Buyer Attitude

Title: The Relationship of Brand Attitudes and Brand Choice
           Behavior

Author: George S. Day

Source: George S. Day, Buyer Attitudes and Brand Choice Behavior,
             New York: The Free Press, 1970 , Ch. 2, 48-50

Summary: Stochastic model measuring the joint effects of an
           underlying latent structure which can be partially
           elicited by attitude measures and the state of the
           environment at the time of the purchase.

Model: For convenience we will assume the relationship is

           linear, although any monotonic function would serve

           just as well. Let

$$P\{K|P\}_i = \alpha - \beta A_i + u_i \qquad i = 1, \ldots, N \qquad (1)$$

   where: $P\{K|P\}_i$ = The estimated probability of person i
                       buying brand K, given the purchase of
                       product P, in the fixed time period considered

           A_i = The initial attitude of person i toward brand
                  A. This measure is scaled so that a very
                  favorable attitude has a low score.

           β = f(inhibiting and facilitating effects of
                  the environment, when estimated over all N
                  buyers of the product).   (2)

           u_i = f(stability of attitude, buying style,
                  extraneous determinants of response, nature
                  of the environment encountered, and
                  random error).   (3)

   Equation (3) provides a specific recognition that buyers
with the same initial evaluation may respond differently to
similar environmental influences (deals, price changes,
promotions, displays, out-of-stocks, and so forth) encountered
during the time period being studied. In part, the
differences in response will depend on the stability of
the initial attitude.

Subject: Buyer Attitude

Title: Television Show Preference: Application of a Choice Model

Author: Donald R. Lehmann

Source: Journal of Marketing Research, Vol. 8, February 1971, 47-55

Summary: A model of preferences based on specific product attributes is examined and found to be more powerful than predictions based on personality measures and demographics.

Model: A concept basic to this model is similar to the Lancaster approach [#34]. Goods are assumed to be multi-dimensional bundles of characteristics. Behavior is then governed by consumer's desires to obtain not the goods but their characteristics.

The model assumes that choice can be viewed as taking place in a space spanned by a small number of dimensions in which competing alternatives are treated as points. An ideal point (occupied by the ideal amount of each attribute) is also assumed. Preference is postulated to be inversely related to the distance of an object from the ideal point.

$$Y_{tj} = \sum_{i=1}^{n} V_{ti} \left| B_{tji} - I_{ti} \right|^K$$

where

Y_{tj} = overall preferences for television show j measured as the distance to the ideal point at time t

n = number of dimensions or attributes

V_{ti} = importance weight attached by the individual to attribute i (action, suspense, humor, personal involvement, quality of production, topical or or educational value) at time t

B_{tji} = belief about show j on attribute i at time t

I_{ti} = ideal position on attribute i at time t

K = integer defining the distance measure.

The best form of the model will produce a Y closest to being inversely monotonic with preference.

References: #25 and 34.

Subject: Buyer Attitude

Title: An Exponential Discrepancy Model for Attitude Evaluation

Authors: Hillel J. Einhorn and Nicholas J. Gonedes

Source: Behavioral Science, Vol. 16, 1971, 152-57

Summary: A model based on an exponential decay function is
            formulated to deal with evaluation based on discrepancy.
            This evaluation concerns discrepancy based on attitudes
            and specifically hypothesizes that as discrepancy from
            one's ideal point increases, evaluation of a multi-
            dimensional stimulus object will decrease. The model
            is applied to experimental data dealing with the
            evaluation of political candidates and is found to give
            a good fit for the obtained data. In addition, the
            parameters of the model are shown to be related to
            the intensity of an attitude, and it is further shown
            that the results could be interpreted in terms of the
            psychological concepts of latitudes of acceptance,
            rejection, and noncommitment.

Model: Let

I_{ij} = individuals j's position with regard to attribute i.

P_{ik} = position of the kth stimulus object with with regard to attribute i.

D_{ijk} = $|I_{ij} - P_{ik}|$ = discrepancy between the position of the jth individual and the position of the kth stimulus object in regard to attribute i. (The absolute value operator is used because we shall be dealing with magnitudes of discrepancies, not directions.)

U_{jk} = the evaluation of the kth stimulus object by the jth individual.

We postulate the following functional form for U_{jk}:

$$U_{jk} = \Pi_{i=1}^{n} \exp [-\gamma_{ij}D_{ijk}]; \qquad (1)$$
$$j = 1,2, \ldots, m$$
$$k = 1,2, \ldots, s$$

where $\gamma_{ij} > 0$ is a parameter to be estimated (for all i).

Observe that: $\exp [- \gamma_{ij}D_{ijk}] > 0$

and

$$\partial U_{jk}/\partial D_{ijk} = -\gamma_{ij} \Pi_{i=1}^{n} \exp [-\gamma_{ij}D_{ijk}] < 0, \quad \text{for all i} \qquad (2)$$

$$\partial^2 U_{jk}/\partial D_{ijk}^2 = (-\gamma_{ij})^2 \prod_{i=1}^{n} \exp\ [-\gamma_{ij}D_{ijk}] > 0 \quad \text{for all} \quad i \quad (3)$$

As the discrepancy with respect to a particular attribute increases, evaluation decreases monotonically (2). (3) indicates that the evaluation decreases at a declining rate as discrepancies increase. Behaviorally, this implies that the initial differences from one's own position will be associated with a larger decrease in evaluation than additional differences.

For given values of D_{ijk}, the decrease in evaluation indicated in (2) and (3) will vary across individuals as a function of the value of γ_{ij} (all i and j). If $|\gamma_{ij}| < |\gamma_{ip}|$, $p \neq j$, then, for given D_{ijk},

$$U_{jk} > U_{pk}$$

Subject: Buyer Attitude

Title: The Vector Model of Preferences: An Alternative to
           the Fishbein Model

Author: Olli T. Ahtola

Source: Journal of Marketing Research, Vol. 12, February 1975,
           52-59

Summary: Based on certain assumptions in the Fishbein model [#25]
           unacceptable to the author a new model of preferences is
           developed and tested.

Model: In Fishbein's model the attitude is a sum of the product of
           the evaluative reaction to a salient property and the
           strength of belief connecting the property to the attitude
           object or event across all salient properties.

$$A_o = \sum_{i=1}^{N} B_i a_i \qquad (1)$$

where:

B_i = the strength of belief i about the attitude
    object o, that is the probability or
    improbability that o is related to some other
    object x_i,

a_i = the evaluative aspect of B_i, that is,
    the evaluation of x_i--its goodness or badness,

N = number of beliefs.

When operationalizing his model Fishbein uses belief
statements which specify only the direction of the belief
(sweet), not the strength of the belief (very sweet,
fairly sweet, slightly sweet).

The purpose here is to develop an attitude model
which distinguishes between belief strength and the content
of the belief and which does not require the utilization
of negative probabilities. Let

$$A_k = \sum_{i=1}^{n} \sum_{j=1}^{g(i)} B_{ijk} a_{ij} \qquad (2)$$

where:

A_k = an individual's attitude toward alternative k,

B_{ijk} = his strength of belief ij about k (on the dimension i), that is, the probability that k is associated with some other concept ij (i.e., j^{th} category on dimension i),

a_{ij} = the evaluative aspect of B_{ij}, that is, the individual's evaluation of ij,

g(i) = number of associated concepts (categories) on dimension i,

n = number of salient dimensions.

Here an individual's attitude toward any alternative in a choice situation is a function of (1) the strength of his beliefs about the alternative and (2) the evaluative aspect of those beliefs, i.e., the evaluative aspect of associated responses .

From model (2) it becomes clear that a person is hypothesized to have an attitude which can be derived mathematically by calculating expected values on each dimension and then by summing these expected values. To calculate the expected affect for each dimension, the affect column vector is premultiplied by the probability row vector. To get the attitude, the dot products of these vectors are summed. This can be expressed as follows:

$$A_k = \sum_{i=1}^{n} B'_{ik} a_i \qquad (3)$$

where:

A_k = an individual's attitude toward alternative k,

B'_{ik} = vector of his probabilities of k's association with categories of i,

a_i = vector of his evaluations of categories of i, that is, his attitude toward each of these categories,

n = number of salient dimensions.

This model includes everything in model (2) which is directly based on the underlying behavioral theory. The prediction using this model is exactly the same as model (2)'s prediction of attitude. However, deriving the zero probabilities and measuring the whole affect curves, model (3) is simpler in its mathematical form and gives useful information to the investigator.

Reference: #25.

Subject: Buyer Attitude

Title: The Importance of Halo Effects in Multi-Attribute Attitude Models

Authors: Neil E. Beckwith and Donald R. Lehmann

Source: Journal of Marketing Research, Vol. 12, No. 3, August 1975, 265-75

Summary: A simultaneous equation model is used to explain both the overall attitude of heterogeneous individuals towards television shows and also their beliefs about the shows on six relevant attributes. The halo effect (individuals bias their indicated beliefs by their overall attitude) may be the primary reason for the usually good descriptive results of the multi-attribute model.

Model: ATTITUDE AS A FUNCTION OF BELIEFS

Let

$$A_i = \sum_{j=1}^{n} \omega_j B_{ij} + y A_i^* + u_o \qquad (1)$$

where:

A_i = individual's attitude toward stimulus i.

B_{ij} = individual's belief about stimulus i on attribute j,

A_i^* = average attitude of all persons toward the stimulus i,

n = number of attributes,

and where the unknown parameters to be estimated are:

ω_j = weight of attribute j,

y = weight of average attitude,

and u_o is a random disturbance. These coefficients may vary between individuals, depending upon the relative importance of the particular variables to him. In the special case where $y = 0$, this equation of the model reduces to the popular linear multi-attribute attitude model:

$$A_i = \sum_{j=1}^{n} \omega_j B_{ij}. \qquad (1A)$$

Beliefs as a Function of Attitudes

Let

$$B_{ij} = \beta_j A_i + y_j B^*_{ij} + u_j \quad \text{for} \quad j = 1, \ldots, n$$

where:

B_{ij} = individual's belief about stimulus i on attribute j,

A_i = individual's attitude toward stimulus i,

B^*_{ij} = average belief about stimulus i on attribute j,

and where the unknown parameters to be estimated are:

β_j = importance of attitude toward stimulus j to belief B_{ij},

y_j = importance of average belief B^*_{ij} to belief B_{ij},

and u_j is a random disturbance.

Summarizing, this model consists of n+1 relationships, one describing the overall attitude A_i and n describing the beliefs B_{ij} of the individual toward stimulus i.

THE DATA

The attitudes of individuals toward television shows were examined. A target sample of 2000 individuals was drawn from a mail panel. The respondents were questioned about 20 shows on six attributes: action, suspense, humor, personal involvement, well produced and directed, and topical or educational value.

In summary, the empirical results

1. explain the beliefs B_{ij} well, and indicate a strong influence of overall attitude on beliefs (particularly for ambiguous attributes 4 and 5);

2. explain overall attitudes A_i very well when the halo effect is allowed to inflate the ordinary least square results;

3. explain overall attitude A_i very poorly when the beneficial halo effect has been purged from the belief variables.

Subject: Buyer Attitude

Title: The Fishbein Extended Model and Consumer Behavior

Authors: Michael J. Ryan and E.H. Bonfield

Source: Journal of Consumer Research, Vol. 2, No. 2, September 1975, 118-36

Summary: The theoretical development and empirical research testing the Fishbein "extended" or "behavioral intentions" model are described and evaluated. Discussion of conceptual and methodological strengths and weaknesses leads to the proposal of a reconceptualized form of the model.

Model: Fishbein's model [#25] is an adaptation of Dulany's theory [Don E. Dulaney, "Awareness, Rules, and Propositional Control: A Confrontation with S-R Behavior Theory." In D. Horton and T. Dixon (eds.), Verbal Behavior and General Behavior Theory, New York: Prentice-Hall, 1968, 340-87].

Dulany's Theory of Propositional Control

An individual's particular verbal response in a given situation is predicted as

$$B \simeq BI = [(RHd)(RSv)]w_0 + [(BH)(MC)]w_1$$

where B = overt behavior; BI = behavioral intention; RHd = hypothesis of the distribution of reinforcement, that is, the degree to which the individual thinks a specific response will lead to a reinforcement or reward; RSv = the subjective value of a reinforcer, that is, the value the individual places on a reward; BH = behavioral hypothesis, that is, the degree to which the individual believes a particular behavior is expected of the individual by some other; MC = motivation to comply, that is, the degree of the individual's desire to conform to a BH; and w_0 and w_1 are empirically determined weights.

The Fishbein Adaptation

Defining RHd · RSv as attitude-toward-the-act,

$$Aact = \sum_{i=1}^{n} B_i a_i$$

where Aact is the attitude toward performance of a specific act, B_i is the belief about the consequences of performing a particular behavior, a_i is the evaluative aspect of B_i and n is the number of relevant consequences. Fishbein's formulation is modeled

$$B \simeq BI = [Aact]w_0 + [NB \cdot MC]w_1 \qquad (1)$$

where NB = a normative belief, that is, the degree of belief that others expect or do not expect the individual to perform a specific act; MC = motivation to comply or not comply with the expectation of others.
Considering a third component, personal normative beliefs (NBp), the modified formulation is:

$$B \simeq [Aact]w_0 + [NBs \cdot MCs]w_1 + [NBp \cdot MCp]w_2 \qquad (2)$$

where p and s refer to personal and social norms respectively.

A third form of the extended Fishbein model results from recognition that NBs may result from several sources.

$$B \simeq BI = [Aact]w_0 + [\sum_{j=1}^{k} NB_j MC_j]w_1 \qquad (3)$$

where NB_j = the degree of belief or disbelief that a specific act is expected of the individual by the jth person or group, and MC_j = the individual's motivation to comply or not to comply with the expectation of the jth person or group.

Model forms (1), (2), and (3) imply the attitudinal and normative constructs are additive, that is, they are independent of one another. However, it is reasonable to suggest a relationship between the attitudinal evaluative component (a_i) and normative belief (NB_j) since it is likely they would be based on the same value system.

Model form (3) has been tested in marketing studies which are summarized.

A Reconceptualization of the Model

The reconceptualization views B as a dependent variable, BI as a moderating variable, Aact and SC as independent, related variables, and $\sum B_i a_i$ and $\sum NB_j MC_j$ as antecedent unrelated variables. The reconceptualization suggests the causal chain shown in Figure 1. Among the implications of utilizing the proposed framework is the development of operationalization procedures for SC which are methodologically distinct from those used to measure Aact.

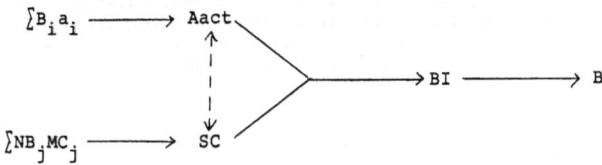

Figure 1

The Suggested Causal Sequence in the Extended
Fishbein Model

Reference: #25.

Subject: Consumer Behavior

Title: A Theory of Demand with Variable Consumer Preferences

Author: R.L. Basmann

Source: Econometrica, Vol. 24, January 1956, 47-58

Summary: A theory of consumer demand with variable preferences. The
           assumption that the individual consumer has a unique ordinal
           utility index function is replaced by the assumption that he
           has a family of ordinal utility functions; advertising
           expenditures by the sellers of commodities are assumed to
           determine which one of these ordinal utility functions is
           to be maximized. From these assumptions are derived a
           number of theoretical relations which measurements defining
           advertising elasticities of demand must satisfy. The rela-
           tions involving shifts in demand and advertising elasticities
           of demand are shown to be analogues of the theorems of
           consumer demand under fixed preferences.

Model:

Assume a utility function of the form

$$\theta_j = \theta_j(a_1, \ldots, a_n / x_1, \ldots, x_n) \quad (j = 1, \ldots, n), \quad (1)$$

where

θ = parameters describing the form of ordinal
utility function

x = quantity of goods and services.

The equilibrium conditions are

$$\sum_{i=1}^{n} p_i x_i = M, \tag{2}$$

$$-\lambda p_i + u_i = 0 \quad (i = 1, \ldots, n), \tag{3}$$

where

p_i = prices

M = money income

λ = marginal utility of money income

and the matrix

$$U = \begin{bmatrix} o & u_j \\ u_i & u_{ij} \end{bmatrix} \quad (i,j = 1, \ldots, n), \tag{4}$$

is negative definite.

The Slutsky-Hicks substitution term, S_{ij}, is obtained by differentiating (3) with the side condition

$$u(x_1, \ldots, x_n; \theta_1, \ldots, \theta_n) = \text{constant}, \qquad (5)$$

partially with respect to p_j and solving the resulting system of equations for

$$S_{ij} = \frac{\lambda |U_{ij}|}{|U|}, \qquad (i,j = 1, \ldots, n), \qquad (6)$$

where $|U_{ij}|$ is the cofactor of u_{ij} in the matrix U. The four major theorems are:

$$S_{ii} < 0, \qquad (7)$$

$$S_{ij} = S_{ji} \qquad (i,j = 1, \ldots, n), \qquad (8)$$

$$\sum_{j=1}^{n} p_j S_{ij} = 0, \qquad (9)$$

$$\sum_{i=1}^{m} \sum_{j=1}^{m} S_{ij} z_i z_j < 0, \qquad (m < n), \qquad (10)$$

where not all $z_i = 0$.

The Tintner-Ichimura Relations are:

$$x_{ia_j} = - \sum_{h=1}^{n} \frac{u_{ha_j}}{\lambda} Sh_i, \qquad (i,j - 1, \ldots, n), \quad (11)$$

and

$$x_{ia_j} = - \sum_{h=1}^{n-1} R_{ha_j} S_{hi} \qquad (12)$$

where $R_{ha_j} = \partial R_h / \partial a_j$, $R_h = u_h / u_n$, and $p_n \equiv 1$, x_n being the numéraire. They satisfy

$$\sum_{i=1}^{n} p_i \, x_{ia_j} = 0. \tag{13}$$

For the market demand

$$X_{ia_j} = \sum_{\mu=1}^{N} x_{ia_j}^{(\mu)} \tag{14}$$

and

$$\sum_{i=1}^{n} p_i \, X_{ia_j} = 0. \tag{15}$$

The individual and market advertising elasticities of demand are

$$e_{ij} = \frac{a_j}{x_i} \, x_{ia_j} \qquad (i,j = 1, \ldots, n), \tag{16}$$

$$E_{ij} = \frac{a_j}{X_i} \, x_{ia_j}. \tag{17}$$

Assume that a_i affects only the marginal rate of substitution of x_n for x_i and let $R_{ia_i} = b_{ii} R_i > 0$, $R_{ja_i} = 0$, for all $j \neq i$. Then (12) becomes

$$x_{ja_i} = -b_{ii} R_i S_{ji} = -b_{ii} S_{ji} p_i, \tag{18}$$

and

$$e_{ji} = w_{ii} z_{ji} \qquad (j = 1, \ldots, n), \tag{19}$$

where

$$z_{ij} = -S_{ij} p_j / x_i \qquad (i,j = 1, \ldots, n) \tag{20}$$

and $w_{hj} = a_j h_{hj}$.

It follows from (18), (7), (14), (20), and the definition of related goods that the advertising elasticity of demand for x_i with respect to a_i is greater than zero, and that the advertising elasticity of demand for x_j with respect to a_i is less than zero if x_i is a substitute for x_j, is greater than zero if x_i is a complement of x_j, and is zero if x_i is independent of x_j.

Theorem analogous to the Leontief-Hicks theorem:

If the marginal rates of substitution are all increased in the same proportion, then the group of goods can be treated logically as a single commodity.

There are the same relations between aggregate elasticities of demand with respect to advertising expenditure and the income-compensated price elasticities of one aggregate commodity for another as there are between elasticities of demand with respect to advertising which affects the marginal rate of substitution of x_n for only one commodity and the income-compensated price elasticities of that one commodity.

Subject: Consumer Behavior

Title: A New Approach to Consumer Theory

Author: Kelvin J. Lancaster

Source: Journal of Political Economy, Vol. 74, April 1966, 132-57

Summary: Activity analysis is extended into consumption theory. It is assumed that goods possess, or give rise to, multiple characteristics in fixed proportions and that it is these characteristics, not goods themselves, on which the consumer's preferences are exercised.

Model: Assume that:

1. The good, per se, does not give utility to the consumer; it possesses characteristics, and these characteristics give rise to utility.

2. In general, a good will possess more than one characteristic, and many characteristics will be shared by more than one good.

3. Goods in combination may possess characteristics different from those pertaining to the goods separately.

We shall regard an individual good or a collection of goods as a consumption activity and associate a scalar (the level of the activity) with it. We shall assume that the relationship between the level of activity k, y_k, and the goods consumed in that activity to be both linear and objective, so that, if x_j is the jth commodity we have

$$x_j = \sum_k a_{jk} y_k, \tag{1}$$

and the vector of total goods required for a given activity vector is given by

$$x = Ay. \tag{2}$$

Since the relationships are assumed objective, the equations are assumed to hold for all individuals, the coefficients a_{jk} being determined by the intrinsic properties

of the goods themselves and possibly the context of
technological knowledge in the society.

We shall assume that each consumption activity
produces a fixed vector of characteristics and that
the relationship is again linear, so that, if z_i is
the amount of the ith characteristic

$$z_i = \sum_k b_{ik} y_k, \qquad (3)$$

or

$$z = By. \qquad (4)$$

Again, we shall assume that the coefficients b_{ik}
are objectively determined for some arbitrary choice of
the units of z_i.

We shall assume that the individual possesses an
ordinal utility function on characteristics $U(z)$ and
that he will choose a situation which maximizes $U(z)$.
$U(z)$ is provisionally assumed to possess the ordinary
convexity properties of a standard utility function.

In this model, the relationship between the collections
of characteristics available to the consumer--the vectors z--
which are the direct ingredients of his preferences and
his welfare, and the collections of goods available to him--the
vectors x--which represent his relationship with the
rest of the economy, is not direct and one-to-one, but
indirect, through the activity vector y.

The standard choice situation becomes

Maximize $U(z)$

subject to $p_x \leq k$

with $\qquad z = By$

$\qquad\qquad x = Ay$

$\qquad\qquad x, y, z \geq 0.$

References: #41 and 113.

Subject: Consumer Behavior

Title: A Geographic Model of an Urban Automobile Market

Authors: Theodore E. Hlavac, Jr. and John D.C. Little

Source: Proceedings of the Fourth International Conference on Operations Research, David B. Hertz and Jacques Melese (eds.), New York: John Wiley & Sons, Inc., 1966, 302-11

Summary: A model is developed in which a customer's probability of purchase at a given dealer is affected by dealer location and customer make preference, as well as the locations and strengths of all other dealers. Aggregation of the customer model gives a dealer market share (penetration) model, which may also be viewed as a model of competitive interaction. Such a model is fit to data for metropolitan Chicago. After fitting, the model permits estimation of the sales of a dealership with specified strength and location. The most obvious practical use of the model relates to market strategy for new dealerships in the automobile industry, but the model appears to be adaptable to site location problems in other fields as well.

Model: Dealer Pull

The attractiveness of a specified dealer, his "pull," is a function of dealer characteristics (such as make of car sold, extent of advertising), buyer characteristics (such as make preference), and distance from dealer to buyer. Pull will not be a directly observable quantity but will be used to develop expressions that are. The number of car purchases is assumed fixed for the time period under consideration.

Buyers are separated into market segments. The pull of a dealer on a buyer in a given segment is broken into two parts: (1) an "intrinsic pull" independent of the make sold by the dealer, and (2) the "make preference" of the buyer. Let

$g(i,j)$ = the pull of dealer j on a buyer in market segment i ($i = 1, \ldots, S$, $j = 1, \ldots, D$)

$h(i,j)$ = the intrinsic pull of dealer j on a buyer in segment i

$q(i,m)$ = the make preference of a buyer in segment i for make m ($m = 1, \ldots, M$)

$$q(i,m) \geq 0 \quad \text{and} \quad \sum_{m=1}^{M} q(i,m) = 1.$$

We stipulate that the above quantities be related by

$$g(i,j) = h(i,j)\, q(i,m(j)). \tag{1}$$

Thus, the pull of a dealer on a buyer is the dealer's intrinsic pull weighted by the buyer's brand preference.

Purchase Probability

The probability that a buyer purchases at a given dealer is taken as the pull of that dealer on the buyer divided by the total pull on the buyer. Let

$p(i,j)$ = the probability that a buyer in market segment i purchases at dealer j

$$p(i,j) = \frac{g(i,j)}{\sum\limits_{k=1}^{s} g(i,k)}. \tag{2}$$

Make preference can be interpreted as the probability of purchase of the make under the conditions that the sum of the intrinsic pulls on the buyer is the same for each make. This result can be deduced from (1) and (2).

Geographic Effect

We hypothesize that pull falls off exponentially with the distance between dealer and buyer.
Let

$x(i,j)$ = distance of the buyers in market segment i to dealer j

$$h(i,j) = a_j d^{-b_j x(i,j)}. \tag{3}$$

Here a_j and b_j are constants specific to dealer j. The constant a_j expresses the dealer's strength in his own immediate neighborhood. The constant b_j tells how fast his sales fall off with distance. Using (1), (2), and (3), we get

$$p(i,j) = \frac{q[i,m(j)] \, a_j e^{-b_j x(i,j)}}{\sum\limits_{k=1}^{D} q[i,m(k)] \, a_k c^{-b_k x(i,k)}}. \tag{4}$$

Dealer Sales and Penetration

Let

$N(i)$ = number of buyers in market segment i (called the potential of the segment) in a given time period

$s(j)$ = expected sales of dealer j in the given time period

$\pi(j)$ = expected penetration of dealer j in the whole city.

Then

$$s(j) = \sum_{i=1}^{s} N(i)\, p(i,j) \qquad (5)$$

$$\pi(j) = \frac{s(j)}{\sum\limits_{i=1}^{s} N(i)}. \qquad (6)$$

Subject: Consumer Behavior

Title: Consumer's Purchase Decision Process: Stochastic Models

Author: Tanniru Rao

Source: Journal of Marketing Research, Vol. 6, August 1969,
           321-29

Summary: Probabilistic analysis of relationships among selected
           elements of a consumer's purchase decision: brand purchased,
           store visited, and size of purchase.

Model:

Three consumer products were selected: A (a paper product),
B (a drug product: toothpaste), and C (a food product:
coffee). The leading brands of Product A are indicated
by A_1 and A_2; of Product B by B_1, B_2, B_3 and B_4; of Product C
by C_1, C_2, C_3, and C_4.

Let $b(n)$ and $s(n)$ indicate respectively brand purchased
and store visited by the consumer for her nth purchase.
The random variables $b_j(n)$ and $s_h(n)$ are defined as follows:

$$b_j(n) = \begin{cases} 1 \text{ if Brand } j \text{ is purchased nth time} \\ 0 \text{ if Brand } j \text{ is not purchased nth time} \end{cases}$$

$$s_h(n) = \begin{cases} 1 \text{ if nth purchase of product is made in Store } h \\ 0 \text{ if nth purchase of product is not made in Store } h. \end{cases}$$

Given the information on a consumer's past store visits
and the store selected for her subsequent purchase, the
purchase history of store visits can be described by a
vector of S's and D's with the following definition. Suppose
the consumer has selected Store h for her nth purchase.
Then,

$s(t)$ = S, if $s_h(t) = 1$ and $s_h(n) = 1$,   t = 1, 2, ..., n-1

      = D, if $s_h(t) = 0$ and $s_h(n) = 1$,   t = 1, 2, ..., n-1.

The following hypothesis were found valid:

Hypothesis 1.

A consumer's selection of a store for the purchase of any product is not completely random and she exhibits bias in her choice of the store. The more recent her purchase experience in a particular store and the more frequent her visits to the store, the more likely she is to repurchase the product in that store.

Hypothesis 2.

A consumer exhibits bias in selecting the kind of retail outlet in which she shops for a particular product.

Hypothesis 3.

Store switching increases brand switching.

Hypothesis 4.

A consumer changes her purchase size when she changes store or brand; generally, she decreases rather than increases purchase size with a change in store or brand.

Subject: Consumer Behavior

Title: An Empirical Test of the Howard-Sheth Model of Buyer Behavior

Authors: John U. Farley and L. Winston Ring

Source: Journal of Marketing Research, Vol. 7, November 1970, 427-38

Summary: The Howard-Sheth buyer behavior model was cast in the form of a multiple-equation regression model for testing data on a grocery product in a specific market. Estimated structural parameters were generally consistent with the model's predictions, but some goodness-of-fit measures were weak. The model was useful for organizing this analysis of consumer behavior, but the test put extreme pressure on the data. Considerably improved data collection techniques and procedures will be needed before the full empirical potential of such models will be realized.

Model: The variables in this econometric model are:

Endogenous		Exogenous	
Y_1	Attention level	X_1	Advertising exposure
Y_2	Perceptual bias	X_2	Level of word of mouth activity
Y_3	Stimulus ambiguity	X_3	Receipt of coupon
Y_4	Motive	X_4	Price
Y_5	Overt search activity	X_5	Various characteristics of the social and organizational setting
Y_6	Attitude		
Y_7	Intention		
Y_8	Brand comprehension		
Y_9	Confidence	.	
Y_{10}	Purchase	.	
Y_{11}	Satisfaction	X_K	

The model is given the form:

$$Y_{(i)} = \sum_{\substack{j=1 \\ (j \neq i)}}^{11} \beta_{i,j} Y_{(j)}$$

$$+ \sum_{k=1}^{K} \gamma_{i,k} X_{(k)} + \gamma_{i,0} + u_{(i)}; \quad i = 1, \ldots, 11. \tag{1}$$

Each $Y_{(j)}$ and $X_{(k)}$ are observed values of endogenous variable j and exogenous variable k. $Y_{i,0}$ is the additive constant in the ith equation; $u_{(i)}$ is an error term related to the model's ability to predict values of the endogenous variables, $Y_{(i)}$.

The linear system (1) has as many equations as there are endogenous variables. These eleven equations are written compactly using matrix notation as:

$$BY + \Gamma X = u. \qquad\qquad (2)$$

B is an 11x11 matrix of coefficients of the endogenous variables, Y is an 11-component column vector of observations of the endogenous variables for one sample point, Γ is an 11x(K+1) matrix of coefficients of the exogenous variables, X is a (K+1)-element column vector of values of the exogenous variables for the same sample point (including X(0) = 1), and u is an 11x1 column vector with elements $u_{(i)}$.

Subject: Consumer Behavior

Title: A Model of Consumer Behavior in a Single Market
          with Incomplete Information

Author: Antonio Bosch-Domenech

Source: Research Report 7312, Department of Economics, The University
          of Western Ontario, London 72, Canada, June 1973

Summary: The model differentiates between consumer behavior (1) with
          knowledge of the price distribution and (2) with imperfect
          knowledge of the price distribution and postulates a maximizing
          behavior. The purchase decision of the consumer is based
          on a comparison between the net utility that he obtains by
          buying at the observed price at one store and the expected
          net utility of searching for another store. The market is a
          single market with an indefinite, fixed number of stores
          and consumers.

Model: Consider a single market with an indefinite, very large but
          fixed number of stores and consumers. At each period of
          time every firm fixes a price and each consumer visits one
          store. The price set by a store can only be modified
          once at the beginning of each period. Every time the
          consumer visits one store he observes a price and decides
          whether he buys at this price or whether he postpones the
          purchase in the hope of finding a store with a cheaper
          price.

Consumer Behaviour with Knowledge of the Price Distribution

Let

c = searching cost per unit of search, considered constant

$U(P)$ = consumer's utility function on the price set, strictly
decreasing in price

$\phi(P)$ = distribution function of prices (finite)

$P_1, P_2 \ldots$ = observed prices, considered as independent,
identically distributed random variables
from $\phi(P)$

$U(P_n) - nc$ = consumer's utility if he buys after n searches.

The consumer has to find a rule for stopping his search

that maximizes $E[U(P_N) - cN]$, where E is the expectation operator and N is the random number of searches that he undertakes under a particular stopping rule.

Maximum expected utility $\alpha^* = E[\max U(P), \alpha^*] - c$.

The optimal rule for the utility maximizer consumer is to buy as soon as he finds a price p such that $U(p)$ is at least as large as α^*, where α^* is the unique solution of

$$\int_\alpha^\infty [U(P) - \alpha] \, dF[U(P)] = c.$$

If the consumer believes that prices are distributed normally with mean μ and variance σ^2 and $U(P)$ is a linear function,

$$\alpha^* = U(\mu) + \overline{\sigma} \; \Psi^{-1}(\frac{c}{\sigma}).$$

Consumer Behaviour with Imperfect Knowledge of the Price Distribution

Let

$U^*(P,\mu,\tau)$ = expected utility.

If the consumer decides to go to i stores, the posterior distribution of M will be $M \sim N(\mu_i, \tau_i)$ where

$$\mu_i = \frac{\tau_0 \mu_0 + \sum_{j=0}^{i-1} p(\mu_j, \tau_j)}{\tau_0 + i} \text{ and } \tau_i = \tau_0 + i \text{ and } p(\mu_j, \tau_j)$$

is price observed after $j-1$ more visits to stores. The marginal distribution of the price observed after visiting i stores will be

$$P(\mu_{i-1}, \tau_{i-1}) \sim N(\mu_{i-1}, \; \frac{\tau_0 + i - 1}{\tau_0 + i}).$$

The consumer will buy if $U(P) \geq E[U^*(P(\mu_i, \tau_i), \mu_{i+1}, \tau_{i+1})] - c$
and will continue searching otherwise. Therefore,

$$U^*(P, \mu_i, \tau_i) = \max \{U(P), E[U^*(P(\mu_i, \tau_i), \mu_{i+1}, \tau_{i+1}] - c\}.$$

Let $E[U^*(p(\mu_i, \tau_i), \mu_{i+1}, \tau_{i+1}] - c \equiv \alpha(\mu_i, \tau_i)$.

Then

$$\alpha(\mu_i, \tau_i) = \alpha(\mu_{i+1}, \tau_{i+1}) + \int_{\alpha(\mu_{i+1}, \tau_{i+1})}^{\infty} [U(P(\mu_i, \tau_i))$$

$$- \alpha(\mu_{i+1}, \tau_{i+1})] \, dF[U(P(\mu_i, \tau_i))] - c$$

$$\equiv \alpha(\mu_{i+1}, \tau_{i+1}) + T_{F[U(P(\mu_i, \tau_i))]}(\alpha(\mu_{i+1}, t_{i+1})) - c.$$

If the utility function is <u>linear</u> on prices,

$U[P(\mu_i, \tau_i)] \sim N(m_i, \pi_i)$ where $m_i = U(\mu_i)$ and $\pi_i = b\tau_i$, $b > 0$.

Therefore,

$$\alpha(\mu_i, \tau_i) = \pi_i^{\frac{1}{2}} \Psi[\pi_i^{\frac{1}{2}} (m_i - \alpha(\mu_{i+1}, \tau_{i+1}))] + m_i - c.$$

Let $\varepsilon(z, m_i, \pi_i) = U^*(p_i, \mu_i, z_i)$. Then

$$\varepsilon(z, m_i, \pi_i) = \max \{z, E[\varepsilon(Z(m_i, \pi_i), m_{i+1}, \pi_{i+1})] - c\}.$$

Suppose that the present and future observed utilities, as
well as the mean of A, are reduced by k.

Calling $E(\varepsilon(Y(\pi_i), o, \pi_{i+1})) - c = \bar{\alpha}(\pi_i)$,

we establish that the optimal rule for the consumer that
has just observed price $p = U^{-1}(z)$ and believes, therefore,
that $\Lambda \sim N(m_i, \pi_i)$ is to buy at this price if $z \geq \bar{\alpha}(\pi_i) + m_i$
and to look for another store otherwise.

In this way, (m_i, π_i) can be expressed as the sum of m_i

plus another term, $\bar{\alpha}(\pi_i)$, <u>independent of m_i</u>. The higher
the expectation of the price mean, the more likely it is
that the number of stores visited will decrease, and the
lower will be the expected utility.   The higher the
precision of the distribution of the <u>price mean</u>, the more
likely it is that the number of stores visited will
increase and the higher will be the expected utility.
These results confirm, therefore, that as higher and higher
prices are observed, <u>ceteris paribus</u>, the cutoff price,

$\bar{p} = U^{-1}(\bar{z})$, $\bar{z} = \bar{\alpha}(\pi_i) + m_i$, will rise.

Subject: Consumer Behavior

Title: Mathematical Programming Models for the Determination of Attribute Weights

Authors: Dov Pekelman and Subrata K. Sen

Source: Management Science, Vol. 20, No. 8, April 1974, 1217-29

Summary: Several versions of a mathematical programming model which determines attribute weights for each consumer are empirically evaluated using data on dry cereals and automobiles. Managerial implications for product repositioning, new product design, and market segmentation are discussed.

Model: The market consists of n brands each of which can be characterized by m attributes. For each attribute, a consumer has a specified preferred position known as his ideal point. The consumer can estimate the "distance" between a brand's location on each attribute and his ideal point. He then computes his "overall distance" from the brand by means of a weighted combination of the individual attribute distances.

Let

d_{ik} = distance of object k from the consumer's ideal point on attribute i $(i = 1, 2, \ldots, m$ and $k = 1, 2, \ldots, n)$;

W_i = weight assigned to attribute i by the consumer;

D_k = overall distance from object k.

Then

$$D_k = \sum_i d_{ik}^2 W_i. \qquad (1)$$

Let S = set of all ordered pairs (c,k) of the n brands where c designates the consumer's preferred brand;

$$D_{kc} = (D_k - D_c) = \sum_i (d_{ik}^2 - d_{ic}^2) W_i. \qquad (2)$$

If brand c is preferred to brand k, D_i should be greater than D_c. However, for a particular set of weights D_c might exceed D_k, indicating a violation of this particular paired preference. The distance between the two objects in the

violated pair can be computed for every violation. Our
objective is to select the weights such that the sum of
these distances is minimized.

Minimize $\sum_{(c,k) \in S} Y_{kc}$ (3)

s.t. $\sum_i (d_{ik}^2 - d_{ic}^2) W_i + Y_{kc} \geq 0$ for all $(c,k) \in S$ (4)

$\sum_i W_i = 1$ (5)

$W_i, Y_{kc} \geq 0$ for all i and $(c,k) \in S$. (6)

The model can be solved by any linear programming code.

Alternative Objective Function

Instead of minimizing the <u>amount</u> of violation, we
could choose the attribute weights such that we minimize
the <u>number</u> of violations of the consumer's $n(n-1)/2$ paired
preferences.

Minimize $\sum_{(c,k) \in S} \delta_{kc}$ (7)

s.t. $\sum_i (d_{ik}^2 - d_{ic}^2) W_i + M\delta_{kc} \geq 0$ for all $(c,k) \in S$ (8)

$\sum_i W_i = 1$ (9)

$W_i \geq 0$ for all i (10)

$\delta_{kc} = 0,1$ for all $(c,k) \in S$ (11)

M is a large number greater than $\max_{i,k} d_{ik}^2$. The above
model can be solved by a mixed integer programming code.

Treatment of Ties

The two models described above cannot prevent ties
between pairs of brands since the weights are not constrained to prevent the occurrence of $D_{kc} = 0$ (the consumer

has an equal preference for brands c and k). Using
the minimization of the amount of violation as the
objective function, the model which prevents ties can be
formulated as the following mixed integer programming
problem:

Minimize $\sum_{(c,k) \varepsilon S} Y_{kc}$ (12)

s.t. $\sum_i (d_{ik}^2 - d_{ic}^2) W_i + Y_{kc} \geq 0$ for all $(c,k) \in S$ (13)

$\sum_i W_i = 1$ (14)

$\sum_i (d_{ik}^2 - d_{ic}^2) W_i + MZ_{kc} \geq m$ for all $(c,k) \in S$ (15)

$\sum_i (d_{ik}^2 - d_{ic}^2) W_i - M(1-Z_{kc}) \leq -m$ for all $(c,k) \in S$ (16)

$W_i, Y_{kc} \geq 0$ for all i   and $(c,k) \in S$ (17)

$Z_{kc} = 0,1$ for all $(c,k) \in S.$ (18)

M is a large number defined as before while m is an
arbitrary small positive number. Constraints (15) and (16)
are designed to select the weights such that $D_{kc} \neq 0$.
Sometimes it is not possible to ensure that $D_{kc} \neq 0$. For
example, if $d_{ik} = d_{ic}$ for all attributes i for a
pair of brands (c,k), D_{kc} will equal zero for any values of
the weights. In such situations, constraints (15) and (16)
can be deleted from the program for the pair (c,k).

Subject: Consumer Behavior

Title: An Interaction Model of Consumer Utility

Authors: Paul E. Green and Michael T. Devita

Source: Journal of Consumer Research, Vol. 2, No. 2, September 1975, 146-53

Summary: The main-effects, additive utility model is extended to include two-factor interaction via a two-stage data collection procedure. The model is applied illustratively to a set of menu preferences data. The article concludes with a discussion of the model's implications for consumer behavior theory.

Model: To date, application of conjoint measurement to preference measurement in consumer behavior has emphasized non-interactive models of either an additive or multiplicative nature. That is, the total utility for a multi-attribute alternative, described as an n-component vector:

$$x = (x_1, x_2, \ldots, x_n) \tag{1}$$

has been defined as either additive:

$$U(x) = \sum_{j=1}^{n} u_j(x_j) \tag{2}$$

or multiplicative:

$$U(x) = \prod_{j=1}^{n} u_j(x_j) \tag{3}$$

where each attribute $j = 1, 2, \ldots, n$ can be viewed as nominal-scaled and each u_j is a real-valued function that is found from the conjoint measurement procedure.

If all utility scale values u_j are positive, the expressions (2) and (3) are formally equivalent since an order-preserving (specifically, a logarithmic) trans-formation of (3), makes it equivalent to (2). However, situations can arise in which we would like to consider interactions among factors.

The model proposed here is designed to be implementable at the individual-subject level and can handle all main effects and all two-factor interactions for a reasonably large number of factors (e.g., nine or ten factors if need be), each at various levels, ranging from two to six.

Assume a four-factor utility model. In conventional experimental design terms, the formal model is:

$$U(x) = \mu + \alpha_i + \beta_j + \gamma_k + \delta_\ell + (\alpha\beta)_{ij} + (\alpha\gamma)_{ik}$$
$$+ (\alpha\delta)_{i\ell} + (\beta\gamma)_{jk} + (\beta\delta)_{j\ell} + (\gamma\delta)_{k\ell} + \varepsilon_{ijk\ell} \quad (4)$$

where: $i = 1,2, \ldots, I$; $j = 1,2, \ldots, J$; $k = 1,2, \ldots, K$; $\ell = 1,2, \ldots, L$; μ denotes the overall mean and $\varepsilon_{ijk\ell}$ is the error term. In addition, the usual zero-sum restrictions are assumed to apply to each main effect and interaction.

Further assume that the original response data $Y(x)$ are measured only at the level of an ordinal scale so that:

$$Y(x) \overset{m}{=} U(x) \quad (5)$$

where $\overset{m}{=}$ is some order-preserving function of $U(x)$ that is found from the analysis. The utility interaction model is applied to menu preference data where the four factors of (4) are: (a) entrees; (b) potatoes; (c) vegetables; and (d) salads.

Subject: Consumer Behavior

Title: Product Attraction, Marketing Effort and Sales: Towards a Utility Model of Market Behavior

Authors: Martin J. Beckmann and Ursula H. Funke

Source: (forthcoming)

Summary: Utility theory is adapted to incorporate marketing behavior in the theory of consumption.

Model: Let

$$u = u(a,x) \qquad (1)$$

where

u = utility function of a household

$x = (x_1, \ldots x_i, \ldots x_n)$ vector of product quantities

$a = (a_1, \ldots a_i, \ldots a_n)$ vector of product attractions.

Attractions are quantifications of product properties, be they physical or psychological; they can also be the result of marketing efforts. Consider the separable utility function

$$u = \sum_{i=1}^{n} \phi_i\,(a_i,x_i). \qquad (2)$$

We obtain

$$u = \sum_{i=1}^{n} a_i\,\phi_i(x_i) \qquad (3)$$

when the attractions are utility changing;

$$u = \sum_{i=1}^{n} \phi_i\,(a_i,x_i) \qquad (4)$$

when the attractions are product changing; and

$$u = \sum_{i=1}^{n} a_i\,\phi_i\,(b_i,x_i) \qquad (5)$$

when the attractions are product as well as utility changing. (3), (4), and (5) coincide when ϕ is a power function. In fact, the specification

$$u = \begin{cases} \displaystyle\sum_{i=1}^{n} a_i x_i^{\alpha} & 0 < \alpha < 1 \qquad (6a) \\[2ex] \displaystyle -\sum_{i=1}^{n} a_i x_i^{\alpha} & \alpha < 0 \qquad (6b) \end{cases}$$

is the utility equivalent of the CES production function. After a monotone transformation,

$$[\sum_i a_i x_i^\alpha]^{\frac{1}{\alpha}} \tag{7}$$

reduces to (6a) and (6b) respectively.

Letting $\alpha \to 0$ we obtain

$$u = \sum_{i=1}^{n} a_i \log x_i. \tag{8}$$

In the conventional consumer analysis we have the budget constraint

$$\sum_{i=1}^{n} p_i x_i = z. \tag{9}$$

Maximizing (8) subject to (9) yields

$$x_i = \frac{a_i}{\lambda p_i}. \tag{10}$$

Using (9) to eliminate λ in (10)

$$x_i = \frac{a_i}{a} \cdot \frac{z}{p_i}. \tag{11}$$

In order to investigate how attractions relate to market shares we indicate that we have so far considered a specific household h with budget z_h and taste or attractions a_{ih}. Rewrite (11) as

$$p_i x_{ih} = \frac{a_{ih}}{\sum_i a_{ih}} \, z_{ih}.$$

Now

$$\frac{a_{ih}}{\sum_i a_{ih}} = a_i + \alpha_{ih} \tag{12}$$

where

$$\sum_i \alpha_{ih} = 0 \quad \sum_h \alpha_{ih} = 0 \quad \sum a_i = 1.$$

Assume that budgets are linear functions of household incomes

$$z_{ih} = \beta y_h + \gamma_h. \qquad (13)$$

Aggregate expenditures on good i is then given by

$$X_i = \sum_h p_i x_{ih} = \sum_h (a_i + \alpha_{ih})(\beta y_h + \gamma_h). \qquad (14)$$

Now

$$X_i = a_i(\beta \bar{y} + \bar{\gamma}) + \beta \, cov \, (\alpha_{ih}, \, y_h) + cov \, (\alpha_{ih}, \, \gamma_h)$$

$$X_i = a_i(\beta \bar{y} + \bar{\gamma}) + b_i^*, \text{ say}. \qquad (15)$$

In view of (12) we have

$$\sum X_i = \beta \bar{y} + \bar{\gamma}.$$

Define market share m_i in terms of relative expenditure

$$m_i = \frac{X_i}{\sum X_i} = a_i + \frac{b_i^*}{\beta \bar{y} + \bar{\gamma}} \qquad (16)$$

$$m_i = a_i + b_i, \qquad \text{say}. \qquad (17)$$

Thus market share is linearly related to attraction. In particular, when the covariances vanish then market share equals relative attraction. In fact, Bell, et. al. [#146] have introduced attractions as relative market shares. Assume that there are two sellers, $i = 1,2$. Net revenue of seller 1 is then

$$R_1 = \frac{a_1}{a_1 + a_2} \, \gamma_1 M \qquad (18)$$

where γ is the profit margin and M is aggregate income or gross revenue. Let advertising represent all marketing activities and let the costs of advertising be $c_i(a_i)$. Profits of firm 1 after advertising outlay are then

$$G_i = \frac{a_1}{a_1 + a_2} \, \gamma_1 M - c_1(a_1). \qquad (19)$$

Assume that each firm maximizes profits on the assumption that the other firm does not change its strategies. Profit maximization by firm 1 yields

$$0 = \frac{\partial G_1}{\partial a_1} = \gamma_1 M \frac{a_2}{(a_1 + a_2)^2} - c_1'(a_1). \tag{20}$$

Now assume that firm 2 uses the same optimal strategy. From (20)

$$a_i = \frac{c_i}{\gamma_i} \frac{(a_1 + a_2)^2}{M}. \tag{21}$$

Write $a_1 + a_2 = a$ and $\frac{c_i}{\gamma_i} = \mu_i$.

Then

$$a = \frac{1}{M} (\mu_1 + \mu_2) \cdot a^2 \quad \text{or} \quad a = \frac{M}{\mu_1 + \mu_2}.$$

Substituting in (21)

$$a_i = \mu_i \frac{M}{(\mu_1 + \mu_2)^2} \quad \text{and} \tag{22}$$

$$\frac{a_i}{a} = \frac{\mu_i}{\mu_1 + \mu_2}. \tag{23}$$

The net revenue is

$$R_i = \frac{\mu_i}{\mu_1 + \mu_2} \gamma_i M. \tag{24}$$

Taking the ratio of advertising outlay $A_i = c_i a_i$ to net revenue R_i we have

$$\frac{A_i}{R_i} = \frac{\mu_i}{\mu_1 + \mu_2}. \tag{25}$$

In particular,

$$\frac{A_1}{R_1} = \frac{1}{1 + \frac{\mu_2}{\mu_1}}.$$

Thus the optimal advertising outlay should be proportionate
to the net revenue (before advertising) for all values of gross
revenue or market size M. In the case of n firms with the
same profit margin γ using the same marketing strategies we
obtain

$$a_i = \frac{n-1}{n^2} \ \mu M. \tag{26}$$

Since each firm's revenue is $\frac{\gamma M}{n}$, this means that each firm
spends $\frac{n-1}{n}$ of its net revenue on marketing activities,
say advertising. Thus the model predicts that (optimal)
advertising outlay as a percentage of net revenue increases
with the number of (monopolistic) competitors in a given
market.

References: #34 and 146.

Subject: Media Selection

Title: Linear Programming in Media Selection

Author: Ralph L. Day

Source: Journal of Advertising Research, Vol. 2, No. 2, June
           1962 , 40-44

Summary: Linear programming applied to advertising media
           selection.

Model: The objective to be maximized is the marketing effective-
           ness of the advertising program.

$$\text{Maximize:} \quad P_1 X_1 + P_2 X_2 + \ldots P_n X_n$$

$$\text{Subject to:} \quad A_{11} X_1 + A_{12} X_2 + \ldots + A_{1n} X_n \leq C_1$$

$$A_{21} X_1 + A_{22} X_2 + \ldots + A_{2n} X_n \leq C_2$$

$$\vdots$$

$$A_{m1} X_1 + A_{m2} X_2 + \ldots + A_{mn} X_n \leq C_m$$

$$X_1, X_2, \ldots, X_n \geq 0$$

where

    X = advertising unit

    P = advertising effectiveness measure

    C = total advertising budget

    A = price of advertising unit.

Subject: Media Selection

Title: Media Selection by Linear Programming

Authors: Douglas B. Brown and Martin R. Warshaw

Source: Journal of Marketing Research, Vol. 2, February 1965, 83-88

Summary: The authors present a general media mix model which assumes linear response and illustrate how the model can be modified to accommodate nonlinearity while still using the revised simplex method as a solution algorithm.

Model: Two models are presented: The first illustrates the general linear programming method. The second shows how a nonlinear objective function can be treated within the framework of a linear model under certain conditions.

THE LINEAR RESPONSE MODEL

Let N_i $(i = 1, 2, \ldots n)$ be the number of times the ith advertising alternative is used per period.

Let the advertising budget be M dollars, and let the cost of one use of media i be c_i. Then the budget constraint is

$$c_1 N_1 + c_2 N_2 + c_3 N_3 + \ldots + c_n N_n \leq M.$$

In addition, subjective constraints can be used which reflect management's conception of the limits to be placed on the media program. The objective is to maximize the number of effective exposures that can be attained given the advertising budget:

$$\max z = s_1 c_1 q_1 e_1 r_1 N_1 + \ldots + s_n c_n q_n e_n r_n N_n,$$

$$\text{with } (0 \leq s_i \leq 1), \ (0 \leq c_i \leq 1), \ (0 \leq q_i \leq 1),$$

$$\text{and } (0 \leq e_i \leq 1)$$

where

s_i = the relative effectiveness of the size of ad i when compared with the largest size ad under consideration, expressed as a decimal;

c_i = the relative effectiveness of the color characteristics of ad i compared with the most effective color ad available, expressed as a decimal;

q_i = qualitative characteristics rating coefficient of alternative i (e.g., appropriateness of editorial climate for the product or past ability to produce successful advertising readership);

e_i = effectiveness rating coefficient (= importance weight times corresponding per cent incidence in magazine readership);

r_i = total readership for medium i.

In general, the response to advertising as a function of the number of insertions in a particular medium is non-linear. (Vidale and Wolfe [#72] use an exponential response function.)

THE NONLINEAR RESPONSE MODEL

The response-to-advertising function is being approximated by straight-line segments. An example is given and it is shown that this method can only be used if the function has a non-increasing first derivative as the amount of promotion increases. Thus, for each medium a nonlinear function of this form can be incorporated into the objective function and the linearity of the model will be maintained.

Reference: #72.

Subject: Media Selection

Title: Media Selection by Decision Programming

Author: Willard I. Zangwill

Source: Journal of Advertising Research, Vol. 5, No. 3, September 1965, 30-36

Summary: An approach for finding an optimal media mix through decision programming, a technique that can take into account discounting, duplication, and availability of the medium, and that will still find an optimum solution.

Model:

Goal: Select a media mix which (1) reaches the most potential customers possible, given the influence of (2) cost, and (3) advertising theme.

Let

selection = specific advertising vehicle, such as specific TV show, magazine, newspaper

choice = sponsor time or space, e.g., half page each week.

Decision Programming constraint equation for selection one:

$$\sum_{j=0}^{3} x_{1j} = 1 \quad x_{1j} = (0 \text{ or } 1) \quad j = 0,1,2,3$$

where

x_{11} = first possible selection, first possible choice

x_{10} = slack variable.

Constraint equation for selection two:

$$\sum_{j=0}^{n_2} x_{2j} = 1 \quad x_{2j} = (0 \text{ or } 1) \quad j = 0,1, \ldots, n_2$$

where

n_2 = total number of possible choices in selection two.

Budget constraint:

$$\sum_{ij} c_{ij} x_{ij} \leq B$$

where B = total budget and the sum is over all possible selections (i) and choices (j).

Other possible constraints:

$$\sum_{ij} a_{ij} x_{ij} \geq A$$

where

a_{ij} = effective exposures for particular choice in the region

A = total number of effective exposures in a given region.

Objective function:

$$\sum_{ij} x_{ij} R_{ij} - \sum_{ijhk} x_{ij} r_{(ij)(hk)} x_{hk}$$

where

R_{ij} = total rated effectiveness units for selection i and choice j

$r_{(ij)(hk)}$ = decrease in effectiveness if selection i and choice j and selection h and choice k are both in the media mix.

The entire program can now be written as:

$$\sum_{ij} x_{ij} R_{ij} - \sum_{ijhk} x_{ij} r_{(ij)(hk)} x_{hk}.$$

Subject to:

$$\sum_{j} x_{ij} = 1 \quad i = 1, \ldots, I \quad \text{where there are I selections}$$

$$\sum_{ij} c_{ij} x_{ij} \leq B \qquad \text{the budget constraint}$$

$$\sum_{ij} x_{ij} a_{ij} \geq A \qquad \text{other constraints including regional, socio-economic, etc.}$$

all $x_{ij} = (0 \text{ or } 1)$.

Subject: Media Selection

Title: Linear Programming and Space-Time Considerations in Media Selection

Author: Stanley F. Stasch

Source: Journal of Advertising Research, Vol. 5, 1965, 40-46

Summary: A linear program is presented that incorporates when and in which markets advertisements should appear. The objective function is modified to assure no multiple counting.

Model: Assume the objectives to be achieved by the media selection to be:

B_1: The total number of effective exposures resulting from the media selection should equal or exceed B_1.

B_2: The total number of effective exposures by readers whose income exceeds \$5,000/year should equal or exceed B_2.

B_3: The total number of effective exposures by readers with a high school education or better should equal or exceed B_3.

B_4: The minimum number of units to be placed in medium one should equal or exceed B_4.

B_5: The maximum number of units to be placed in medium one should be less than or equal to B_5.

B_6: The maximum number of units to be placed in medium two should be less than or equal to B_6.

The problem can now be stated as:

Minimize:

$$c_{1,1}X_{1,1} + c_{1,2}X_{1,2} + c_{2,1}X_{2,1} + c_{2,2}X_{2,2} \qquad (1)$$

Subject to:

$$a_1 f_{1,1}X_{1,1} + a_1 f_{1,2}X_{1,2} + a_2 f_{2,1}X_{2,1} + a_2 f_{2,2}X_{2,2} \geq B_1 \quad (2a)$$

$$d_1 f_{1,1}X_{1,1} + d_1 f_{1,2}X_{1,2} + d_2 f_{2,1}X_{2,1} + d_2 f_{2,2}X_{2,2} \geq B_2 \quad (2b)$$

$$e_1 f_{1,1}X_{1,1} + e_1 f_{1,2}X_{1,2} + e_2 f_{2,1}X_{2,1} + e_2 f_{2,2}X_{2,2} \geq B_3 \quad (2c)$$

$$X_{1,1} + X_{1,2} \geq B_4 \quad (2d)$$

$$X_{1,1} + X_{1,2} \leq B_5 \quad (2e)$$

$$X_{2,1} + X_{2,2} \leq B_6 \quad (2f)$$

The first subscript identifies the medium and the second the color or black and white option. In order to incorporate time into the model, a third subscript is added, and a fourth subscript will be used to identify the particular market area of interest. A two week example is:

Minimize:

$$c_{1,1}X_{1,1,1,s} + c_{1,2}X_{1,2,1,s} + c_{2,1}X_{2,1,1,s} + c_{2,2}X_{2,2,1,s}$$
$$+ c_{1,1}X_{1,1,2,s} + c_{1,2}X_{1,2,2,s}$$
$$\vdots \qquad \vdots$$
$$+ c_{1,1}X_{1,1,1,n} + c_{1,2}X_{1,2,1,n} + c_{2,1}X_{2,1,1,n} + c_{2,2}X_{2,2,1,n}$$
$$+ c_{1,1}X_{1,1,2,n} + c_{1,2}X_{1,2,2,n} \qquad (3)$$

where the subscript s means south market and n means north market.

(3) includes some inaccuracies. If the advertiser purchases one black and white unit of medium one during period one, he pays a price of $c_{1,1}$. For this price he gets one such unit in the north market and one such unit in the south market. If there were also an east market and a west market, the cost value should prefix only one of the four unknown variables; the other three should be prefixed with a zero coefficient. The linear program has a tendency to assign large values to the variables associated with zero cost coefficients. This can be overcome by writing constraints which are internal to the model.

If both an east and a west market were also reached by medium one, the proper internal constraint could be written as:

$$X_{1,1,1,n} = X_{1,1,1,s} = X_{1,1,1,e} = X_{1,1,1,w} \qquad (4)$$

Similar expressions could be written for other media which reach more than one market, independent of the number of markets they reach. An entire set of internal constraints--such as expression (5)--would have to be written.

$$
\left.
\begin{aligned}
X_{1,1,1,n} &= X_{1,1,1,s} \\
X_{1,2,1,n} &= X_{1,2,1,s} \\
X_{2,1,1,n} &= X_{2,1,1,s} \\
\vdots \quad & \quad \vdots
\end{aligned}
\right\}
\qquad (5)
$$

Subject: Media Selection

Title: An Exploration of Linear Programming in Media Selection

Authors: Frank M. Bass and Ronald T. Lonsdale

Source: Journal of Marketing Research, Vol. 3, No. 2, May 1966, 179-88

Summary: Operational details of the application of linear pro-gramming to advertising media selection. The influence of weighting systems used to adjust audience data and various restraint systems is examined with actual data on an oral hygiene product as inputs.

Model:

Linear Models Examined in the Analysis

In the first set of programs, the model was

$$\text{Max} \sum_j c_j x_j,$$

subject to:

$$\sum_j \theta_j x_j < B.$$

Where c_j = number of exposures per unit of vehicle j, x_j = number of units of vehicle j, and θ_j = cost per unit of vehicle j. An appropriate set of c_j values corresponded to each of the three weighting systems. In the first set of programs, additional restraints were introduced sequentially over the x_j values. The process of restraining the variables appearing in the previous solution was carried out for each of the three different weighting systems through six solutions.

In the second set of programs, restraints were imposed over variables, in some cases individual restraints; in others collective restraints, simultaneously. Thus, the program was

$$\text{Max} \sum_j c_j x_j,$$

subject to:

$$\sum_j \theta_j x_j \leq B$$

$$x_1 \leq b_1$$

$$x_2 \leq b_2$$

$$\vdots \qquad \vdots$$

$$x_{15} \leq b_{15}$$

$$x_{16} + x_{61} \leq b_0 \qquad \text{etc.}$$

In the third set of programs, the program was

$$\text{Max} \sum_j C_j x_j ,$$

subject to:

$$\sum_j \theta_j x_j \leq B$$

$$x_1 \leq b_1$$

$$x_2 \leq b_2$$

$$\vdots \qquad \vdots$$

$$x_{15} \leq b_{15}$$

$$x_{16} + x_{61} \leq b_0$$

$$\sum_j \varepsilon_{j1*} \theta_j x_j \leq B_1$$

$$\vdots \qquad \vdots$$

$$\sum_j \varepsilon_{jk*} \theta_j x_j \leq B_k \qquad \text{etc.}$$

Where in the final set of restraints, the number of dollars spent in a certain type of medium is restrained. Thus, if B_1 is an upper bound on the dollars to be spent in magazines, $j \in j_1^*$ indicates that the j index is to run over those vehicles which are magazines.

In the final set of programs, in which restraints are placed over the number of exposures going to different segments of the market, we have:

$$\text{Max} \sum_j c_j x_j,$$

subject to:

$$\sum_j \theta_j x_j \leq B$$

$$x_1 \leq b_1$$

$$x_2 \leq b_2$$

$$\cdot \qquad \cdot$$
$$\cdot \qquad \cdot$$
$$\cdot \qquad \cdot$$

$$x_{15} \leq b_{15}$$

$$x_{16} + x_{61} \leq b_0$$

$$\sum_j \varepsilon_{j1} * \theta_j x_j \leq B_1$$

$$\cdot \qquad \cdot$$
$$\cdot \qquad \cdot$$
$$\cdot \qquad \cdot$$

$$\sum_j \varepsilon_{jk} * \theta_j x_j \leq B_k$$

$$\sum_j a_{1j} x_j \geq \lambda_1$$

$$\sum_j a_{2j} x_j \geq \lambda_2$$

$$\sum_j a_{3j} x_j \geq \lambda_3 \qquad \text{etc.}$$

Where in the final set of constraints, the number of dollars spent in a certain type of medium is restrained and, in addition, the number of exposures going to different segments of the market are also restrained, λ_i is the minimum number of exposures to be allocated to market segment i, and a_{ij} is the number of exposures in segment i per unit of vehicle j.

Subject: Media Selection

Title: A Goal Programming Model for Media Planning

Authors: A. Charnes, W.W. Cooper, J.K. DeVoe, D.B. Learner, W. Reinecke

Source: Management Science, Vol. 14, No. 8, April 1968, B-423-430

Summary: A goal programming model for selecting media is presented which accounts for cumulative duplicating audiences over a variety of time periods. This permits detailed control of the distribution of message frequencies directed at each of numerous marketing targets over a sequence of interrelated periods. This is accomplished via a logarithmic nonreach device and a continuous lognormal generation of the discrete message frequencies.

Model:

Let

$d_{kij}(t)$ = gross kth audience segment obtained by the jth cumulative purchase of medium i in period t,

$x_{ij}(t)$ = the jth cumulative purchase of medium i in period t

where $x_{ij}(t) = 0$ or 1. We approximate this by requiring

$$\sum_j x_{ij}(t) \leqq 1, \quad x_{ij}(t) \geqq 0. \tag{1}$$

Thus,

$$\sum_i \sum_j d_{kij}(t) x_{ij}(t). \tag{2}$$

Considering the net audience or "reach," let

$r_{kij}(t)$ = proportion of the kth net audience segment obtained by jth cumulative purchase of medium i in period t (3)

and $R_k(t)$ = proportion of the net kth audience segment obtained by media purchases in period t.

Then

$$\ln(1 - R_k(t)) = \sum_i \sum_j \ln(1 - r_{kij}(t) x_{ij}(t) \tag{4}$$

since

$$1 - R_k(t) = \Pi_{i,j}(1 - r_{kij}(t))^{x_{ij}(t)}. \tag{5}$$

Estimating the distribution of frequencies for the net k^{th} audience segment, let

$H_{ks}(t)$ = proportion of the net k^{th} audience (6) segment which is reached s or more times in period t.

Replacing the relation between means and variances of normal and log-normal distributions by the following approximations:

$$\mu_k(t) = A + B \sum_i \sum_j P_{ki}(t)x_{ij}(t) + C \sum_i \sum_j jx_{ij}(t) \quad (7)$$

$$\sigma_k(t) = D + E \sum_i \sum_j jx_{ij}(t)$$

with $\qquad P_{ki}(t) = \dfrac{d_{ki1}(t)}{U_k}$

and U_k = total number in universe for the kth audience segment,

where μ_k and σ_k are, respectively, the mean and standard deviation of the associated normal distribution and the A, B, C, D and E are constants determined empirically.

The correspondence with the discrete distributions is made via

$$\frac{\ln(s-0.5) - \mu_k(t)}{\sigma_k(t)} = Z_{(1-H_{ks}(t))} \quad (8)$$

where Z is the studentized normal variate--viz., $N(0,1)$. Employing (7)

$$Z = \frac{\ln(s-0.5) - A - B \sum_{i,j} P_{ki}(t)x_{ij}(t) - C \sum_{i,j} jx_{ij}(t)}{D + E \sum_{i,j} jx_{ij}(t)} \quad (9)$$

where Z is the fractile associated with $1 - H_{ks}(t)$ for $N(0,1)$.

The constraints on media choices are of various types such as, for instance, those on gross k^{th} audience segments at specified times--viz.,

$$\sum_i \sum_j d_{kij}(t) \, x_{ij}(t) \geq D_k(t) \quad (10)$$

as well as over selected time intervals,

$$\sum_{t \varepsilon T} \sum_i \sum_j d_{kij}(t) \; x_{ij}(t) \geq D_k(T).\tag{11}$$

Similarly, for the net kth audience segment we might require

$$R_k(t) \geq N_k(t)\tag{12}$$

or

$$\ln \; (1 - R_k(t) \leq \ln \; (1 - N_k(t))\tag{13}$$

so that, via (4),

$$\sum_i \sum_j \ln \; (1 - r_{kij}(t)) x_{ij}(t) \leq \ln \; (1 - N_k(t))\tag{14}$$

where $N_k(t)$ is some prescribed fraction of the kth audience segment at time t.

We may wish to specify constraints such as, "at least 80% of the kth audience segment must be reached at least twice at specified times, t." Then

$$Z_{1-H_{k,2}}(t) \leq Z_{1-.8} = Z_{.2} = N^{-1}(.2)\tag{15}$$

where N is the distribution function for $N(0,1)$. Thus, from (8) typical constraints of this type may be represented by

$$\frac{\ln \; (s - 0.5) - \mu_k(t)}{\sigma_k(t)} \geq Q\tag{16}$$

or, from (9),

$$(C + EQ) \sum_{i,j} j x_{ij}(t) + B \sum_{i,j} P_{ki}(t) \; x_{ij}(t)\tag{17}$$

$$\leq \ln \; (s - 0.5) - A - DQ.$$

From the constraints a specific subset may be selected to be fulfilled as close to equality as possible in the sense of "goal programming." For instance, it may be desired to come as close as possible to reaching, say, 85% of the kth audience segment in time t_1. Thus,

$$\sum_i \sum_j \ln \; (1 - r_{kij}(t_1)) x_{ij}(t_1) + u^+ - u^- = \ln(1 - .85)\tag{18}$$

where $u^+ + u^-$ would also appear in the functional to be minimized. Similarly if it were desired to establish a goal for the frequency constraint, from (17)　we obtain

$$(C + EQ) \sum_{i,j} j x_{ij}(t_1) + B \sum_{i,j} P_{ki}(t_1) x_{ij}(t_1) + v^+ - v^- \quad (19)$$

$$= \ln (s - 0.5) - A - DQ$$

and, $v^+ + v^-$ would appear in the functional.
If the goals were to have relative weights W_1, W_2 the functional to be minimized would be

$$z \equiv W_1 (u^+ + u^-) + W_2 (v^+ + v^-). \quad (20)$$

Subject: Media Selection

Title: A Probabilistic Approach to Industrial Media Selection

Author: David A. Aaker

Source: Journal of Advertising Research, Vol. 8, September 1968,
          46-54

Summary: The Probabilistic Optimizing Model for Selecting Insertion
          Schedules (POMSIS) is an approach to the following problem:
          Within a given budget, which journal insertion schedule
          will obtain the greatest impact for an advertising campaign
          of given length? By disaggregating to the level of a
          potential exposure to individuals in a sample population,
          the model permits to assign appropriate weights to
          successive exposures.

Model: The objective is to select an insertion schedule subject
          to budget limitations (and other constraints) that will
          provide the maximum number of "effective exposures" within
          a given time period. There are four dimensions to
          determine the value of the effective exposure. An individual
          will receive one with a relative value of 1.0, if:

   1. This exposure is the first for the particular individual.

   2. The individual is a member of the customer class
      regarded as the most important target of the advertise-
      ment.

   3. The journal involved provides the best possible
      environment for the advertisement.

   4. The individual is actually exposed to the advertisement
      in that journal.

POMSIS

Select the Insertion Schedule that will maximize
Total Effective Exposures:

$$TEE = \sum_k (N_k/n_k) \sum_{i \varepsilon k} \sum_j x_j (y_i w_k v_j P_{ij})$$

Subject to:

$$\sum_j c_j x_j \leq b \quad \text{budget constraint}$$

$$x_j \leq max_j \quad \text{insertion constraint}$$

$$\sum_{i\epsilon k}\sum_{j} x_j v_j P_{ij} \geq m_k \quad \text{minimum exposure constraint}$$

where:

k = index referencing the individual;

i = index referencing the individual in the sample;

j = index referencing the journal;

N_k = total size of segment k;

n_k = sample size from segment k;

x_j = number of insertions in journal j;

y_i = weight reflecting the multiple exposure effect on individual i;

w_k = relative weight attached to segment k;

v_j = relative weight attached to journal j;

P_{ij} = probability that individual i is exposed to an advertisement in journal j;

c_j = cost of the marginal insertion in journal j (subscript denoting the insertion number suppressed);

b = allowable dollar expenditure;

$\max_j$ = maximum number of insertions allowed in journal j;

m_k = minimum exposure level for segment k.

The selection heuristic adds insertions incrementally until the budget constraint is reached.

Subject: Media Selection

Title: A Media Planning Calculus

Authors: John D.C. Little and Leonard M. Lodish

Source: Operations Research, Vol. 17, January-February 1969,
          1-35

Summary: Heuristic problem solving. The model incorporates non-
          linear response, market segmentation, and forgetting,
          and is optimized by dynamic programming. The response
          is expanded as a power series in exposure level.
          The expected response then becomes a weighted sum of
          the moments of the exposure-level distribution.

Model:

1.1 Media, Exposure Levels, and Forgetting

A media class will be a general means of communication, such as television, magazines, or newspapers. A
media vehicle will be a cohesive grouping of advertising
opportunities within a class, such as a particular TV
show, magazine, or newspaper. A media option will be a
detailed, purchasable unit within a vehicle.

It is assumed that a media option: (1) is available
exactly once in every time period, (2) has substantial
continuity of audience, and (3) has continuity in outward
format.

Let

M = number of media options under consideration;

T = number of time periods in the planning horizon;

S = number of market segments;

$$x_{jt} = \begin{cases} 1, \text{ if an insertion is made in option } j \text{ in time} \\ \quad \text{period } t, \\ 0, \text{ if not.} \end{cases}$$

$$z_{ijt} = \begin{cases} 1, \text{ if the person in segment } i \text{ is exposed to an} \\ \quad \text{insertion in media option } j \text{ in period } t, \\ 0, \text{ if not.} \end{cases}$$

e_{ij} = exposure value (weight) for an exposure in media option j going to a person in market segment i (exposure value/exposure).

$\sum_{j=1}^{j=M} e_{ij} z_{ijt}$ = increase in exposure level of a particular individual in market segment i in time period t (exposure value/capita).

We suppose that the effect of advertising wears off because of forgetting. Specifically, it is assumed that, in the absence of new input, exposure level decreases by a constant fraction each time period. Let

y_{it} = exposure level of a particular individual in market segment i in time period t (exposure value/capita).

α = memory constant: the fraction of y_{it} retained from one time period to the next, $0 \leq \alpha < 1$.

Then

$$y_{it} = \alpha y_{i,t-1} + \sum_{j=1}^{j=M} e_{ij} z_{ijt} \qquad (1)$$

(1) can be rewritten as

$$y_{it} = \sum_{s=-\infty}^{s=t} \sum_{j=1}^{j=M} \alpha^{t-s} e_{ij} z_{ijs} \qquad (2)$$

1.2 Market Response

Let

n_i = number of people in market segment i,

w_{it} = sales potential (weight) of a person in segment i in time period t (potential units/capita/time period),

$r(y_{it})$ = response function: the fraction of potential realized when a person has exposure level y_{it},

$f_{it}(\cdot)$ = probability density of y_{it}.

Let E denote the taking of expected values. Then $w_{it}E\{r(y_{it})\}$ is the average realized sales potential per person in market segment i at time t. Summing, we obtain

$$R = \sum_{i=1}^{i=s} \sum_{t=1}^{t=T} n_i w_{it} E\{r(y_{it})\}$$

$$= \text{total market response (potential units)} \quad (3)$$

A function with only diminishing return is

$$r(y) = r_o + a(1 - e^{-by}) \quad (0 \leq y < \infty) \quad (4)$$

where r_o, a, and b are nonnegative constants specific to the product at hand.

The expected response $E\{r(y_{it})\}$ for a given market segment and time period can be expressed in terms of the moments of the distribution $f_{it}(it)$. For notational simplicity, we drop the subscripts i and t for the present. Let

$\mu = E\{y\} = $ mean of y,

$\mu_n = E\{(y-\mu)^n\} = $ nth moment of y about the mean, $n > 1$.

We can expand $r(y)$ in a Taylor series about μ:

$$r(y) = r(\mu) + \sum_{k=1}^{n-1} (1/k!) r^{(k)}(\mu)(y-\mu)^k$$

$$+ (1/n!) r^{(n)}(y_1)(y-\mu)^n \quad (5)$$

where $r^{(k)}(\mu)$ is the kth derivative of $r(y)$ evaluated at $y = \mu$ and y_1 is some value between y and μ. Taking expectations:

$$E(r) = r(\mu) + \sum_{k=2}^{n-1} (1/k!) r^{(k)}(\mu)\mu_k$$

$$+ (1/n!) E\{r^{(n)}(y_1)(y-\mu)^n\}. \quad (6)$$

1.3 Exposure Arithmetic

Let

y = exposure level of a particular individual,

$$z_j = \begin{cases} 1, & \text{if the individual is exposed to option } j, \\ \\ 0, & \text{if not,} \end{cases} \tag{7}$$

$$y = \sum_{j=1}^{j=M} e_j z_j$$

$p_j = P(z_j = 1) = P$ (a person is exposed to option j),

$p_{jk} = P(z_j = 1, z_k = 1) = P$ (a person is exposed to both option j and option k).

The mean of y is

$$E(y) = \sum_{j=1}^{j=M} e_j p_j \tag{8}$$

The second moment of y is

$$E\{y^2\} = E\{(\sum_{j=1}^{j=M} e_j z_j)^2\} = \sum_{j=1}^{j=M} \sum_{k=1}^{k=M} e_j e_k E\{z_j z_k\}$$

$$= \sum_{j=1}^{j=M} e_j^2 p_j + 2 \sum_{j=1}^{M-1} \sum_{k=j+1}^{k=M} e_j e_k p_{jk} \tag{9}$$

Generally,

$$\mu_\eta = \mu_\eta(\mu_2, \mu) \tag{10}$$

Then, using (8), (9), and (10), we obtain moments

$$\mu_{it} = E(y_{it}) = \sum_{s=-\infty}^{s=t} \sum_{j=1}^{j=M} \alpha^{t-s} e_{ij} p_{j|is} \tag{11}$$

Let

$g_{j|i}$ <u>market coverage</u> of the media vehicle of option j in segment i, defined as the fraction of people in segment i who are in the audiences of the vehicle of option j, averaged over a year.

s_{jt} = <u>audience seasonality</u>, the seasonal index for the vehicle of option j in time period t. Average value over a year is 1.0.

h_j = <u>exposure probability for audience member</u>. The probability a person is exposed to an insertion in option j given that he is in the audience of the vehicle of j.

Then

$$P_{j|it} = g_{j|i} h_j s_{jt} x_{jt} \tag{12}$$

The duplication probabilities are

$$P_{jk|its} = g_{jk|i} h_j h_k s_{jt} s_{ks} x_{jt} x_{ks} \tag{13}$$

where $g_{jk|i}$ = <u>segment duplication</u>.

1.4 Budget Constraint

Let

c_{jt} = cost of an insertion in media option j in time t (dollars/insertion).

B = total budget for the planning period (dollars).

Then

$$\sum_{j=1}^{j=M} \sum_{t=1}^{t=T} c_{jt} x_{jt} \leq B \tag{14}$$

1.5 Mathematical Program

Find x_{jt} $(j = 1, \ldots, M; \; t = 1, \ldots, T)$ and maximal R subject to

$$R = \sum_{i=1}^{i=s} \sum_{t=1}^{T+E} n_i \cdot w_{it} \{ r(\mu_{it})$$

$$+ \sum_{m=2}^{m=n} (1/m!) r^{(m)} (\mu_{it}) \mu_{mit} \} \sum_{j=1}^{j=M} \sum_{t=1}^{t=T} c_{jt} x_{jt} \leq B$$

$$\mu_{it} = \sum_{s=-K}^{s=t} \sum_{j=1}^{j=M} \alpha^{t-s} e_{ij} h_j g_{j|i}{}^s_{js} x_{js}$$

$$\mu_{2it} = \sum_{s=-K}^{s=t} \sum_{j=1}^{j=M} (\alpha^{t-s} e_{ij})^2 g_{j|i} h_j{}^s_{js} x_{js}$$

$$+ 2 \sum_{s=-K}^{t-1} \sum_{r=s+1}^{r=t} \sum_{j=1}^{j=M} \sum_{k=1}^{k=M} \alpha^{2t-r-s} e_{ij} e_{ik} g_{jk|i} h_j h_k{}^s_{js}{}^s_{kr} x_{js} x_{kr}$$

$$+ 2 \sum_{s=-K}^{s=t} \sum_{j=1}^{M-1} \sum_{k=j+1}^{k=M} \alpha^{2t-2s} e_{ij} e_{ik} g_{jk|i} h_j h_k{}^s_{jt}{}^s_{ks} x_{jt} x_{ks} \; - \mu_{it}^2$$

$$\mu_{mit} = \mu_m (\mu_{it}, \mu_{2it}) \; m = 3, \ldots, n \quad i = 1, \ldots, S \quad t = 1, \ldots, T+E$$

$x_{jt} \varepsilon \{0,1\}$ for all (j,t), $x_{jt} = 1$ for $(j,t) \varepsilon I_1$, $x_{jt} = 0$ for $(j,t) \varepsilon I_2$

Reference: #50.

Subject: Media Selection

Title: Considering Competition in Media Planning

Author: Leonard M. Lodish

Source: Management Science, Vol. 17, No. 6, February 1971, B-293-306

Summary: A normative mathematical procedure for the media planning
problem is proposed which explicitly considers the effect
of competitors' media schedules. A predictive model is
developed to evaluate expected market response due to an
advertising media schedule considering the anticipated
schedules of competitors as well as other major advertising
phenomena. Heuristic search routines are used to select
and schedule media with the objective of maximizing market
response subject to budget limitations. The procedure
has been applied on real problems.

The market response model first divides people into
market segments which are characterized by product class
sales potential. Ads placed by the competing firms cause
people in segments to be exposed to this advertising and
thereby create a level of exposure value which decays
over time in the absence of new exposures. The individual's
response during a time period is a function of his
retained exposure value for each competing firm and his
market segment. Summing over individuals and over time
to obtain total market response is approximated analytically
using media coverage and overlap data that is relatively
easy to gather and store.

The model represents refinements and additions to #49.

Model: Let

F = number of firms in the industry or product class

M = number of media options under consideration

T = number of time periods in the planning horizon

S = number of market segments

x_{jt} = 1 if an insertion is made in media option j in
time period t

= 0 if not

z_{ijt} = number of times the person in segment i is exposed
during period t to an insertion in media option j
in period t

e_{ij} = exposure value conveyed by one exposure in option j
to a person in segment i (exposure value/exposure)

J_c = set of options of company c

$\sum_{j \in J_c} e_{ij} z_{ijt}$ = increase in exposure level of an individual
in market segment i in time period t from
insertions in options of company c (exposure
value/capita)

y_{cit} = retained exposure value of a particular individual in market segment i in time period t due to insertions in options of company c

α = memory constant: the fraction of y_{cit} retained from one time period to the next. $0 \leq \alpha \leq 1$.

Then

$$y_{c,i,t} = \alpha y_{c,i,t-1} + \sum_{j \in J_c} e_{ij} z_{ijt} \tag{1}$$

(1) can be rewritten as:

$$y_{cit} = \alpha^t y_{c,i,0} + \sum_{s=1}^{t} \sum_{j \in J_c} \alpha^{t-s} e_{ij} z_{ijs} \tag{2}$$

or, going back indefinitely, as

$$y_{cit} = \sum_{s=-\infty}^{t} \sum_{j \in J_c} \alpha^{t-s} e_{ij} z_{ijs} \tag{3}$$

Considering a single market segment and a single time period, let

N = number of firms in the industry

y_f = retained exposure value of a person due to firm f's advertising

$Y_I = \sum_{f=1}^{N} y_f$ = retained exposure value of a person due to advertising of all firms in the industry

w = average sales potential (weight) of a person for products of all the firms in the industry at a given industry price and distribution level

$\Gamma_I(Y_I)$ = industry response function: the average fraction of industry potential realized when a person has industry retained exposure value of Y_I

K_f = relative sales effects for firm f of other marketing variables such as the expected price and distribution levels

$\Gamma_f(y_f)$ = firm f relative response function.

Market share for firm f, MS_f, is then modeled as follows:

$$MS_f = K_f \Gamma_f(y_f) / \sum_{f=1}^{N} K_f \Gamma_f(y_f) \quad \text{(dimensionless)}. \tag{4}$$

Expected sales of firm f to the individual, SLS_f, are then

$$SLS_f = w\Gamma_I(\textstyle\sum_{f=1}^{N}Y_f)\,(K_f\Gamma_f(y_f)/\textstyle\sum_{f=1}^{N}K_f\Gamma_f(y_f)) \quad \text{(potential} \tag{5}$$
$$\text{units/capita)}.$$

Taking expected values in (5) we can obtain the average sales per person of firm f as:

$$E(SLS) = wE\{\Gamma_f(\textstyle\sum_{f=1}^{N}Y_f)\,(K_f\Gamma_f(y_f)/\textstyle\sum_{f=1}^{N}K_f\Gamma_f(y_f))\} \tag{6}$$
$$\text{(potential unit/capita)}$$

We can now restore segment and time period subscripts i and t respectively.

Let

S = the number of segments,

T = the number of time periods,

n_i = number of people in segment i.

If we sum the average sales of firm f over all segments and periods, we obtain:

$$R_j = \sum_{i=1}^{S}\sum_{t=1}^{T} n_i w_{it} E\{\Gamma_I(\textstyle\sum_{j=1}^{F} y_{fit})\,(K_{fi}\Gamma_f(y_{fit})/\textstyle\sum_{f=1}^{F}K_{fi}\Gamma_f(y_{fit}))\} \tag{7}$$
$$= \text{total expected sales for firm } f \text{ (potential units).}$$

This is the objective function of the mathematical program. Now let

$$r(\mathbf{y}_{it}) = r(y_{1it},\ \ldots,\ y_{fit}) = \Gamma_I(\textstyle\sum_{f=1}^{F}y_{fit})\,(K_{1i}\Gamma_1(y_{1it})/$$
$$\textstyle\sum_{f=1}^{F} K_{fi}\Gamma_f(y_{fit}))$$

$$u_{fit} = E(y_{fit}),$$

$$\mathbf{u}_{it} = (u_{1it},\ u_{2it},\ \ldots,\ u_{fit}), \tag{8}$$

$D_{jk}r(\mathbf{y}_{it}) = D_r(D_kr(\mathbf{y}_{it}))$ the second partial of r with respect to y_{kit} and y_{jit} at $\mathbf{y}_{it}$,

$D^kr(\mathbf{y}_{it},\mathbf{h}) = $ the kth order differential of r at $\mathbf{y}_{it}$ with respect to h.

If we assume r has continuous partial derivatives of order 3, we can expand $r(\mathbf{y}_{it})$ about $\mathbf{u}_{it}$ as follows:

$$r(\mathbf{y}_{it}) = r(\mathbf{u}_{it}) + \sum_{k=1}^{2} (1/k!) d^k r(\mathbf{u}_{it}; \mathbf{y}_{it} - \mathbf{u}_{it})$$

$$+ \frac{1}{3}! \, d^3 r(\mathbf{z}_{it}; \mathbf{y}_{it} - \mathbf{u}_{it})$$

where $\mathbf{z}_{it}$ lies between $\mathbf{y}_{it}$ and $\mathbf{u}_{it}$.

Considering a specific segment and time period, and expressing the above explicitly as sums of differentials we obtain:

$$r(y_1, y_2, \ldots, y_N)$$

$$= r(u_1, u_2, \ldots, u_N) + \frac{1}{1} \left(\sum_{i=1}^{N} D_i r(u_1, \ldots, u_N)(y_i - u_i) \right)$$

$$+ \frac{1}{2} \left(\sum_{i=1}^{N} \sum_{j=1}^{N} D_{ij} r(u_1, \ldots, u_N)(y_i - u_i)(y_j - u_j) \right)$$

$$+ \frac{1}{6} \left(\sum_{i=1}^{N} \sum_{j=1}^{N} \sum_{k=1}^{N} D_{ijk} r(z, \ldots, z_N)(y_i - u_i)(y_j - u_j)(y_k - u_k) \right),$$

where z_i is some value between y_i and u_i for $i = 1, N$. Taking expectations:

$$E\{r\} = r(\mathbf{u}) + \frac{1}{2} \left(\sum_{i=1}^{N} \sum_{j=1}^{N} D_{ij} r(\mathbf{u}) u_{2ij} \right) + \frac{1}{6} \left(\sum_{i=1}^{N} \sum_{j=1}^{N} \sum_{k=1}^{N} D_{ijk} r(\mathbf{z}) u_{3ijk} \right)$$

$$(8a)$$

where

$$u_{2ij} = E\{(y_i - u_i)(y_j - u_j)\} = \text{the covariance of } y_i \text{ and } y_j$$

and

$$u_{3ijk} = E\{(y_i - u_i)(y_j - u_j)(y_k - u_k)\}.$$

If we observe that all differentials, u_{2ij}, and u_{3ijk}, are constant under permutation of subscripts (e.g., $D_{ij} = D_{ji}$) when the function r has all of its partials continuous at all points under consideration, then (8a) can be simplified to

$$E(r) = r(\mathbf{u}) + \frac{1}{2} \left(\sum_{i=1}^{N} D_{ii} r(\mathbf{u}) u_{2ii} + 2 \sum_{i=1}^{N-1} \sum_{j=i+1}^{N} D_{ij} r(\mathbf{u}) u_{2ij} \right)$$

$$+ \frac{1}{6} \left(\sum_{i=1}^{N} \sum_{j=1}^{N} \sum_{k=1}^{N} D_{ijk} r(\mathbf{z}) u_{3ijk} \right). \tag{9}$$

In this paper, we will take the first two terms above as our approximation. Restoring the i and t subscripts, we can approximate $E\{r_i(\mathbf{y}_{it})\}$ as follows:

$$E\{r_i(\mathbf{y}_{it})\}$$

$$= r_i(\mathbf{u}_{it}) + \left(\frac{1}{2}\right) \left(\sum_{j=1}^{F} D_{jj} r_i(\mathbf{u}_{it}) u_{2itjj} \right.$$

$$+ 2 \sum_{j=1}^{F-1} \sum_{k=j+1}^{F} D_{jk} r(\mathbf{u}_{it}) u_{2itjk} \right) \tag{10}$$

where

$$u_{2itjk} = E\{(y_{jit} - u_{jit})(y_{kit} - u_{kit})\} = \text{the covariance}$$

of y_{jit} and y_{kit}.

Exposure Subgroups.

Deleting i and t subscripts, let

$V_c = 1$ if an individual has been exposed to at least one option of firm c,

$\quad = 0$ otherwise.

$V(k_1, k_2, \ldots, k_N) = $ the even $V_1 = k_1$ and $V_2 = k_2$, and $V_N = k_N$ where $k_i = 0$ or 1 for $i = 1, \ldots, N$

$P(k_1, \ldots, k_N) = $ probability of the event $V(k_1, \ldots, k_N)$

$E(r | V(k_1, \ldots, k_N)) = $ the conditional expectation of r given that $V_1 = k_1, \ldots, V_N = k_N$

$\mathbf{u}(k_1, \ldots, k_N) = E(\mathbf{y} | V(k_1, \ldots, k_N))$

$u_{2ij}(k_1, \ldots, k_N) = E\{y_i - u_i)(y_j - u_j) | V(k_1, \ldots, k_N)\}$

$\quad = $ the conditional means and second moments about the mean of $\mathbf{y}$ given that $V_1 = k_1, \ldots, V_N = k_N$.

The approximation of (10) then becomes

$$E(r) = \sum_{k_1=0}^{1} \sum_{k_2=0}^{1} \cdots, \sum_{k_N=0}^{1} P(k_1, \ldots, k_N) \tag{11}$$

$$\cdot [r(u(k_1,\ldots,k_N)) + \frac{1}{2} \sum_{i=1}^{N} D_{ii} r(u(k_1,\ldots,k_N)) u_{2ii}(k_i,\ldots,k_N)$$

$$+ 2 \sum_{i=1}^{N} \sum_{j=i+1}^{N} D_{ij} r(u(k_1,\ldots,k_N)) u_{2ij}(k_1,\ldots,k_N)].$$

Let

y_c = exposure level of a particular individual due to insertions in options of company c.

z_j = the number of times an individual is exposed to option j.

$$y_c = \sum_{j \in J_c} e_j z_j. \tag{12}$$

Let

$$P_{rj} = p(z_j = r) = p(\text{a person is exposed } r \text{ times to option j}),$$

$P_{rv;jk} = p(z_j = r, z_k = v) = p$ (a person is exposed both r times to option j and v times to option k).

We assume that the person sees the insertion during the same time period in which the insertion was made. The mean of y_c is simply

$$E(y_c) = \sum_{j \in J_c} e_j \sum_{r=1}^{\infty} r p_{rj} = u_c. \tag{13}$$

The second moment of y_c is

$$E\{y_c^2\} = E\{(\sum_{j \in J_c} e_j z_j)^2\} = \sum_{j \in J_c} \sum_{k \in J_c} e_j e_k E\{z_j z_k\} \tag{14}$$

$$= \sum_{j \in J_c} e_j^2 \sum_{r=1}^{\infty} r^2 p_{rj}$$

$$+ \sum_{j \in J_c} \sum_{k \in J_c; k \neq j} e_j e_k (\sum_{r=1}^{\infty} \sum_{v=1}^{\infty} r v p_{r,v;j,k}).$$

Then the variance of y_c is

$$u_{2cc} = E\{y_c^2\} - u_c^2. \tag{15}$$

To determine the covariance between y_c and y_d, u_{2cd} we need $E\{y_c y_d\}$ which is

$$E\{y_c y_d\} = E\{\sum_{j \in J_c} e_j z_j\}(\sum_{k \in J_d} e_k z_k)\} = \sum_{j \in J_c} \sum_{k \in J_d} e_j e_k E\{z_j z_k\}$$

$$= \sum_{j \in J_c} \sum_{k \in J_d} e_j e_k \ (\sum_{r=1}^{\infty} \sum_{v=1}^{\infty} rv p_{r,v;j,k}). \quad (16)$$

The covariance of y_c and y_d is

$$u_{2cd} = E\{y_c y_d\} - u_c u_d.$$

Let

g_j = market coverage of the media vehicle of option j in the segment,

s_j = audience seasonality,

h_j = average exposures per audience member,

g_{jk} = segment duplication.

Then we take

$$g_{jk} h_j h_k = \sum_{r=1}^{\infty} \sum_{v=1}^{\infty} rv p_{r,v;j,k} = E(z_j z_k). \quad (17)$$

Let

$q_j = h_j$ if the medium of option j is a broadcast medium

$\quad = h_j + h_j^2$ if the medium of option j is a print medium.

Then

$$\sum_{r=1}^{\infty} r^2 p_{rj} = g_j q_j. \quad (18)$$

Using the definitions in (17) and (18), equation (14) becomes

$$E\{y_c^2\} = \sum_{j \in J_c} e_j^2 \ s_j g_j q_j + \sum_{j \in J_c} \sum_{k \in J_c; k \neq j} e_j e_k g_{jk} h_j h_k. \quad (19)$$

(13) becomes

$$E(y_c) = u_c = \sum_{j \in J_c} e_j g_j s_j h_j. \quad (20)$$

(16) becomes

$$E\{y_c y_d\} = \sum_{j \in J_c} \sum_{k \in J_d} e_j e_k g_{jk} h_j h_k. \tag{21}$$

Time period t and market segment i subscripts and relations to the decision variables are now restored by the following correspondences.

$$z_j \longleftrightarrow z_{ijt}, \qquad g_j \longleftrightarrow g_{ij},$$

$$e_j \longleftrightarrow \alpha^{t-s} e_{ij}, \qquad s_j \longleftrightarrow s_{jt},$$

$$y_c = \sum_{j \in J_c} e_j z_j \longleftrightarrow y_{cit} = \sum_{s=-\infty}^{t} \sum_{j \in J_c} \alpha^{t-s} e_{ij} z_{ijs},$$

$$p_{rj} \longleftrightarrow p_{rijt} = p(z_{ijt} = r).$$

$$q_j \longleftrightarrow q_j,$$

$$g_{ji} \longleftrightarrow g_{ijt;ks} = \text{segment duplication.}$$

We redefine J_c as the set of all available options of company c.

$$u_{cit} = E(y_{cit}) = \sum_{s=-\infty}^{t} \sum_{j \in J_c} \alpha^{t-s} e_{ij} g_{ij} s_{js} h_j x_{js}. \tag{22}$$

$$u_{2itcc} = V_{ar}(y_{cit}) = \sum_{s=-\infty}^{t} \sum_{j \in J_c} (\alpha^{t-s} e_{ij})^2 s_{js} g_{ij} q_j x_{js} \tag{23}$$

$$+ \sum_{j \in J_c} \sum_{k \in J_c; k \neq j} \sum_{s=-\infty}^{t} \sum_{r=-\infty}^{t} [\alpha^{t-s} e_{ij} \alpha^{t-r} e_{ik} h_j h_k g_{ijt;ks} x_{js} x_{kr}]$$

$$- u_{cit}^2.$$

$$u_{2itcd} = \text{Cov}(y_{cit}, y_{dit}) \tag{24}$$

$$= \sum_{j \in J_c} \sum_{k \in J_d} \sum_{s=-\infty}^{t} \sum_{r=-\infty}^{t} [\alpha^{t-s} e_{ij} \alpha^{t-r} e_{ik} h_j h_k g_{ijt;ks} x_{js} x_{kr}]$$

$$- u_{cit} u_{dit}.$$

It is shown that the conditional moments are simple functions of the unconditional moments above and the probability that a person is in a particular exposure subgroup. The various ways of estimating the required probability are discussed.

Finally, the budget constraint is

$$\sum_{j=1}^{M} \sum_{t=1}^{M} c_{jt} x_{jt} \leqq B,$$

where

c_{jt} = cost of an insertion in media option j in time t (dollars),

B = total budget for the planning period (dollars).

Reference: #49.

Subject: Media Selection

Title: ADMOD: An Advertising Decision Model

Author: David A. Aaker

Source: Journal of Marketing Research, Vol. 12, February 1975, 34-45

Summary: ADMOD is an advertising decision model designed to address simultaneously the budget decision, the copy decision, and the media-allocation decision. The model focuses upon specific consumer decisions that advertising is attempting to precipitate and is illustrated with data on dishes.

Model: The goal of ADMOD is to select a budget level, a copy approach, and a media insertion schedule to maximize the objective function. The task of the objective function is to assign a value to a given media insertion schedule. A selection heuristic has the task of generating the various insertion schedules to be evaluated.

The value of a given insertion schedule, the objective function, can be written:

$$V = \sum_s \frac{N_s}{n_s} \sum_{i \varepsilon s} w_s \sum_{z_i=0}^{\infty} a_{ci}(z_i) f_{ci}(z_i) - \sum_j k_j x_{cj},$$

where:

c = index referencing the copy alternative,

i = index referencing the individual in the sample population,

j = index referencing the media vehicle (Time, NBC News, etc.),

s = index referencing the market segment,

N_s = the size of segment s,

n_s = the size of the sample from segment s,

w_s = the value to the firm of the consumer action (i.e., a trial purchase) by a member of market segment s,

z_i = the number of exposures received by individual i, given the insertion schedule,

$a_{ci}(z_i)$ = the probability that the desired consumer action (i.e., a trial purchase) will occur, given the fact that z_i exposures occurred,

$f_{ci}(z_i)$ = the probability that individual i will receive exactly z_i exposures, given the insertion schedule,

k_j = the cost of an insertion in vehicle j,

x_{cj} = the insertion of copy alternative c into vehicle $j(x_{cj} = 0,1)$.

There is a wide variety of constraints that could be introduced. The number of insertion options considered for a given vehicle may be limited by the frequency with which the vehicle appears during the time period, for example.

Because of the irregular nature of the objective function, it is necessary to use a heuristic to search systematically through media insertion schedules to select one with a high value. To implement the constraint that any one schedule will include one copy alternative, the heuristic will be employed for each copy alternative. The one with the highest value will be the one selected.

Subject: Advertising and Promotion Expenditure

Title: The Determination of Advertising Expenditure

Author: Arne Rasmussen

Source: Journal of Marketing, Vol. 16, No. 4, April 1952, 439-46

Summary: The optimal advertising expenditure in the short-run
is determined and compared to the Percentage-of-Sales
method.

Model: Let q = quantity, p = price, and A = advertising dollars.
Then the price elasticity of demand is dq/q: dp/p and the
advertising elasticity of demand is dq/q: dA/A.
Assume that price and unit cost are constant and that the
advertising dollars are spent in the most efficient way.
Let r = gross revenue per unit (defined as price minus
all variable costs per unit apart from advertising); then

$$dqr = dA, \tag{1}$$

$$dq/dA = 1/r, \qquad \text{and} \tag{2}$$

$$\frac{dq}{dA}\frac{A}{q} = \frac{A}{qr}. \tag{3}$$

Thus, the advertising expenditure which maximizes profit
in the short-run is reached when the (expected) advertising
elasticity of demand equals the relation between total
advertising expenditure and total gross revenue.

The content of the Percentage-of-Sales method is that
advertising expenditure is determined as a given percentage
of the expected sales in dollars. If we call this percentage
k, we obtain

$$A = k(pq). \tag{4}$$

However, since price and gross revenue per unit are constant,
we have

$$\frac{A}{rq} = k\ \frac{p}{r}. \tag{5}$$

Thus, if a firm spends a given percentage of the sales in
dollars of advertising, then it spends a given percentage of
the total gross revenue on advertising.

Spending the same percentage of sales in dollars on advertising means that the firm "assumes" the advertising elasticity of demand to be constant under all conditions. This cannot be true, since the advertising elasticity must fluctuate according to the size of advertising expenditure, cyclical position, competitors' reactions, and so on. Advertising elasticity must be <u>shifting</u> and not be constant.

Subject: Advertising and Promotion Expenditure

Title: Optimal Advertising and Optimal Quality

Authors: Robert Dorfman and Peter O. Steiner

Source: American Economic Review, Vol. 44, No. 5, December 1954,
         826-36

Summary: Equilibrium conditions for a firm marking decisions
         with respect to price, quality, and advertising
         expenditure.   Theorems on optimal advertising and
         optimal quality with fixed prices.

Model:

1. Joint Optimization of Advertising Budget and Price

A firm which can influence the demand for its product
by advertising will, in order to maximize its profits,
choose the advertising budget and price such that the
increase in gross revenue resulting from a one dollar
increase in advertising expenditure is equal to the ordinary
elasticity of demand for the firm's product.

$$\mu = \eta \quad \text{if} \quad s > 0,$$
$$\mu \leq \eta \quad \text{if} \quad s = 0$$

where

   μ = marginal value product of advertising

   η = elasticity of demand and

   s = advertising budget.

2. Joint Optimization of Quality and Price

$$\eta = \frac{p}{c} \ \eta_c$$

where

   p = product price

   c = average cost of production

η_c = elasticity of demand with respect to quality variation.

3. Joint Optimization of Advertising, Quality, and Price

$$\mu = \eta = \frac{p}{c} \ \eta_c.$$

4. Optimal Advertising with Fixed Prices

If the price which a firm can charge is predetermined by conventional, oligopolistic, legal or other considerations, and if the firm can influence its demand curve by advertising, it will, in order to maximize its profits, choose that advertising budget and the resulting rate of sales such that

$$MC = p(1 - \frac{1}{\mu})$$

where MC = marginal cost.

5. Optimal Quality with Fixed Prices

If the price which a firm can charge is predetermined and if the firm can influence its demand curve by altering its product, it will, in order to maximize its profits, choose the quality such that the ratio of price to average cost multiplied by the elasticity of demand with respect to quality expenditure equals the reciprocal of the mark-up on the marginal unit:

$$\frac{p}{c} \ \eta_c = \frac{p}{p - MC}.$$

References: #54, 56, 86.

Subject: Advertising and Promotion Expenditure

Title: Optimal Advertising Policy Under Dynamic Conditions

Authors: Marc Nerlove and Kenneth Arrow

Source: Economica, Vol. 29, May 1962, 129-42

Summary: The Dorfman-Steiner model [#53] is extended to cover
           the situation in which present advertising expenditures
           affect the future demand for the product.

Model:

Let a stock, called goodwill and denoted by $A(t)$,
summarize the effects of current and past advertising
outlays on demand. Goodwill depreciates at a constant
proportional rate δ. Then we have

$$\frac{dA}{dt} + \delta A = a \geq 0 \tag{1}$$

where a = current advertising expenditure
and t = time.
The following demand, cost, and profit relations hold:

$$q = f(p, A, z)$$
$$c(t) = C(q)$$
$$r(t) = pf(p, A, z) - C(q) - a$$

where

- q = unit sales
- p = price
- z = other variables not under the control of the
     firm, e.g., consumer incomes, population, prices
     of related products
- c = rate at which total production costs are incurred
- r = rate at which revenue net of production costs
     and current advertising outlays accrues to the firm
- R = revenue net of production expenses.

The optimal price policy is

$$p = \frac{C'\eta}{(\eta-1)} \tag{2}$$

where η = price elasticity.

The dynamic counterpart to the Dorfman-Steiner [#53] price-advertising theorem can be summarized as

$$\frac{A^*}{pq} = \frac{\beta}{\eta(\alpha+\delta)} \tag{3}$$

where

A^* = optimal advertising stock

β = elasticity of demand with respect to goodwill

α = interest rate at which funds are borrowed.

For the case that z is not held constant over time, assume that the demand function is linear in logarithms

$$f(p,A,z) = kp^{-\eta} A^{\beta} z^{\xi} \tag{4}$$

(where ξ = elasticity of demand with respect to z) and

$$C' = \gamma, \quad \text{a constant.} \tag{5}$$

Then

$$A^* = \left[\frac{k\,\beta\,\gamma^{1-\eta}}{(\eta-1)(\alpha+\beta)}\right]^{1/(1-\beta)} (\frac{\eta}{\eta-1})^{-\eta/(1-\beta)} z^{\xi/(1-\beta)} \tag{6}$$

where $k = 1$ without loss of generality. The optimal current advertising expenditure is

$$a^*(t) = [\delta + \frac{\xi}{1-\beta}(\frac{dz}{dt}/z)]\, A^*. \tag{7}$$

If we suppose that z is expected to change at a constant proportional rate, ρ, then a^*/pq or

$$\sigma = \frac{\beta}{\eta(\alpha+\delta)}\quad (\delta + \frac{\xi\rho}{1-\beta}). \tag{8}$$

Thus firms should try to keep a constant ratio of sales to advertising.

References: #53, 55, 56.

Subject: Advertising and Promotion Expenditure

Title: Diffusion Processes and Optimal Advertising Policy

Author: John P. Gould

Source: Edmund S. Phelps, et. al., Microeconomic Foundations of
           Employment and Inflation Theory, New York: W.W. Norton & Co.,
           Inc., 1970, 338-68

Summary: This model extends the Nerlove-Arrow model [#54] of the
           dynamic optimal advertising policy for the firm by relaxing
           their assumption of a linear advertising cost function and
           by recognizing that there exist alternative hypotheses about
           the way in which information spreads through a market.

Model: The Nerlove and Arrow Model

Let

$A(t)$ = stock of goodwill summarizing the effects of
current and past advertising outlays on demand

$a(t)$ = advertising expenditure at time t.

Assuming that goodwill depreciates at a constant proportional
rate, we have

$$\dot{A}(t) + \delta A(t) = a(t). \qquad (1)$$

(A dot over a variable denotes its derivative with respect
to time.)

Let

$q(t)$ = rate at which sales are made at time t

$g(t) = C(q(t))$ = total manufacturing cost of $q(t)$.

Then the cash flow (net of manufacturing expenses only) is

$$R(p(t), A(t)) = p(t)q(t) - g(t). \qquad (2)$$

The problem is

$$\max_{a,p} \int_o^\infty e^{-rt}[R(p(t),A(t)) - a(t)]\, dt \qquad (3)$$

subject to $a(t) \geq 0$, $p(t) \geq 0$,

$A(0) = A_o$ (i.e., initial level of goodwill is given)

and

$$\dot{A} + \delta A = A(t),$$

where r is the fixed rate of interest.

The derivative $\dot{p}(t)$ is unrestricted, so that the optimal price policy is to choose $p(t)$ to maximize the integrand of (3) at each point in time for given $A(t)$ and $a(t)$. The first-order condition for the optimal price, p^*, is

$$\frac{\partial R(p,A)}{\partial p}\bigg|_{p=p^*} = q + p\frac{\partial f}{\partial p} - \frac{dq}{dq}\frac{\partial f}{\partial p} = 0. \tag{4}$$

Define $p^*(A(t))$ to be that p for which (4) holds when goodwill is $A(t)$ and let

$$\pi(A(t)) = R(p^*,A). \tag{5}$$

Substituting (5) into (3), the problem becomes

$$\max_{a\geq0} \int_o^\infty e^{-rt}[\pi(A(t)) - a(t)]\, dt \tag{3a}$$

subject to

$$A(0) = A_o,$$

$$\dot{A} + \delta A = a.$$

Nerlove and Arrow show, under certain regularity assumptions, that the optimal policy is to jump instantaneously from A_o to A^* (assuming $A^* > A_o$), where

$$A^* = \frac{\beta pq}{\eta(r + \delta)}$$

and where β and η are the elasticities of demand with respect to goodwill and price, respectively. For $t > 0$, the optimal policy is $a^* = \delta A^*$. If $A^* < A_o$, the optimal policy is to set $a^* = 0$ until the stock of goodwill depreciates to the level A^* and then set $a^* = \delta A^*$ from that point on.

The Stigler and Ozga Models of Information Spread

In the Stigler model [#86] let

$K(t)$ = number of individuals who know of the firm at time t

    N = total number of individuals in the market $N \geq K(t)$

    b = forgetfulness coefficient, that is, the instantaneous proportional rate at which individuals forget the message

$u(t)$ = contact coefficient or the instantaneous proportional rate at which individuals become aware of the firm.

Then

$$K(t+dt) = K(t)(1-b\ dt) + u(t)dt(N-K(t)). \qquad (6)$$

Rewriting,

$$\frac{K(t+dt) - K(t)}{dt} = u(t)N - (b+u(t))K(t). \qquad (6a)$$

The limit of (6a) as dt approaches zero yields

$$\dot{K}(t) = u(t)N - (b+u(t))K(t). \qquad (6b)$$

In Ozga's model [#85] information spreads by word of mouth. Let $K(t)$, N and b be the same as above and let $c(t)$ be the contact coefficient. The $K(t)(1-b\ dt)$ people who remember the message will contact and inform a total of $K(t)(1-b\ dt)c(t)dt$ individuals during dt of which the proportion

$$1 - \frac{K(t)(1-b\ dt)}{N}$$

will not know of the message at the time of contact. Combining these results we have

$$K(t+dt) = K(t)(1-b\ dt)$$

$$+ K(t)(1-b\ dt)c(t)dt \left[1 - \frac{K(t)(1-b\ dt)}{N}\right]. \qquad (7)$$

Dividing both sides of (7) by dt and rewriting,

$$\frac{K(t+dt) - K(t)}{dt}$$

$$= -bK(t) + K(t)(1-b\ dt)c(t)\left[1 - \frac{K(t)(1-b\ dt)}{N}\right]. \qquad (7a)$$

The limit of (7a) as dt approaches zero yields the differential equation

$$\dot{K}(t) = -bK(t) + c(t)K(t)\left[1 - \frac{K(t)}{N}\right]. \qquad (7b)$$

The Optimal Policy for the Nerlove-Arrow Model with Nonlinear Costs for Adding to Goodwill

We introduce the twice continuously differentiable cost function $W(a)$ where for $a \geq 0$,

$$w(a) > 0, \qquad\qquad (8a)$$

$$w'(a) > 0, \qquad\qquad (8b)$$

$$w''(a) > 0, \qquad\qquad (8c)$$

It is also assumed that $\pi(A)$ is twice continuously differentiable and that for $A \geq 0$ $\quad \pi'(A) > 0$, $\pi''(A) \leq 0$. The problem is

$$\max_{0 \leq a(t)} \int_o^\infty e^{-rt}[\pi(A) - w(a)] \, dt \qquad\qquad (9)$$

subject to

$$\dot{A} = a - \delta A,$$

$$A(0) = A_o.$$

It is shown that there exists a path which satisfies the conditions of Pontryagin's maximum principle with $A(0) = A_o$ and a finite stationary point (a^*, A^*). The optimal policy for $A_o < A^*$ is to advertise most heavily in the initial periods and continually decrease the level of advertising expenditures as A increases toward the equilibrium level A^*.

The Optimal Policy for the Stigler Diffusion Model

One obtains an operating profit function $\pi(K)$, where K is the number of persons in the market who know of the firm's advertising message. It is assumed that $\pi'(K) > 0$ and $\pi''(K) \leq 0$ for $K \geq 0$.

The contact coefficient, u, is assumed to be related to advertising expenditures by the cost function $w(u)$, where $w'(u) > 0$ and $w''(u) > 0$. It is also assumed that $\pi'(0) > [(b+r)/N]w'(0)$. The formal problem, using the diffusion process given by (6b), thus becomes

$$\max_{u(t) \geq 0} \int_o^\infty e^{-rt}[\pi(K) - w(u)] \, dt \qquad\qquad (14)$$

subject to

$$\dot{K} = u(t)N - (b + u(t)K(t),$$

$$K(0) = K_o.$$

Since $\lim_{u \to \infty} u(t) = \infty$,

(14) has a finite upper bound over all feasible paths.
It is shown that there exists a saddle point (u*,K*) and a
unique path approaching it. The general qualitative
properties of the optimal path are similar to those found
for the Nerlove-Arrow model with nonlinear costs of adding
to goodwill. In particular, for K_o < K* the optimum policy
is to advertise most heavily at the start of the campaign
and continually decrease advertising expenditures as K
approaches K*. In contrast to the Nerlove-Arrow model,
this general pattern of advertising expenditures is maintained even when it is assumed that π'(k) is constant.

Optimal Advertising Policy for the Ozga Diffusion Model

If information spread is assumed to occur according to
the model (7b), where the firm affects the parameter c(t)
through its advertising policy, then the formal problem
becomes

$$\max_{0 \le c(t)} \int_o^\infty e^{-rt} [\pi(K) - w(c)]\, dt \tag{15}$$

subject to

$$\dot{K}(t) = -bK(t) + c(t)K(t)\left[1 - \frac{K(t)}{N}\right].$$

The diffusion process is such that if K_o is zero, there is
no way of achieving any positive level of K no matter how much
the firm spends on advertising. This means that for certain
values of K_o the path to the saddle point is not optimal, because
for small enough K_o it is presumably optimal to keep c at
zero indefinitely. Assuming that K_o is large enough to put
the firm on the unique path which leads to the saddle point,
the optimum path differs significantly from that found in the
previous two models. In the earlier models, the optimum path
of advertising expenditures always requires the heaviest outlays
in the early periods with continuous reductions in expenditures

as K (or A) approaches its equilibrium level. In contrast, the optimum path for the present model may begin with a low level of expenditure, build up to a maximum of c_m, which is greater than the equilibrium level c^*, and then cut back toward c^* as K approaches K^*. Thus, while K increases all along the optimum path, c first increases and then decreases over time.

References: #54, 85, 86.

Subject: Advertising and Promotion Expenditure

Title: Monopoly Advertising: Dynamic Conditions

Author: Richard Schmalensee

Source: R. Schmalensee, The Economics of Advertising, Amsterdam-London: North-Holland, 1972, 26-32

Summary: The model examines optimal advertising by a monopolist under dynamic conditions.

Model: Assume that at every moment of time there is an equilibrium demand given by $Q^* = Q^*(A,P,t)$ where Q = quantity demanded, P = price, A = number of advertising messages. Q moves towards Q^* at all times. The goodwill of a product is determined by past prices and incomes, not just by past advertising.

The simplest sort of adjustment mechanism would be

$$\frac{dQ}{dt} = \lambda(Q^* - Q). \tag{1}$$

Assuming that $Q(t)$ is a weighted average of all past values of Q^*, with $Q^*(t-\tau)$ having a weight $h(\tau)$,

$$Q(t) = \int_{-\infty}^{t} Q^*(\tau)h(t-\tau)d\tau, \text{ with } \int_{o}^{\infty} h(\tau)d\tau = 1. \tag{2}$$

We retain time-invariance and drop the restriction of linearity

$$Q(t) = f\left\{\int_{-\infty}^{t} h[Q^*(\tau), t-\tau]\, d\tau\right\}, \text{ with} \tag{3}$$

$$K = f\left\{\int_{-\infty}^{t} h[K, t-\tau]\, d\tau\right\}, \text{ for all } t \text{ and } K \geq 0.$$

Differentiating,

$$\frac{dQ(t)}{dt} = f'\, h[Q^*(t), 0]. \tag{4}$$

When $f(\cdot)$ is monotone,

$$\frac{dQ(t)}{dt} = F[Q^*(t), Q(t)]. \tag{5}$$

Assuming

$$\frac{\partial F}{\partial Q^*} \equiv F_1 > 0, \quad \text{and} \quad \frac{\partial F}{\partial Q} \equiv F_2 < 0. \tag{6}$$

The flow of profit may be written as

$$\Pi(t) = PQ - C(Q) - AT, \tag{7}$$

where C = total production cost, T = message costs, and AT = advertising spending.

The monopolist faces the problem of choosing the time-paths of A and P so as to maximize the present value of the stream of profits,

$$PV = \int_{0}^{\infty} \Pi(t) \, e^{-rt} \, dt, \tag{8}$$

where r is the relevant discount rate. The problem is solved using optimal control theory. The result is

$$\frac{AT}{PQ} = \frac{a^*}{E^*} \tag{9}$$

where the long-run elasticities of demand with respect to advertising messages and price are

$$a^* = \frac{\partial Q^*}{\partial A} \left(\frac{A}{Q^*}\right), \quad \text{and} \quad E^* = -\frac{\partial Q^*}{\partial P} \left(\frac{P}{Q^*}\right).$$

(9) holds at all points along an optimal path, regardless of the particular form of $F(\cdot)$ or of how Q^* varies with t.

If the long-run elasticities in (9) are constant, neither changes in the interest rate nor fluctuations in the price of advertising messages will have any impact on the dollar advertising dollar sales ratio. We thus have the Dorfman-Steiner formula [#53], except that we now distinguish between short-run and long-run elasticities. We also have a strong defense for policies involving constant advertising-sales ratios.

Equation (9) does not reduce in any case to the Nerlove-Arrow equilibrium relation [#54]. The fundamental difference arises because we have assumed that all demand-determining variables have the same dynamic impact. This formulation is in accord with most empirical determinations of distributed lag mechanisms.

References: #53 and 54.

Subject: Advertising and Promotion Expenditure

Title: An Optimum Geographical Distribution of Publicity Expenditure
           in a Private Organization

Authors: A.P. Zentler and Dorothy Ryde

Source: Management Science, Vol. 2, No. 4, July 1956, 337-52

Summary: The problem is: Given an international organization with
           branches in n different countries and concerned with
           publicity for a commodity X--in competition with a sub-
           stitute X'--how should such an organization allocate its
           expenditure among them so as to get the maximum overall
           return. First, the general form of a curve representing
           response to advertising in favor of one commodity only
           is developed. Response is defined as the increase in the
           consumption of X above its natural level, by which is meant
           the consumption of X that would exist if there were no
           promotion for either X or X'. It is assumed that inter-
           country differences in response depend on the following
           parameters: population, relative publicity and other costs,
           per caput consumption of (X + X'), and a psychological
           factor θ (which measures the "X-mindedness" of the popula-
           tion); these are introduced into the general equation of
           the response curve. Allowance is also made for a time-lag
           in the effects of promotion. The "interaction" effect of
           simultaneous publicity in favor of both X and X' is then
           considered and a formula is developed for the resultant
           response $\mathcal{R}$ in this case.

Model: Assume an individual's response function, $R(\xi)$, which is

           a continuous, increasing function of ξ in the interval

           $(0,\infty)$ with a minimum at $R(0) = 0$ and a maximum A as

           $\xi \to \infty$:

$$R(\xi) = A \ \frac{c_2 \xi^2}{1 + d_1 \xi + d_2 \xi^2}. \tag{1}$$

where

           R = response function

           ξ = real expenditure on promotion

           A = limiting value of response function as $\xi \to \infty$.

Imposing the condition that the point where average response

equals marginal response shall correspond to an expenditure

β, we have $c_2 = 1/\beta^2$ and obtain

$$R = A \ \frac{(\xi/\beta)^2}{1 + d_1 \xi + (\xi/\beta)^2} \tag{2}$$

For $d_1' = 1$ we get as basic response curve

$$R = A \ \frac{(\xi/\beta)^2}{1 + (\xi/\beta) + (\xi/\beta)^2}. \tag{3}$$

We assume that the following parameters are responsible for the variation between countries:

1. The total consumption per head of $(X + X')$.
2. A psychological factor which we have called "X-mindedness" measuring the extent to which people are favorably disposed towards X or resistant to the subsitution of X by X'.
3. Cost of promotion in terms of sterling.
4. Population.

Let "X-mindedness" = θ and "X'-mindedness" = θ', then we take L'/W and L/W as our measures of θ and θ' so that

$$\theta = L'/W \quad \theta' = L/W \quad \theta + \theta' = 1.$$

If "response" is defined as the increase in consumption of X above the natural or unpromoted level L', then L represents a physical upper limit to R and should be identified with A of the equation: we then have

$$R = L \ \frac{(\xi/\beta)^2}{1 + (\xi/\beta) + (\xi/\beta)^2}. \tag{4}$$

Assume β to be inversely proportional to θ and let R = response per capita, P = promotional power per unit sterling expenditure, x = promotional expenditure per capita in sterling, θ = "X-mindedness," L = maximum physically possible response, b = constant scale factor. Then we may take the curve representing per caput response to varying amounts of promotional expenditure in favor of one commodity only, to be represented by the following formula

$$R = L \ \frac{(\frac{\theta P_x}{b})^2}{1 + (\frac{\theta P_x}{b}) + (\frac{\theta P_x}{b})^2} \equiv L\phi \ (\frac{\theta P_x}{b}). \tag{5}$$

Introducing time lags we write

$$\phi \ (\frac{\theta P_x}{b}) \equiv \phi (\xi).$$ (6)

Then,

$$R_t = L\phi (\xi) \ \psi(t)$$ (7)

where $\psi(t)$ is an increasing function of t, which approaches unity for large t. We may suppose $\psi(t)$ to be generated in the following way:

$$\Delta R_t = L\phi (\xi)\alpha[1 - \psi(t)]$$ (8)

$$R_{t+1} = \alpha L\phi (\xi) + (1 - \alpha)R_t.$$ (9)

By repeated substitution:

$$R_t = L\phi (\xi) [1 - \alpha)^t].$$ (10)

Varying ξ,

$$R_t = \alpha L\phi (\xi_t) + (1 - \alpha)R_{t-1}$$ (11)

which leads, by successive substitution, to

$$R_t = \alpha L\{\phi (\xi_t) + (1- \alpha)\phi (\xi_{t-1}) + (1 - \alpha)^2\phi (\xi_{t-2}) \cdots$$

$$+ (1 - \alpha)^{t-1}\phi (\xi_1)\}$$ (12)

which reduces to (10) if all the ξ are made equal.

For promotion in favor of commodity X only, response will be

$$R = L\phi (\xi).$$ (13)

Hence,

$$R' = L'\phi (\xi').$$ (14)

Assume that the X' promoters have got in a long time ahead and have established a response R' before the X promoters start, then the latter will be confronted with the "regainable" space L + R'.

When the two potential responses are in equilibrium, it follows from condition (1) that

$$R* = (L + R'*)\phi (\xi)$$ (15)

$$R'* = (L' + R*)\phi (\xi')$$ (16)

and from condition (2) that the resultant responses $\mathcal{R}$ and $\mathcal{R}'$ in favor of X and X' respectively are

$$\mathcal{R} = R* - R'*$$

$$\mathcal{R}' = R'* - R*.$$

Solving (15) and (16) we get

$$R* = \frac{\phi(\xi)[L + L'\phi(\xi')]}{1 - \phi(\xi)\phi(\xi')} \tag{17}$$

$$R'* = \frac{\phi(\xi')[L' + L\phi(\xi)]}{1 - \phi(\xi)\phi(\xi')}. \tag{18}$$

The net response, from the point of view of commodity X, is then

$$\mathcal{R} \equiv R* - R'* = \frac{L\phi(\xi)[1 - \phi(\xi')] - L'\phi(\xi')[1 - \phi(\xi)]}{1 - \phi(\xi)\phi(\xi')} \tag{19}$$

$$\equiv \Phi(\xi,\xi').$$

If we now introduce time lags and variable ξ and ξ' the arguments already used will apply, with $\Phi(\xi,\xi')$ replacing $\phi(\xi)$ everywhere.

The final equation for net response is then

$$\mathcal{R}_t = \alpha\{\Phi(\xi_t,\xi'_t) + (1-\alpha)\Phi(\xi_{t-1},\xi'_{t-1}) + \ldots + (1-\alpha)^{t-1}\Phi(\xi_1,\xi'_1)\}. \tag{20}$$

Write

$$\rho_t \equiv \frac{\mathcal{R}_t}{W}$$

and

$$\Psi(\xi,\xi') \equiv \frac{\theta'\phi(\xi)[1 - \phi(\xi')] - \theta\phi(\xi')[1 - \phi(\xi)]}{1 - \phi(\xi)\phi(\xi')}$$

so that

$$\rho_t = \alpha\{\Psi(\xi_t,\xi'_t) + (1-\alpha)\Psi(\xi_{t-1},\xi'_{t-1}) \ldots + (1-\alpha)^{t-1}\Psi(\xi_1,\xi'_1)\}. \tag{21}$$

For purposes of computation we have replaced (21) by the expression

$$\rho_t = 0.6\Psi(\xi_t,\xi'_t) + 0.25\Psi(\xi_{t-1},\xi'_{t-1}) + 0.15\Psi(\xi_{t-2},\xi'_{t-2}) \tag{22}$$

which is roughly equivalent to putting a= 0.6 and ignoring changes in the ξ that took place further back than (t-2).

Determining θ,

$$\frac{X'}{W} = \theta' - \rho$$

$$= 0.6G(\xi_t, \xi'_t) + 0.25G(\xi_{t-1}, \xi'_{t-1}) + 0.15G(\xi_{t-2}, \xi'_{t-2}) \quad (23)$$

where

$$G(\xi, \xi') = \frac{[\theta' + \theta\phi(\xi')][1 - \phi(\xi)]}{1 - \phi(\xi)\phi(\xi')}.$$

Given n countries $1, 2, \ldots n$, with populations N_r and a total budget S to be allocated among them, and given that the per capita response curves for the various countries are

$$\mathcal{R}_t = \psi \left(\frac{P_r \theta_r x_r}{b}\right) \qquad r = 1, 2, \ldots n \quad (24)$$

we have to find the set of x_r which will given the maximum total return for S. We have

$$\mathcal{R} = W \frac{k_1 \phi \left(\frac{P\theta x}{b}\right) - k_2}{1 - k_3 \phi \left(\frac{P\theta x}{b}\right)} + B \quad (25)$$

where

$$\phi(z) \equiv \frac{z^2}{1 + z + z^2},$$

by definition and

$$k_1 = 0.6\{\theta'[1 - \phi(\xi'_t)] + \theta\phi(\xi'_t)\}$$

$$k_2 = 0.6\theta\phi(\xi'_t)$$

$$k_3 = \phi(\xi'_t).$$

B is a constant involving promotional expenditure in years $(t-1)$ and $(t-2)$. Then

$$\frac{\partial \mathcal{R}}{\partial x} = \frac{WP\theta}{b} (k_1 - k_2 k_3) \frac{\phi' \left(\frac{P\theta x}{b}\right)}{[1 - k_3 \phi \left(\frac{P\theta x}{b}\right)]^2} \quad (26)$$

where

$$\phi'(z) = \frac{2z + z^2}{(1 + z + z^2)^2}.$$

A graphical solution is indicated.

Subject: Advertising and Promotion Expenditure

Title: Game-Theory Models in the Allocation of Advertising Expenditures

Author: Lawrence Friedman

Source: Operations Research, Vol. 6, No. 5, September-October 1958, 699-709

Summary: Five game-theory models are discussed relating to the allocation of advertising expenditures under assumptions in which the major factor governing advertising allocation is competitive expenditures.

Model:

Model I

Let two competitors A and B controlling an industry have equal price, quality, reputation, service, and fixed advertising budgets. Both companies have divided their total market into marketing areas.

Let S_i = total potential sales influenced by advertising in the ith area and $S = \sum_{i=1}^{n} S_i$.

If

x_i = the amount of advertising expenditure by company A in area i,

y_i = the amount of advertising expenditure by company B in area i,

the total advertising influenced sales by company A will be

$$\sum_{i=1}^{n} \left(\frac{x_i}{x_i + y_i}\right) S_i,$$

while the total advertising influenced sales by company B will be

$$\sum_{i=1}^{n} \left(\frac{y_i}{x_i + y_i}\right) S_i.$$

The two-person zero-sum game: Maximizing

$$D = \sum_{i=1}^{x} (\frac{x_i - y_i}{x_i + y_i}) \, S_i , \tag{1}$$

subject to

$$\sum_{i=1}^{i=n} x_i = A, \tag{2}$$

$$\sum_{i=1}^{i=n} y_i = B \tag{3}$$

has the following solution:

$$x_i = S_i \, A/S, \qquad (i = 1,2,\ldots, n) \tag{4}$$

$$y_i = S_i \, B/S. \qquad (i = 1,2,\ldots, n) \tag{5}$$

Thus, the optimal allocation of funds in each area will be proportional to the sales potential in the area. The expected difference in sales, if both companies use their optimum strategies, will be

$$D = (\frac{A-B}{A+B}) \, S. \tag{6}$$

Model II

Company B's advertising in each of n areas is known and may not be optimal. A is to choose a set of x_i, which maximizes

$$\sum_{i=1}^{n} (\frac{x_i}{x_i + y_i}) \, S_i ,$$

subject to $\sum_{i=1}^{n} x_i = A$.

The solution is

$$x_i = (A+B) \, \frac{\sqrt{S_i y_i}}{\sum_{i=1}^{i=n} \sqrt{S_i y_i}} - y_i .$$

Model III

Assume two competitors A and B, each of whom is trying to obtain the business from a finite number of customers. Each competitor has a fixed advertising budget that must be allocated among the potential customers. Each customer's business, S_i, all other things being equal, will go completely to the company directing the most advertising and promotion in his direction. Thus company A wishes to maximize and company B wishes to minimize the difference, D, between A's and B's sales,

$$D = \sum_{i=1}^{i=n} S_i \, \text{sign} \, (x_i - y_i), \tag{7}$$

subject to

$$\sum_{i=1}^{i=n} x_i = A, \tag{8}$$

$$\sum_{i=1}^{i=n} y_i = B. \tag{9}$$

Assume that the advertising budgets are equal (A = B). The solution lies in mixed strategies for the two competitors:

$$x_1 : x_2 : \ldots : x_n = A_1 : A_2 : \ldots : A_n. \tag{10}$$

Model IV

Extension of the previous problem, with the exception that both parties do not have the same amount of money available. Let A > B.

The optimum strategy for company A is to allocate an amount x_i to customer i chosen randomly from a rectangular distribution in the interval

$$(0, \; 2S_i A / \sum_{i=1}^{i=n} S_i).$$

Competitor B does not always advertise to each customer. He should use a mixed strategy, which would assign a probability of B/A of advertising to any given customer. When he does advertise to a customer he uses the same allocation strategy as competitor A.

The expected difference in sales, D, if both companies use their optimum strategies will be

$$D = S(1 - B/A). \tag{11}$$

Model V

Let the cost of producing and selling an amount of sales, N, exclusive of advertising costs, A, be given by

$$\text{total costs} = C_1 + C_2 N, \tag{12}$$

where C_1 is the fixed cost and C_2 is the variable cost per unit. If C_3 is the gross return per sale, then the total profits as a function of the advertising budget, A, will be

$$\text{profit} = C_3 [A/(A+B)]S - \{C_1 + C_2 [A/(A+B)]S\} - A. \tag{13}$$

Maximizing with respect to A, one finds as the advertising expenditure that produces the most profit:

$$A_{max} = \sqrt{(C_3 - C_2)SB} - B. \tag{14}$$

Subject: Advertising and Promotion Expenditure

Title: Advertising Without Supply Control: Some Implications of a Study of the Advertising of Oranges

Authors: Marc Nerlove and Frederick V. Waugh

Source: Journal of Farm Economics, November 1961, 813-37

Summary: Static analysis concerned with the economically optimal expenditure on advertising in the long run. When supplies are uncontrolled, the variables are: the price elasticity of demand, the long-run effects of advertising expenditures on demand, the price elasticity of industry supply, the nature and extent of external economies or diseconomies of scale to the industry, and the rate of return on alternative forms of investment.

Model:

Let

p = price of the commodity

q = its quantity

a = advertising expenditures

$v = pq$

π = producers' surplus

$R = \pi - a$, cooperative returns to advertising

ρ = rate of return on other investments open to firms

η = elasticity of demand with respect to price

ε = elasticity of industry supply with respect to price.

The industry demand and supply functions are, respectively,

$$q = D(p,a) \tag{1}$$

$$q = S(p) \tag{2}$$

Using the stability condition

$$\frac{\partial S}{\partial p} < \frac{\partial D}{\partial p}$$

the effect of an increase in advertising expenditures on the dollar value of marketings (or total industry revenue) is:

$$\frac{\partial v}{\partial a} = p \; \frac{\partial q}{\partial a} + q \; \frac{\partial p}{\partial a} = \frac{p \; \frac{\partial S}{\partial p} + q}{\frac{\partial S}{\partial p} - \frac{\partial D}{\partial p}} \quad \frac{\partial D}{\partial a} = \frac{1+\varepsilon}{\varepsilon - \eta} \; \alpha \qquad (3) \text{ and } (4)$$

where α = marginal gross revenue from advertising, holding prices constant.

In cooperative advertising (= promotional expenditures by any group which exercises little or no control over supplies) a is determined by

$$\frac{\partial R}{\partial a} = \rho. \qquad (5)$$

If $\partial R/\partial a$ exceeds ρ, the cooperative should increase advertising expenditures; and, conversely, if $\partial R/\partial a$ is less than ρ, it should reduce expenditures.

When there are no external economies or diseconomies, the industry supply curve corresponds to the $\sum$- curve, which is unique. The producers' surplus is

$$\pi = v - \int_0^q S^{-1}(t)dt, \qquad (6)$$

so that

$$\frac{\partial R}{\partial a} = \frac{\partial v}{\partial a} - S^{-1}(q) \; \frac{\partial q}{\partial a} - 1 = p \; \frac{\partial q}{\partial a} + q \; \frac{\partial p}{\partial a} - p \; \frac{\partial q}{\partial a} - 1$$

$$= \frac{\alpha}{\varepsilon - \eta} - 1. \qquad (7) \text{ and } (8)$$

Thus, the optimal condition for cooperative advertising policy is

$$\frac{\alpha}{\varepsilon - \eta} = 1 + \rho. \qquad (9)$$

When external economies or diseconomies are present:

$$\frac{\alpha}{\varepsilon - \eta} \; \left(\frac{1+\varepsilon}{1+\sigma}\right) = 1 + \rho. \qquad (10)$$

Subject: Advertising and Promotion Expenditure

Title: A Model for Budgeting Advertising

Author: Alfred A. Kuehn

Source: F.M. Bass, et. al. (eds.), Mathematical Models and Methods in Marketing, Homewood, Illinois: Richard D. Irwin, Inc., 1961, 315-53

Summary: A profit-maximizing decision rule for the budgeting of consumer-directed advertising is developed. It is based upon a dynamic, empirically supported model of consumer brand shifting in their purchases of grocery products, coupled with mechanisms of consumer influence. The decision rule model is examined under conditions of competitive equilibrium and disequilibrium and compared with sales-oriented rules of thumb widely used in business practice.

Model: Gain Operator: $P_{i,t} = P_{i,t-1} + g(U_i - P_{i,t-1})$

Loss Operator: $P_{i,t} = P_{i,t-1} - \ell(P_{i,t-1} - L_i)$

where

$$0 \leq L_i \leq P_{i,t-1} \leq U_i \leq 1; \quad 0 \leq g, \ell \leq 1;$$

and

$P_{i,t}$ = probability of the consumer purchasing Brand i on the tth purchase occasion.

g = gain parameter, the fraction of maximum possible gain in purchase probability $(U_i - P_{i,t-1})$ which is realized when the brand is purchased.

ℓ = loss parameter, the fraction of maximum possible loss in purchase probability $(P_{i,t-1} - L_i)$ which is realized when the brand is rejected. In general, g and ℓ bear subscripts much as do U and L for brand identification. For established brands in homogeneous product-price classes with widespread retail availability, however, these parameters appear to be a constant.

U_i = upper limit of probability of purchase of Brand i attained by consumers. $1-U_i$ is the extent of incomplete adjustment or learning in the limit and, in a closed-market system, is equal to $\sum_{j \neq i} L_j$.

L_i = lower limit of probability of purchase of Brand i, approached if a consumer repeatedly rejects the brand. There is incomplete extinction of the purchase response if L_i is greater than 0.

The gain and loss operators can be combined, each weighted by its probability of being operative, to yield the expected value purchase probability relationship:

$$P_{i,t} = (\ell - g)P_{i,t-1}^2 + (1 + gU_i - \ell - \ell L_i)P_{i,t-1} + \ell L_i. \quad (1)$$

If the g_i and ℓ_i are each constant for all brands, then, in a closed-market system g also equals ℓ and we obtain

$$P_{i,t} = P_{i,t-1}(1 - g[1 - U_i + L_i]) + gL_i. \quad (2)$$

For the aggregate brand shifting in the market we have

$$S_{i,t} = r_i S_{i,t-1} + I_{t-1}(1 - \overline{r}_t)Z_{i,T}, \quad (3)$$

where:

$S_{i,t}$ = unit sales of Brand i in time period t,

r_i = repeat probability of purchase for Brand i (1 minus the rate of decay of brand loyalty), assumed to be constant over time,

$I_{t-1} = \sum S_{i,t-1}$, the unit sales of all brands (industry sales) in time period t-1, assumed to be constant over time. This restriction will be relaxed in (7).

$\overline{r}_t$ = weighted average of decay rate of brand loyalty for the industry, not constant over time (unless the r_i's are all equal) because of shifts in relative sales volume of individual brands.

$Z_{i,T}$ = fraction of "potential brand shifters" attracted to Brand i in t as a result of the brand's retail price and availability relative to competition at time t and the advertising in its behalf during T. The distinction between t and T takes cognizance of the fact that some time lag may exist between the advertising expenditures and the sales results.

$$z_i = b_p \frac{P_i}{\sum P_i} + b_d \frac{D_i}{\sum D_i} + b_a \frac{A_i}{\sum A_i} + b_{pd} \frac{(PD)_i}{\sum (PD)_i} + b_{pa} \frac{(PA)_i}{\sum (PA)_i}$$
$$+ b_{da} \frac{(DA)_i}{\sum (DA)_i} + b_{pda} \frac{(PDA)_i}{\sum (PDA)_i}, \quad (4)$$

where:

$$\sum b_k = 1$$
$$\sum P_i = \sum D_i = \sum A_i = 1$$

$(PD)_i = P_i D_i$; $(PA)_i = P_i A_i$; $(DA)_i = D_i A_i$; $(PDA)_i = P_i D_i A_i$.

P_i = the share of brand shifters which would be attracted by Brand i if product characteristics and price were the only merchandizing variables (i.e., $b_p = 1$, $b_d, b_a, \ldots, b_{pda} = 0$). P_i is also the equilibrium share of market if it is constant over time, $b_p = 1$, and all the r_i's are equal.

Similarly, D_i and A_i represent the share of potential brand shifters that would be attracted if <u>distribution-display space</u> and <u>advertising</u>, respectively, were the only influential merchandising variables.

$\dfrac{(PD)_i}{\sum (PD)_i}$ = the share of brand shifters which would be attracted by Brand i if the interaction (joint) effect of product characteristics-price and distribution-display were the only influence upon consumer choice. The other interaction effects are defined in similar fashion.

Setting b_p, b_a and b_{pa} as well as b_d, b_a and b_{da} equal to zero and $\sum (PD)_i$ equal to 1 we obtain

$$S_{i,t} = r_i S_{i,t-1} + I_{t-1}(1 - \bar{r}_t) b_{pd} (PD)_i$$

$$+ I_{t-1}(1 - \bar{r}_t) b_{pda} \frac{(PDA)_i}{\sum (PDA)_i}. \qquad (5)$$

If we assume a constant and equal rate of exit for customers of each brand and assume that entries (newcomers) are distributed among brands as are the potential brand shifters, we obtain

$$S_{i,t} = r_i e S_{i,t-1} + I_{t-1}[g + (1 - \bar{r}_t)e] b_{pd} (PD)_i$$

$$+ I_{t-1}[g + (1 - \bar{r}_t)e] b_{pda} \frac{(PDA)_i}{\sum (PDA)_i}, \qquad (6)$$

where:

e = probability of <u>survival</u> of past customers of the product, and

g = entry of potential consumers to the market for the product as a fraction of the size of the market in the previous time period.

Let $k = e + g$, the net growth of the industry sales per period. Then

$$S_{i,t} = r_i e S_{i,t-1} + I_o k^{t-1}(k - \bar{r}_t e) b_{pd}(PD)_i$$
$$+ I_o k^{t-1}(k - \bar{r}_t e) b_{pda} \frac{(PDA)_i}{\sum (PDA)_i}. \quad (7)$$

Since

$$\frac{(PDA)_i}{\sum (PDA)_i} = \frac{(PD)_i A_i}{(PD)_c + (PD)_i A_i - (PD)_c A_i}, \quad (8)$$

we have

$$S_{i,t} = r_i e S_{i,t-1} + I_o k^{t-1}(k - \bar{r}_t e)(PD)_i [b_{pd}$$
$$+ \frac{b_{pda} A_{iT}}{(PD)_c + (PD)_i A_{iT} - (PD)_c A_{iT}}], \quad (9)$$

which is shown to become the general expression

$$S_{i,t} = (r_i e)^t S_{i,0} + I_o b_{pd}(PD)_i \sum_{T=0}^{t-1} (r_i e)^{t-T-1} k^T (k - \bar{r}_{T+1} e)$$
$$+ I_o b_{pda} \sum_{T=1}^{t-L} (r_i e)^{t-L-T} k^{L+T-1}(k - \bar{r}_{L+T} e) \frac{(PD)_i A_{i,T}}{(PD)_c + [(PD)_i - (PD)_c] A_{i,T}} \quad (10)$$

for any t considered. Maximizing profits with respect to advertising we obtain for Brand i

$$\Pi_i = \sum_{i=L+1}^{\infty} \rho^t m_i S_{i,t} - \sum_{T=1}^{\infty} \rho^T C_{i,T}, \quad (11)$$

where

ρ = discount factor = $1/(1 + \text{rate of interest})$

m_i = dollars profit margin per unit sale apart from advertising costs

$C_{i,T}$ = advertising expenditures for Brand i at Time T.

We define advertising effectiveness as

$$\frac{A_{i,T}}{C_{i,T} E_{i,T}} = \frac{A_{j,T}}{C_{j,T} E_{j,T}} = \frac{\sum_{j \neq i} A_{i,T}}{\sum_{j \neq i} C_{j,T} E_{j,T}}, \quad (12)$$

where

$C_{i,T}$ = advertising expenditures in behalf of Brand i during period T

$E_{i,T}$ = effectiveness of advertising in behalf of Brand i during period T.

Then

$$C_{i,T} = C_{c,T} \; \frac{E_{c,T}}{E_{i,T}} \; \frac{A_{i,T}}{1 - A_{i,T}}, \tag{13}$$

where

$C_{c,T} = \sum\limits_{j \neq i} C_{j,T}$, total advertising expenditures by competitors at time T

$E_{c,T} = \dfrac{\sum\limits_{j \neq i} C_{j,T} E_{j,T}}{\sum\limits_{j \neq i} C_{j,T}}$, the average weighted effectiveness of advertising in behalf of competitive brands in T.

It is now possible to define optimal advertising expenditures in each time period from $T = 1$ until $T = \infty$ on the basis of (11) and (13). It is shown that

$$C_{i,T} = C_{c,T} \left[\sqrt{\frac{m_i I_o b_{pda} (k - \overline{r}_{L+T} e)(pk) L_k^{T-1}}{C_{c,T}(1 - \rho r_i e) E_R (PD)_R}} - \frac{1}{E_R (PD)_R} \right]. \tag{14}$$

Competitive Equilibrium

In a two-brand market (Brands i and j)

$$C_{i,\text{equilibrium}} = \frac{W m_i Q_R}{(1 - \rho r_j e)[(1 - \rho re) R + Q_R]^2}, \tag{15}$$

where $W = I_o b_{pda} (k - \overline{r}_{L+T} e)(\rho k) L_k^{T-1}$

$Q_R = m_R E_R (PD)_R$ and

$(1 - \rho re)_R = (1 - \rho r_i e)/(1 - \rho r_i e).$

$$C_{j,\text{equilibrium}} = \frac{W m_j Q_R (1 - \rho re)_R^2}{(1 - \rho r_i e)[(1 - \rho re) R + Q_R]^2}, \tag{16}$$

and total industry advertising expenditures are

$$C_{i,\text{eq.}} + C_{j,\text{eq.}} = \frac{W Q_R}{[(1 - \rho re) R + Q_R]^2} \left[\frac{m_i}{(1 - \rho r_j e)} + \frac{m_j (1 - \rho re)_R^2}{(1 - \rho r_i e)} \right]. \tag{17}$$

The ratio of advertising expenditures for Brand i relative

to Brand j is

$$\frac{C_{i,eq.}}{C_{j,eq.}} = C_R = \frac{m_i(1 - \rho r_i e)}{m_j(1 - \rho r_i e)} = \frac{m_R}{(1 - \rho re)_R}. \tag{18}$$

The equilibrium market share for Brand i is

$$\frac{S_{i,t}}{I_0 k^t} = \frac{k - \bar{r}_t e}{k - r_i e} \left[\frac{b_{pd}(PD)_R}{1 + (PD)_R} + \frac{b_{pda}Q_R}{(1 - \rho re)_R + Q_R}\right]. \tag{19}$$

For Brand j it is

$$\frac{S_{j,t}}{I_0 k^t} = \frac{k - \bar{r}_t e}{k - r_j e} \left[\frac{b_{pd}}{1 + (PD)_R} + \frac{b_{pda}(1 - \rho re)_R}{(1 - \rho re)_R + Q_R}\right]. \tag{20}$$

Dividing (19) by (20) we obtain

$$S_R = \frac{S_{i,eq.}}{S_{j,eq.}} = \frac{1}{(k - re)_R}$$

$$\left[\frac{b_{pd}[(PD)_R Y_R + (PD)_R Q_R] + b_{pda}[Q_R + (PD)_R Q_R]}{b_{pd}[Y_R + Q_R] + b_{pda}[Y_R + Y_R(PD)_R]}\right], \tag{21}$$

where $Y_R = (1 - \rho re)_R$.

The profit for Brand i in the two-brand market in t is

$$\Pi_{i,t} = m_i S_{i,t,eq.} - C_{i,T+L,eq.}$$

$$= m_i I_0 k^t \frac{k - \bar{r}_t e}{k - r_i e} \left[b_{pd}(PD)_i + \frac{b_{pda}Q_R}{(1 - \rho re)_R + Q_R}\right] \tag{22}$$

$$- \frac{I_0 b_{pda}(k - \bar{r}_t e)(\rho k)^L k^{t-1} m_i Q_R}{(1 - \rho r_j e)[(1 - \rho re)_R + Q_R]^2}.$$

Subject: Advertising and Promotion Expenditure

Title: Optimal Advertising Appropriation

Author: Kristian S. Palda

Source: Economic Analysis for Marketing Decisions, Englewood Cliffs, New Jersey: Prentice-Hall, Inc., 1969, 189-95

Summary: Assuming that the parameters of the distributed lag model of the geometric progression form representing the impact of advertising outlays on sales over time can be estimated, it is shown how to use these estimates to determine the optimal advertising appropriation.

Model: Advertising has a lagged effect on sales, of the specific form of convergent geometric series:

$$S_t = k + \alpha A_t + \alpha \lambda A_{t-1} + \alpha \lambda^2 A_{t-2} + \ldots \qquad 0 < \lambda < 1.$$

This can be reduced to

$$S_t = c + \alpha A_t + \lambda S_{t-1}.$$

The equilibrium level is then

$$S_e = c + \alpha A_t + \lambda S_e.$$

The long-run effect of advertising is the derivative of sales with respect to advertising:

$$\frac{dS_e}{dA_t} = \frac{\alpha}{1-\lambda}.$$

Using a stock of goodwill, the advertising-sales relationship can be expressed as

$$S_t = k + \alpha a_t^*$$

where a_t^* is advertising capital or goodwill, accumulated over the past. In turn, this capital can be viewed as capital accumulated up to the beginning of the current period, suitably depreciated over this current period and comprising this period's advertising expenditure as well:

$$a_t^* = (1 - r) a_{t-1}^* + A_t.$$

This we multiply by $1-r$, lag one period, and subtract from the original equation to get

$$S_t = kr + (1-r)S_{t-1} + \alpha[a_t^* - (1-r)a_{t-1}^*]$$

with the rate of depreciation being

$$r = 1 - \lambda.$$

In the optimal advertising appropriation the present value of the last dollar invested in advertising should be equal to the present value of revenue stimulated by advertising net of all costs, except advertising.

$$\$1 = c\alpha + \frac{c\alpha\lambda}{1+r} + \frac{c\alpha\lambda^2}{(1+r)^2} + \ldots$$

$$= c\alpha \left[\frac{1}{1-(\lambda/1+r)} \right] \qquad 0 < c < 1.$$

The marginal rate of return on advertising investment, r, is

$$r = \frac{c\alpha + \lambda - 1}{1 - c\alpha}.$$

The firm, in equilibrium, equalizes the marginal rates of return on the investment possibilities open to it (such as in sales force, product development, manufacturing) and the rate at which it can borrow capital, ρ. The optimal advertising appropriation then yields a rate of return r_A, which is equal to other internal rates of return, r_1, r_2, ... and to ρ:

$$r_A = r_1 = r_2 = \ldots = r_n = \rho.$$

Given that advertising returns are in the diminishing-returns stage, this rule implies that advertising expenditures should be increased when $r_A > r_n$ or $r_A > \rho$, and vice versa.

Subject: Advertising and Promotion Expenditure

Title: Advertising Expenditures in Coupled Markets--A Game-Theory
          Approach

Author: Melvin F. Shakun

Source: Management Science, Vol. 11, No. 4, February 1965,
          B-42-47

Summary: A game-theory static approach to advertising expenditures
          in coupled markets. Noncooperative equilibrium solutions
          are obtained for the case of two competing companies each
          selling two products in coupled markets.

Model:

X_i = advertising expenditures of Company 1

$\overline{X}_i$ = advertising expenditures of Company 2

i = product ($i = 1,2$)

Y_i = sales in units of products

X_i = "effective" advertising dollars spent

c_i = saturation level

a_i = constant representing slope of curve at origin

Then

$$dY_i/dX_i = a_i(c_i - Y_i)/c_i$$

For the ith class of product as a whole, the rate of change
of sales Y_i in units of products with respect to "effective"
advertising dollars X_i spent is proportional to the
difference between a saturation level c_i and Y_i expressed
as a fraction of c_i.

$$Y_i = c_i B_i \tag{1}$$

where

$$B_i = 1 - \exp(-\alpha_i X_i) \quad \text{and} \quad \alpha_i = a_i/c_i \tag{2}$$

$$X_i = x_i + \overline{x}_i - k_j(x_j + \overline{x}_j) \quad \text{for } i = 1,2, \tag{3}$$
$$j = 1,2, \text{ with } i \neq j$$

where k_1 and k_2 are coupling coefficients which express the degree to which advertising dollars spent by both companies in attempting to generate sales of one product interact (additively) with the advertising dollars for the other product. The profits P and $\overline{P}$ earned by Company 1 and 2, respectively, are

$$P = \textstyle\sum_{i=1}^{2} (m_i x_i (x_i + \overline{x}_i)^{-1} c_i B_i - x_i) \tag{4}$$

$$\overline{P} = \textstyle\sum_{i=1}^{2} (\overline{m}_i \overline{x}_i (x_i + \overline{x}_i)^{-1} c_i B_i - \overline{x}_i) \tag{5}$$

where combining (2) and (3)

$$B_i = 1 - \exp \{ -\alpha_i [x_i + \overline{x}_i - k_j (x_j + \overline{x}_j)] \} \tag{6}$$

$$i = 1,2, \ j = 1,2 \ \text{with} \ i \neq j,$$

and m_i and $\overline{m}_i$ represent unit profit margins of Company 1 and 2, respectively.

Noncooperative equilibrium solutions are obtained by partially differentiating P with respect to x_1 and a_2 in equation (4) and $\overline{P}$ with respect to $\overline{x}_1$ and $\overline{x}_2$ in equation (5) and setting the derivatives equal to zero. The following four simultaneous nonlinear equations result [(7) from (4) and (8) from (4)]:

$$m_i [a_i (1 - B_i) - c_i B_i / A_i] x_i + m_i c_i B_i$$
$$- A_i - k_i m_j a_j (1 - B_j) (A_i / A_j) x_j = 0 \tag{7}$$

$$\overline{m}_i [a_i (1 - B_i) - c_i B_i / A_i] \overline{x}_i + \overline{m}_i c_i B_i$$
$$- A_i - k_i \overline{m}_j a_j (1 - B_j) (A_i / A_j) \overline{x}_j = 0 \tag{8}$$

where $i = 1,2$, $j = 1,2$ with $i \neq j$ and $A_i = x_i + \overline{x}_i$.

Subject: Advertising and Promotion Expenditure

Title: A Simple Model for Determining Advertising Appropriations

Author: Julian L. Simon

Source: Journal of Marketing Research, Vol. 2, August 1965, 285-92

Summary: Profit-maximizing model that requires knowledge of (a) the sales effect of advertising in the current period of different possible levels of advertising; (b) sales in the prior period; (c) the rate at which sales decline in the absence of advertising; and (d) the interest rate for capital for the firm. The applicability of the model to general advertisers with and without monopoly power, established and new products, small and large market shares, and in oligopoly is examined.

Model:

Model for general advertisers with small market shares and a relatively undifferentiated product:

Let

T = advertising period ($T = 1, 2, \ldots, m$)

t = revenue period ($t = 1, 2, \ldots, n$)

$T = t$

$R_{T,t}$ = net revenue caused by advertising in period T, realized in period t, undiscounted (gross revenue less all production costs)

$V_{T,t}$ = present value net revenue

A_T = advertising expenditure in period T

P_T = profit from advertising in period T

b = retention rate, equal to 1 minus the decay rate of customer purchases from period to period (in absence of further advertising)

ρ = discount rate of the cost of money to the firm.

$\sum\limits_{T=1}^{t} R_{T,t}$ = sum of sales in period t, caused by all prior advertising in periods $1, 2, \ldots, t$.

Then

$$\sum_{T=1}^{t} R_{T,t} = \sum_{T=1}^{t-1} R_{T,t} + R_{T=t,t} \tag{1}$$

$$\sum_{T=1}^{t-1} R_{T,t} = b \sum_{T=1}^{t-1} R_{T,t-1} \tag{2}$$

$$\sum_{T=1}^{t} R_{T,t} = b \sum_{T=1}^{t-1} R_{T,t-1} + R_{T=t,t} \tag{3}$$

The basic result of this model is:

$$\sum_{t}^{\infty} R_{T=t,t} = \frac{1}{1-b \cdot \rho} \left(\sum_{T=1}^{t} R_{T,t} - b \sum_{T=1}^{t-1} R_{T,t-1} \right). \tag{4}$$

The profit from advertising in the present period is:

$$P_T = \sum_{t}^{\infty} V_{T=t,t} - A_{T=t}. \tag{5}$$

The profit-maximizing rule can be expressed thus: advertise until

$$\Delta A_{T=t} = \Delta \sum_{t}^{\infty} V_{T=t,t}. \tag{6}$$

Subject: Advertising and Promotion Expenditure

Title: Theory Versus Practice in Allocating Advertising Money

Author: Donald C. Marschner

Source: Journal of Business, Vol. 40, 1967, 286-302

Summary: A normative model about the best way to appropriate an advertising budget among geographical markets is outlined and compared with the practices of two large, sophisticated, and profitable corporations selling products directly to the consumer (gasoline and coffee).

Model:

$$X_i = X_j \left\{ \frac{D_{bi} \cdot V_{bi} + U_{bi}(V_{bi} - V_{bi})}{D_{bj} \cdot V_{bj} + U_{bj}(V_{pj} - V_{bj})} \right\} \cdot N_p \left(\frac{Q_{pi}}{V_{pi}} \div \frac{Q_{pj}}{V_{pj}} \right) \cdot F_p \left(\frac{E_{pi}}{E_{pi}} \div \frac{E_{pj}}{E_{pj}} \right) \cdot \frac{C_i}{C_j}$$

Key

X = Budget

D = Rate of Sales Decay, established by market research

V = Sales Volume, during time period t-1

U = Sales Opportunity for Brand "b," expressed as a decimal fraction. (This may be a specified "share of the available market," representing a practical estimate of the growth in sales volume to be aimed at as a sales goal; or, it may be the entire available market, less only such volume as has already been achieved, in which case the value of U would be 1, or 100 per cent.)

Q = Advertising Dollar Weight, during time period t-1

N = Conversion Factor, indicating the importance of dollars spent on advertising as a determinant of Sales Response, in relation to all other ingredients in the marketing mix; the values and nature of N are determined through market research

E = Copy Effectiveness Indicator, determined through market research or motivational research

F = Conversion Factor indicating the importance of advertising copy effectiveness as a determinant of Sales Response, in relation to all other ingredients in the marketing mix; the values and nature of F are determined through market research

C = Average cost of reaching one prospect with one satisfactory message by means of the advertising medium or media normally used

i (subscript) = Any specific market

b (subscript) = Brand "b": our brand

p (subscript) = All competing brands plus Brand "b" in a specific product category--that is, "the industry"

j (subscript) = All markets, or the total of all markets-- that is, $j = \sum i_{1,2,3...n}$.

Reference: #72.

Subject: Advertising and Promotion Expenditure

Title: A Dynamic Model for Competitive Marketing in Coupled Markets

Author: Melvin F. Shakun

Source: Management Science, Vol. 12, No. 12, August 1966, B-525-530

Summary: A dynamic model involving advertising expenditures over time for coupled markets is outlined as a game-theory approach with noncooperative equilibrium solutions for advertising expenditures. Effectiveness is measured by weighted future profits. The model treats any number of competitors and products.

Model: Parameters

x_{hin} advertising expenditure of company h on product i during time period n ($h = 1,2, \ldots, H$; $i = 1,2, \ldots, I$; $n = t, t+1, \ldots, t+T$)

Y_{in} sales in units of product i during time period n

c_i sales saturation level for product i

a_i slope of curve at origin

X_i "effective" advertising dollars

k_{ij} coupling components (positive or negative) which express the degree to which advertising dollars spent by the H companies in attempting to generate sales of product j interact (additively) with the advertising dollars spent for product i

m_{hi} unit profit margin for company h on product i

β_i smoothing constants

$$Y_{in} = c_i B_{in} \tag{1}$$

where $B_{in} = 1 - \exp(-\alpha_i X_{in})$ and $\alpha_i = a_i/c_i$ (2)

$$X_{in} = \beta_i \left(\sum_h x_{hin} - \sum_{j \neq i} k_{ji} \sum_h x_{hjn} \right) + (1 - \beta_i) X_{i,n-1} \tag{3}$$

The fraction of the total market for product i which goes to company h in time period n is taken as

$$\phi_{hin} = x_{hin} \phi_{hi,n-1} \Big/ \sum_h x_{hin} \phi_{hi,n-1}. \tag{4}$$

The weighted future profit earned by company h over the next T+1 time periods is expressed as

$$P_{ht} = \sum_{i=1}^{I} \sum_{n=t}^{t+T} w_{hn}(m_{hi} \phi_{hin} Y_{in} - x_{hin}) \tag{5}$$

where w_{hn} is a weight assigned by company h to profits realized in time period n; it can reflect short-term vs. long-term profit preferences as well as the usual discounting. In the latter case $w_{hn} = v_h^{n-t}$ where v_h is the one-period discount factor for company h.

Noncooperative equilibrium solutions may be obtained by partially differentiating the P_{ht} with respect to the x_{hin} and setting the derivatives equal to zero. The following simultaneous nonlinear equations result:

$$
\sum_{g=n}^{t+T} w_{hg} \; [-m_{hi}x_{hin}^{-1}D_{hitg}^2 c_i \; (\sum_h D_{hitg})^{-2} B_{itg}
$$

$$
+ \; m_{hi}x_{hin}^{-1}D_{hitg}c_i \; (\sum_h D_{hitg})^{-1} B_{itg}
$$

$$
+ \; m_{hi}D_{hitg} \; (\sum_h D_{hitg})^{-1} a_i\beta_i(1-\beta_i)^{g-n}(1 - B_{itg})
$$

$$
- \; \sum_{j\neq i} k_{ij}m_{hj}D_{hjtg}(\sum_h D_{hjtg})^{-1}a_j\beta_j(1-\beta_j)^{g-n}(1-B_{jtg})]
$$

$$
- \; w_{hn} = 0 \qquad\qquad (6)
$$

where
$$
D_{hitg} = \phi_{hi,t-1} \prod_{p=t}^{g} x_{hip} \qquad\qquad (7)
$$

$$
B_{itg} = 1 - \exp \; \{-a_i [\sum_{f=t}^{g} \beta_i(1-\beta_i)^{g-f}(\sum_h x_{hif} - \sum_{j\neq i} k_{ji} \sum_h x_{hjf})
$$

$$
+ \; (1-\beta_i)^{g-t+1} x_{i,t-1}]\}. \qquad\qquad (8)
$$

Iterative techniques for the solution for x_{hin} are indicated.

Subject: Advertising and Promotion Expenditure

Title: A Model of Adaptive Control of Promotional Spending

Author: John D.C. Little

Source: Operations Research, Vol. 14, November-December 1966, 1075-97

Summary: Information about sales response is collected by performing an experiment. The experimental results are used to update a sales response model. The promotion rate is chosen to maximize expected profit in the next time period. The cycle is repeated. The model employs a quadratic sales response function with a parameter that changes according to a first order, autoregressive process. The optimal adaptive system involves exponential smoothing of the experimental results.

Model:

Let

s = sales rate (dol/hh yr)

x = promotion rate (dol/hh yr)

p = profit rate (dol/hh yr)

c = fixed cost rate (dol/hh yr)

m = gross margin, the incremental profit as a fraction of sales, taken to be constant.

Company profit is

$$p = ms - x - c. \qquad (1)$$

For a given fixed time period, the average sales rate over the period can be represented by

$$s = \alpha + \beta x - \gamma x^2, \qquad (2)$$

where α, β, and γ are constants.

The value of x^* that maximizes profit is

$$x^* = (m\beta - 1)/2m\gamma. \qquad (3)$$

The loss rate, ℓ, relative to maximum profit, is

$$\ell(x) = m\gamma(x-x^*). \qquad (4)$$

At t, the national sales rate for the product is

$$\tilde{s} = \tilde{\alpha} + \tilde{\beta}x - \gamma x^2 \ (\text{dol/hh yr}), \qquad (5)$$

where

$\tilde{\alpha}(t)$ has relatively high variance from time period
  to time period
$\tilde{\beta}(t)$ is dependent on $\tilde{\beta}(t-1)$
$\tilde{\alpha}$ and $\tilde{\beta}$ are independent
γ is constant
$\sim$ means quantity is being viewed as random variable.

Let

β^{o} = long-run average value of $\beta(t)$
 k = constant, $0 \leq k \leq 1$
$\tilde{\epsilon}_\beta(t)$ = random variable.

The model of changing β is

$$\tilde{\beta}(t) = k\tilde{\beta}(t-1) + (1-k)\beta^{\text{o}} + \tilde{\epsilon}_\beta(t). \qquad (6)$$

Suppose that at t, a promotion rate of $x_{\text{o}}(t)$ is thought to be best. This is used everywhere except in n markets where a deliberately low rate x_1 is used, and in another n where a deliberately high rate x_2 is used. We take

$$x_1 = x_{\text{o}}(t) - (\Delta/2),$$
$$x_2 = x_{\text{o}}(t) - (\Delta/2), \qquad (7)$$

where Δ is a design constant.
We assume sales in a market are given by

$$\tilde{s} = s(x) + \tilde{\epsilon}, \qquad (8)$$

where

$\tilde{s}$ = sales rate in the market (dol/hh yr)

$s(x) = \alpha + \beta x - \gamma x^2$ = national sales rate (dol/hh yr)

$\tilde{\epsilon}$ = random variable.

Let $\bar{s}_1$ and $\bar{s}_2$ be the observed mean sales rates in the groups of markets at x_1 and x_2 respectively. The experimental mean for $\beta(t)$ is then

$$\hat{\beta}(t) = (1/\Delta)(\bar{s}_2 + \gamma x_2^2 - \bar{s}_1 - \gamma x_1^2). \tag{9}$$

Let

$v = V[\hat{\beta}(t)].$

$$E[\hat{\beta}(t)] = \beta(t) \tag{10}$$

$$v = 2\sigma^2/n\Delta^2. \tag{11}$$

Let $\beta'(t)$ be an estimate of β at t. We choose a number, a, such that $0 \leq a \leq 1$ and update β' by

$$\beta'(t+1) = a\beta'(t) + (1-a)\hat{\beta}(t). \tag{DR1}$$

Then we choose as the promotion rate at $t+1$

$$x_o(t+1) = [m\beta'(t+1) - 1]/2m\gamma. \tag{DR2}$$

Define

$$\hat{x}_o(t) = [m\hat{\beta}(t) - 1]/2m\gamma. \tag{12}$$

Then DR1 and DR2 combine to

$$x_o(t+1) = ax_o(t) + (1-a)\hat{x}_o(t). \tag{DR3}$$

Let $\beta''(t)$ and v'' denote the mean and variance of the posterior distribution of $\tilde{\beta}(t)$. The posterior distribution is normal with

$$\beta''(t) = [v/(v+v')]\beta'(t) + [v'/(v+v')]\hat{\beta}(t) \tag{13}$$

$$v'' = vv'/(v+v'). \tag{14}$$

The prior distribution of $\beta(t+1)$ is normal with mean and variance:

$$\beta'(t+1) = k\beta''(t) + (1-k)\beta^O \tag{15}$$

$$V'[\tilde{\beta}(t+1)] = k^2 v'' + \sigma_\beta^2. \tag{16}$$

From (13) and (15) we obtain

$$\beta'(t+1) = k[v/(v+v')]\beta'(t) + k[v'/(v+v')]\hat{\beta}(t)$$
$$+ (1-k)\beta^O. \tag{DR1a}$$

However, let

$$a = v/(v+v'). \tag{17}$$

If $k = 1$, then DR1a becomes the exponential smoothing rule DR1. In steady state, $V'[\tilde{\beta}(t+1)] = v'$. Then

$$v' = (\tfrac{1}{2})[\sigma_\beta^2 - (1-k^2)v] + (\tfrac{1}{2})\{[\sigma_\beta^2 - (1-k^2)v]^2$$
$$+ 4\sigma_\beta^2 v\}^{1/2}. \tag{18}$$

When $k \to 1$,

$$v' = (\tfrac{1}{2})\sigma_\beta^2\{1 + [1 + (4v/\sigma_\beta^2)]^{1/2}\}. \tag{19}$$

Substitution of v and v' into (17) gives a in terms of σ, n, Δ, k, and σ_β.

With imperfect information the promotion rate is

$$x_o(t) = [m\beta'(t) - 1]/2m\gamma. \tag{20}$$

Consider a market at x_1 with its loss rate relative to perfect information $\tilde{\ell}_1$.

$$\tilde{\ell}_1 = m\gamma[(x_o - \tilde{x}^*)^2 - \Delta(x_o - \tilde{x}^*) + (\Delta^2/4)].$$

Then

$$E'[\tilde{\ell}_1] = E'[\tilde{\ell}] + m\gamma\Delta^2/4. \tag{21}$$

Similarly we find

that $E'[\tilde{\ell}_2] = E'[\tilde{\ell}_1]$.

To compute the total expected loss rate let

N = total number of markets in the country,

$2n$ = the number of experimental markets,

H = the average number of households in a market,

T = total expected loss rate (dol/yr).

For simplicity, all markets are assumed to have the same size. Then

$$T = NH \ mv'/4\gamma + (\frac{1}{2}) H \ m\gamma n\Delta^2.$$

Let

α^o = long-run average of $\alpha(t)$

β^o = long-run average of $\beta(t)$

$$x^o = (m\beta^o - 1)/2m\gamma \qquad (22)$$

$$s^o = \alpha^o + \beta^o x^o - \gamma(x^o)^2 \qquad (23)$$

L = steady-state expected loss rate (relative to perfect information and no experiment) as a fraction of the long-run average promotion rate, x^o (dimensionless).

Then

$$L = mv'/4\gamma x^o + m\gamma n \ \Delta^2/2 \ N \ x^o. \qquad (24)$$

In the case of the $k = 1$ we minimize with respect to the dimensionless quantity, z, defined by

$$n\Delta^2 = 8\sigma^2/\sigma_\beta^2 \ 2. \qquad (25)$$

Using (25), (11), (19), (24), we obtain L in terms of z:

$$L = (m\sigma_\beta^2/8\gamma x^o)[1 + (1+z)^{1/2}] + 4m\gamma\sigma^2/N\sigma_\beta^2 x^o z. \qquad (26)$$

Setting $dL/dz = 0$, we obtain an equation for optimal z:

$$z/(1+z)^{1/4} = 8\gamma\sigma/\alpha_\beta\sqrt{N}. \qquad (27)$$

The equation can be solved for z graphically, or simply by trial and error.

Subject: Advertising and Promotion Expenditure

Title: Models and Managers: The Concept of a Decision Calculus

Author: John D.C. Little

Source: Management Science, Vol. 16, No. 8, April 1970, B-466-484

Summary: An on-line decision calculus model consisting of numerical procedures for processing data and judgment to assist decision making is described for use by product managers on advertising budgeting questions.

Model: The Advertising Budgeting Model (ADBUDG)

s_t = brand sales rate in period t (sales units/period)

h_t = brand share in period t

c_t = product class sales rate in period t (sales units/period

$$s_t = h_t c_t \tag{1}$$

$\overline{h}_t$ = unadjusted brand share in t

n_t = nonadvertising effects index in t

$$h_t = n_t \overline{h}_t \tag{2}$$

α = persistence constant for unadjusted brand share

β = affectable range of unadjusted brand share

γ = advertising response function exponent for brand

δ = advertising response function denominator constant for brand

λ = long-run minimum brand share

w_t = weighted, normalized brand advertising in t

$$\overline{h}_t = \lambda + \alpha(\overline{h}_t - \lambda) + \beta w_t^{\gamma}/(\delta + w_t^{\gamma}) \tag{3}$$

e_{1t} = brand media efficiency in t

e_1^* = brand media efficiency reference value

e_{2t} = brand copy effectiveness in t

e_2^* = brand copy effectiveness reference value

x_t = brand advertising rate in t (dol/period)

x^* = brand maintenance advertising rate (dol/period)

$$w_t = e_{1t}e_{2t}x_t/e_1^* e_2^* x^*$$ (4)

d_t = product class sales rate index in t

$\overline{c}_t$ = unadjusted product class sales rate in t (sales units/period)

$$c_t = \overline{c}_t d_t$$ (5)

α' = persistence constant for unadjusted product class sales

β' = affectable range of product class sales rate (sales units/period)

γ' = advertising response function exponent for product class

δ' = advertising response function denominator constant for product class

λ' = long-run minimum product class sales (sales units/period)

v_t = normalized product class advertising rate in t

$$\overline{c}_t = \lambda' + \alpha'(\overline{c}_{t-1} - \lambda') + \beta' v_t^{\gamma'}/(\delta' + v_t^{\gamma'})$$ (6)

v^* = maintenance advertising rate for product class sales (dol/period)

$$v_t = (v^* - x^* + x_t)/v^*$$ (7)

m_t = brand contribution per unit in t (dol/sales unit)

p_t = brand contribution rate after advertising in t (dol/period)

$$p_t = m_t s_t - x_t$$ (8)

σ_t = cumulative contribution after advertising for periods 1 to t (dol)

T = number of periods considered

$$\sigma_t = \sum_{s=1}^{t} p_s$$ (9)

$\sigma_T(x_t)$ = value of σ_T as a function of x_t

η_t = the rate of change of σ_T with x_t, called SLOPE, and calculated by:

$$\eta_T = [\sigma_T(x_t + .05x^*) - \sigma_T(x_t)]/.05x^*$$ (10)

References: #70, 137, 148.

Subject: Advertising and Promotion Expenditure

Title: Optimal Advertising- Expenditure Implications of a Simultaneous Equation Regression Analysis

Authors: Leonard J. Parsons and Frank M. Bass

Source: Operations Research, Vol. 19, No. 3, May-June 1971, 822-31

Summary: A simultaneous-equation regression model is developed and tested against aggregative time-series data. The optimal advertising strategies for two brands in the industry are developed.

Model: The simultaneous-equation regression model contains two demand equations, one for the brand being studied, and one for all other brands (remainder) and two advertising equations, one for the advertising of the brand and one for the advertising of all other brands.

The dynamic structural model of the four-equation system is:

$$-S_{B,t} + \beta_{12}A_{B,t} + \beta_{14}A_{R,t} + \gamma_{11}S_{B,t-1} + \gamma_{13}S_{R,t-1} + \gamma_{15}S_{N,t}$$
$$+ \gamma_{16}A_{N,t} + \gamma_{17}D_{7,t} + \gamma_{18}D_{8,t} + \gamma_{19}D_{9,t} + \gamma_{110}D_{10,t}$$
$$+\gamma_{111}D_{11,t} + \gamma_{112} + u_t = 0, \qquad (1a)$$

$$-A_{B,t} + \gamma_{21}S_{B,t-1} + \gamma_{22}A_{B,t-1} + \gamma_{23}S_{R,t-1} + \gamma_{24}A_{R,t-1}$$
$$+ \gamma_{27}D_{7,t} + \gamma_{28}D_{8,t} + \gamma_{29}D_{9,t} + \gamma_{210}D_{10,t} + \gamma_{211}D_{11,t}$$
$$+ \gamma_{212} + v_t = 0, \qquad (1b)$$

$$\beta_{32}A_{B,t} - S_{R,t} + \beta_{34}A_{R,t} + \gamma_{31}S_{B,t-1} + \gamma_{33}S_{R,t-1} + \gamma_{35}S_{N,t}$$
$$+ \gamma_{36}A_{N,t} + \gamma_{37}D_{7,t} + \gamma_{38}D_{8,t} + \gamma_{39}D_{9,t} + \gamma_{310}D_{10,t}$$
$$+ \gamma_{311}z_{11,t} + \gamma_{312} + w_t = 0, \qquad (1c)$$

$$-A_{R,t} + \gamma_{41}S_{B,t-1} + \gamma_{42}A_{B,t-1} + \gamma_{43}S_{R,t-1} + \gamma_{44}A_{R,t-1}$$
$$+ \gamma_{47}D_{7,t} + \gamma_{48}D_{8,t} + \gamma_{49}D_{9,t} + \gamma_{410}D_{10,t} + \gamma_{411}D_{11,t}$$
$$+ \gamma_{412} + x_t = 0, \qquad (1d)$$

where:

$S_{B,t}$ = log of per-capita unit sales of the brand,

$A_{B,t}$ = log of per-capita advertising of the brand,

$S_{R,t}$ = log of per-capita unit sales of the remainder,

$A_{R,t}$ = log of per-capita advertising of the remainder,

$S_{N,t}$ = log of per-capita unit sales of the new brands.

New brands are operationally defined as those that have been on the market for less than one year.

$A_{N,t}$ = log of per-capita advertising of the new brands,

$D_{i,t}$, i = 7, ..., 11, = dummy variables for seasonality.

In studying the optimal advertising strategy for a brand, we have removed the advertising equation for the brand from the system, substituting alternative strategies for it and have then calculated the impact of these strategies upon the system. In this way we are able to calculate the profit consequences of alternative strategies. We define profit for a single period, t, as:

$$\Pi_t(A_1, A_2, \ldots, A_t) = mS_t(A_1, A_2, \ldots, A_t) - A_t, \quad (2)$$

where m = the gross margin per unit for a brand exclusive of advertising costs, S_t = unit sales of the brand in period t, and A_t = advertising expenditures for the brand in period t.

For fixed initial conditions and with a finite n-period planning horizon, the discounted total profit associated with an advertising strategy $(A_1, A_2, \ldots, A_n)$ is:

$$\Pi = \Pi_1(A_1)\rho + \Pi_2(A_1,A_2)\rho^2 + \ldots + \Pi_n(A_1,A_2, \ldots, A_n)\rho^n. \quad (3)$$

The profit function Π will be the criterion used to evaluate alternative strategies.

In developing an advertising strategy, the criterion was discounted total profit over an n-period planning horizon,

since it is clear that the optimal solution is sensitive
to the length of the planning horizon. We have chosen
a planning horizon of 40 bimonthly periods. Although we
have studied the sensitivity of the optimal solution to
the length of the planning horizon, rather than present
the alternative sets of solutions, we have chosen to
present an illustration of the profit per period as a
function of the advertising strategy, since this illustration,
we believe, demonstrates the character of the issue.

For a single-period profit criterion the optimal
advertising to sales ratio is:

$$A^*/S^* = m\eta, \tag{4}$$

where η is advertising elasticity. It follows that
the optimal single-period profit to sales ratio is:

$$\pi^*/S^* = m(1-\eta). \tag{5}$$

Subject: Advertising and Promotion Expenditure

Title: Optimal Advertising Expenditure

Author: Maurice W. Sasieni

Source: Management Science, Vol. 18, No. 4, Part II, December 1971, 64-72

Summary: The optimal rate of advertising expenditure given the relationship between the rate of change of sales and the rate of expenditure is discussed. It is shown that we may assume that the marginal return of increased expenditure is never increasing.

Provided it is profitable to advertise, there exists an overall optimal sales rate and an expenditure level, just sufficient to maintain it, with the following properties with respect to long-run discounted profits:

(1) If sales even reach this level it is optimal to keep them there.

(2) Starting from any other level, the optimal policy is to spend in such a way as to drive sales towards this level.

The only requirements for these results are that the cost of achieving a given change in the sales rate be an increasing function of the sales rate and the rate of change of sales rate. It is also shown that the optimal sales rate to be maintained in the long-run is not the rate which maximizes the rate of gaining profit after advertising, unless the discount rate is zero.

Model: Vidale and Wolfe [#72] assume that

$$\dot{s} = ra(1 - s/m) - \lambda s, \qquad 0 \leq s \leq m, \qquad (1)$$

where $s = s(t)$ is the sales rate at time t, $a = a(t)$ is the rate of advertising expenditure, m, r, λ are parameters, m represents the maximum sales rate attainable.

$$\dot{s} = g(s,a,t), \qquad (2)$$

where g is a known function which increases with respect to both s and a. Assume that the present value of a sales rate s(T) at a time T is $b\Phi(T)s(T)$. The problem is to find $a(u) \geq 0$ to maximize

$$\int_t^T [p(u)s(u) - a(u)]\Phi(u)du + b\Phi(T)s(T).$$

If $s(t) = c$ then $s(T) = \int_t^T g(s,a,u)du + c$ so that for fixed c the problem is equivalent to maximizing the functional

$$J(a) = \int_t^T \{ [p(u)s(u) - a(u)]\phi(u) + b\phi(T)g(s,a,u) \}\, du. \quad (3)$$

Let us write x in place of $\dot{s}$ in (2) solving for a; suppose we obtain the cost function

$$a = h(s,x,t). \qquad (4)$$

The advertising rate cannot be negative; this implies that there is a minimum value for x corresponding to $h = 0$. We can write, using (2),

$$x_{min} = g(s,0,t). \qquad (5)$$

To obtain an equation for x, we note that if we select a value for $x \geqq x_{min}$ during the interval t to $t+w$, then for small $w > 0$, the contribution to J in this interval is

$$w\{\phi(t)[cp(t) - h(c,x,t)] + bx\phi(T)\}.$$

By time $t+w$ the sales rate has become $c+wx$ so that in the interval $t+w$ to T we have a similar maximization problem with $c+wx$ in place of c and $t+w$ in place of t. The result will be $f(c+wx, t+w)$. Hence,

$$f(c,t) = \underset{x \geqq x_{min}}{\text{Max}} \{ w\phi(t)(cp-h) + wbx\phi(T) + f(c+wx, t+w) \}. \quad (6)$$

Now for small w

$$f(c+wx, t+w) \simeq f(c,t) + wxf_c + wf_t.$$

Inserting this expression in (6), subtracting $f(c,t)$ from both sides, dividing by w and allowing w to tend to zero results in

$$0 = \underset{x \geqq x_{min}}{\text{Max}} \{ \phi(t)(cp-h) + x[b\phi(T) + f_c] + f_t \}. \quad (7)$$

Let $\psi(x)$ denote the quantity to be maximized. Provided the maximum occurs for finite $x > x_{min}$ we have

$$f_c + b\phi(T) = \phi(t)h_x \qquad (8)$$

from which we obtain

$$f_{ct} = \phi'(t)h_x + \phi(t)h_{xx} \frac{\partial x}{\partial t} + \phi(t)h_{xt}. \tag{9}$$

We may also insert (8) into (7) to obtain

$$f_{tc} = -\phi(t)(p - h_c - h_x \frac{\partial x}{\partial c} + h_x \frac{\partial x}{\partial c} + xh_{xx} \frac{\partial x}{\partial c} + xh_{xc}). \tag{10}$$

Equating f_{ct} to f_{tc} yields

$$h_{xx} (\frac{\partial x}{\partial t} + \frac{x\partial x}{\partial c}) = h_c + vh_x - p - xh_{xc} - h_{xt},$$

where $v = -\phi'(t)/\phi(t)$.

Since

$$x = \frac{dc}{dt} \tag{11}$$

we have

$$\dot{x}h_{xx} = h_c + vh_x - p - xh_{xc} - h_{xt}. \tag{12}$$

Equations (11) and (12) serve to determine c and x as functions of time. When $h_{xx} \neq 0$,

$$\dot{c} = x, \tag{11}$$

$$\dot{x} = (h_c + vh_x - p - xh_{xc} - h_{xt})/h_{xx}. \tag{13}$$

Optimal policies consist of intervals in which x and c satisfy (11) and (13) interspersed with intervals in which $x = x_{min}$. Assume that neither h, p, nor v depend on time. Then

$$\frac{dx}{dc} = \frac{h_c + vh_x - p - sh_{xc}}{xh_{xx}}. \tag{14}$$

Suppose we start when $c = \alpha$ and move along $x = x_1(c) > 0$ until $c = \beta$. We then move along $x = x_2(c) < 0$ until c is again equal to α; the cycle repeats indefinitely. As $T \to \infty$

$$0 = \underset{x \geq x_{min}}{\text{Max}} \{cp - h + x\theta' - v\theta\}. \tag{15}$$

Now $x_1 > 0 > x_{min}$, so we may differentiate and obtain $-h_{x_1} + \theta' = 0$, where $h_{x_1} = \partial h/\partial x$ evaluated at $x = x_1$.

Insert this in (15),

$$h_2 - h_1 = (x_2 - x_1)h_{x_1}, \qquad (16)$$

where h_i is the value of h when $x = x_i$.

As $T \to \infty$ there is a unique value of x corresponding to each value of c, and consequently no cyclic policy can be optimal. Assume $x = 0$, $c = c^* > 0$, $h_{xx} > 0$. From (11) and (13) we have

$$h_c + vh_x - p = 0 = x. \qquad (17)$$

The solution to (17) will yield a value of c^* of c and provided $h_{xx}(c^*,0) > 0$ we see that the long-run optimal policy is to allow c to approach c^* along the solution curve of (14) which passes through $(c^*,0)$.

To compute numerical values of x along the solution curve through $(c^*,0)$ we need the value of dx/dc at this point. Expanding (13), for points (c,x) near to $(c^*,0)$ we obtain:

$$\dot{x} = [xh_{xx} + (c - c^*)(h_{cc} + vh_{xc})]/h_{xx}, \qquad (13a)$$

where all h-functions are evaluated at $(c^*,0)$.

Now assume that near $(c^*,0)$, $c - c^* = Ae^{\lambda t}$ so that $x = \dot{c} = \lambda Ae^{\lambda t}$ and substitute in (13a); divide by A

$$\lambda^2 - v\lambda - (h_{cc} + vh_{xc})/h_{xx} = 0. \qquad (18)$$

If (18) is satisfied by a suitable value of $\lambda < 0$ we can use (14) starting at $(c^*,0)$ to compute the complete solution.

References: #72 and 73.

Subject: Advertising and Promotion Expenditure

Title: Dynamic Correction in Marketing Planning Models

Author: Charles B. Weinberg

Source: Management Science, Vol. 22, No. 6, February 1976, 677-87

Summary: Most marketing planning models have carry-over effects in which one period's decisions influence the results obtained in future periods. It is shown that failure to allow for the carry-over effect beyond the planning horizon can result in underallocation of resources and in biases of the timing pattern of resource expenditure. For a wide class of market planning models, a procedure is developed to take into account this long-term effect. The distortion and the procedure are illustrated in an example.

Model: The problem is how to define an objective function when a firm must set a budget over k periods and the response in each period is dependent on the expenditures in previous periods (or, equivalently, current expenditures influence market response in future periods). Formally, we have $M_t = G_t(x_t, x_{t-1}, \ldots, x_{t-j})$ where M_t is the market response in period t and x_t is the expenditure in period t. If profit in period t from a market response of M_t is pM_t, an appropriate objective function is

$$\sum_{i=1}^{k} (pM_{t+i} - x_{t+i}) + CpM_{t+k} \qquad (1)$$

where C, the "dynamic correction factor," is independent of the values of the x_{t+i}.

Two main types of dynamic models of carry-over effects can be distinguished. One, the delayed or <u>lagged response</u> effect, is described by

$$M_t = H_t(y_t) \quad \text{where} \quad y_t = x_t + \beta y_{t-1}. \qquad (2)$$

The second type which describes the <u>customer holdover</u> effect is

$$M_t = \alpha_t M_{t-1} + F_t(x_t). \qquad (3)$$

(3) can be extended to the form

$$M_t = \alpha_t M_{t-1} + \sum_{i=0}^{\ell} F_{t-i}(x_{t-i}) \qquad (4)$$

include a delayed response effect as well as a customer holdover effect.

Dynamic Correction for the Customer Holdover Case

Recalling (3), the following response model for the customer holdover case is postulated:

$$M_t = \alpha_t M_{t-1} + F_t(x_t) \qquad (5)$$

where

M_t = market response at time t (e.g., sales, share),

F_t = response function at time t,

x_t = marketing decision variable (or vector) at time t (e.g., firm's advertising expenditure),

α_t = carryover parameter ($\alpha_t > 0$),

t = time period.

The firm's marketing planning objective is to maximize the total net returns from its marketing expenditures in periods $t+1$ to $t+k$, i.e.,

$$\text{Max } Z_t = \sum_{i=1}^{\infty} pM_{t+i} - \sum_{i=1}^{k} x_{t+i}. \qquad (6)$$

When M_t is given by (5) it is shown that many of the terms in (6) are independent of x_{t+i} for $i = 1, \ldots, k$ and consequently (6) can be replaced by (1) where

$$C = \sum_{i=1}^{\infty} \prod_{\ell=1}^{i} \alpha_{t+k+\ell}. \qquad (7)$$

We obtain

$$\text{Max } Z_2 = \sum_{i=1}^{k} pM_{t+i} + pCM_{t+k} - \sum_{i=1}^{k} x_{t+i} \qquad (8)$$

where C is given by (7). Thus, the modification (8) needed to allow for the dynamic effects is to place an additional weighting on the market response in the last period of the horizon.

This dynamic correction procedure is applied to the optimal budget for the ADBUDG model [#67].

Dynamic Correction for the Lagged Response Model

Recalling (2),

$$M_t = H_t(y_t) \tag{9}$$

where $\quad y_t = x_t + \beta y_{t-1}. \tag{10}$

The variable y_t is often called "goodwill" or "accumulated marketing effort."

It is not possible for (9) to build a dynamic correction factor which equalizes the impact of advertising over all periods. However, a dynamic correction factor can be built which will equalize the weight of the $x_{t+1}, x_{t+2}, \ldots, x_{t+k}$ in the objective function. Define W_i = the weight of x_{t+i} in the objective function. Now, consider the objective function

$$\text{Max } Z_3 = \sum_{i=1}^{k} pM_{t+i} + C_1 pM_{t+k} - \sum_{i=1}^{k} x_{t+i} \quad \text{where} \quad (11)$$

$$C_1 = \beta/(1-\beta). \tag{12}$$

Then

$$W_i = W_j = 1 \quad \text{for} \quad i,j = 1,2, \ldots, k. \tag{13}$$

C_1 is the dynamic correction for the lagged response model and is analogous to the dynamic correction factor C in (7).

Reference: #67.

Subject: Sales Response to Advertising and Promotion

Title: The Effect of Promotional Effort on Sales

Author: John F. Magee

Source: Operations Research, Vol. 1, No. 2, February 1953, 64-74

Summary: A static model is presented which develops the functional relationship between sales and promotional effort based upon empirical data on coffee distribution to retail grocery stores.

Model: The company studied distributes coffee in cases to retail grocery stores throughout the country. Promotional salesmen travel from store to store, distributing point-of-sale advertising and displays. Let the distribution of actual number of cases ordered be approximated by the Poisson distribution

$$E(n) = \frac{e^{-c}c^{n}}{n!}, \tag{1}$$

where $E(n)$ is the probability that a dealer with an expected frequency of ordering or characteristic order, c, will actually order n cases in a month.

The fraction of dealers who will order n cases will be given by

$$f(n) = \int_{0}^{\infty} \frac{e^{-c}c^{n}}{n!} \ Y(c)dc \tag{2}$$

where $Y(c)$ is the probability density under the condition that all dealers are given the promotional aids. Assume the exponential form

$$Y(c) = \frac{1}{s} \ e^{-c/s} \tag{3}$$

where s is the average number of cases ordered by one dealer per month for the group of dealers as a whole. Substituting (3) in (2) yields

$$f(n) = \frac{s^{n}}{(s+1)^{n+1}}. \tag{4}$$

After ranking the dealers, the top 40% are given promotion. Approximation yields

$$Y_p(c) = \frac{(1 - e^{-g(c/s)})e^{-c/s}}{s}, \tag{5}$$

where a is the fraction of the dealers promoted, and $g = (a/1-a)$. The values of $f_p(n)$ would then be given by

$$f_p(n) = \int_0^\infty E(n)Y_p(c)dc$$

$$= s^n \left\{\frac{1}{(s+1)^{n+1}} - \frac{1}{(s+g+1)^{n+1}}\right\}. \tag{6}$$

$Y_{np}(c)$, the distribution of characteristic orders of dealers normally not promoted (the lower 60 per cent) is approximated by

$$Y_{np}(c) = \frac{e^{-((g+1)/s)c}}{s}, \tag{7}$$

and the fraction of these dealers ordering n cases, by

$$f_{np}(n) = \frac{s^n}{(s+g+1)^{n+1}}. \tag{8}$$

Comparison of the observed fractions of nonpromoted dealers ordering a given number of cases with the values calculated from (8) therefore gives a basis for estimating the effect of the promotional effort. Two differences were found:

1. According to (8), the ratio of $f_{np}(c)$ to $f_{np}(n+1)$ is $(s+g+1)/s$. When the observed values of these ratios for $n \geq 1$ were used to determine s, the average number of cases ordered by one dealer per month, the value found was 0.71 of the value expected from study of the promoted dealers.

2. When this value of s was substituted in (8) to calculate the values of the fractions ordering n cases, $f_{np}(n)$, the observed values of $f_{np}(n)$ for $n \geq 1$ were found to be 0.7 times the calculated values.

Adjustment of (8) to account for these two effects yields (8a):

$$f_{np}(n) = \frac{0.7(0.71s)^n}{(0.71s + g + 1)^n}, \quad n \geq 1, \tag{8a}$$

$$f_{np}(0) = (1-a) \{1 - \sum_{n=1}^{\infty} f_{np}(n)\}.$$

These two effects can be summarized as follows: When a dealer is given no promotion, the probability is 0.3 that he will act as if his characteristic order, c, were zero; moreover, the probability is 0.7 that he will act as if c were 0.71 of his frequency if promoted. The net effect is a reduction in his expected business of 50 per cent.

Suppose there are N dealers, of which Na are to be selected for promotion. Then the resulting number of cases ordered will be

$$O(a) = N \{a \int_0^{\infty} cY_p(c)dc + 0.5(1-a) \int_0^{\infty} cY_{np}(c)dc\}$$

$$= \frac{N_s}{2} (1 + 2a - a^2). \tag{9}$$

Since the value of a case is fixed, the resulting business will be

$$B(a) = \frac{Nsv}{2} (1 + 2a - a^2), \tag{10}$$

where v is the value per case. The relative efficiency of distribution of promotion is

$$E(a) = \frac{B(a) - B''(a)}{B'(a) - B''(a)}, \tag{11}$$

or, in this case,

$$E(a) = -\frac{(1-a)}{\ln a}; \quad 0 < a < 1. \tag{12}$$

The relative potential gain from improvement of the selection of dealers is given by

$$G(a) = \frac{B'(a) - B(a)}{B(a)} = \frac{a(a - 1 - \ln a)}{1 + 2a - a^2}. \tag{13}$$

Subject: Sales Response to Advertising and Promotion

Title: An Operations-Research Study of Sales Response to Advertising

Authors: M.L. Vidale and H.B. Wolfe

Source: Operations Research, Vol. 5, No. 3, June 1957, 370-81

Summary: Model of the interaction of advertising and sales based on three parameters: sales decay constant, saturation level, and response constant.

Model:

Let

$$dS/dt = rA(t)(M-S)/M - \lambda S \tag{1}$$

where

S = rate of sales

r = response constant

A = rate of advertising expenditure

M = saturation level

λ = exponential sales decay constant.

For a constant A

$$S(t) = [M/(1 + \lambda M/rA)][1 - e^{-(rA/M+\lambda)t}]$$
$$+ S_0 e^{-(rA/M+\lambda)t} \qquad (t < T) \tag{2}$$

where S_0 is the rate of sales at $t = 0$, the start of the advertising campaign. After advertising has stopped

$$S(t) = S(T)e^{-\lambda(t-T)} \qquad (t > T). \tag{3}$$

For an advertising pulse

$$S(t) = M e^{-\lambda t} - (M - S_0)e^{-(ra/M+\lambda)t} \tag{4}$$

where S_0 = rate of sales immediately preceding the promotion and a = total advertising expenditure. The immediate sales increase resulting from the promotion is

$$S(0) - S_0 = (M - S_0)(1 - e^{-ra/M}). \tag{5}$$

The total additional sales generated by this campaign are

$$\int_0^\alpha [S(0) - S_0]e^{-\lambda t}dt = \frac{M - S_0}{\lambda}(1 - e^{-ra/M})t$$

which reduces to $(ra/\lambda)(M - S_0)/M$ for sales well below saturation.

Advertising Budget

Determining the rate of return on capital invested in the promotion of product k:

$$\int_0^\infty C_k(t)e^{-Ikt}dt = \int_0^\infty R_k(t)e^{-Ikt}dt \qquad (6)$$

where

$R_k(t)$ = additional sales resulting from the advertising campaign

$C_k(t)$ = rate of additional expenditures resulting from the advertising campaign

I_k = return on capital invested in advertising product k.

Under the assumption that production and distribution costs are proportional to sales, we have

$$C_k(t) = f_k R_k(t) + a_k \qquad (7)$$

where f_k = the ratio of production and distribution costs to selling price, and a_k = total cost of the proposed advertising campaign for product k.

Assuming that the rate of sales of the unpromoted product decays exponentially at the rate λ_k, we have

$$R_k(t) = R_{0k}e^{-\lambda kt} \qquad (8)$$

where R_{0k} is the instantaneous sales increase resulting from the campaign. Substituting (7) and (8) into equation (6), we obtain

$$I_k = (R_{0k}/a_k)(1 - f_k) - \lambda_k. \tag{9}$$

The rate of return I_k is a function of the intensity of the advertising campaign.

References: #2, 43, 64, 69, 73, 82.

Subject: Sales Response to Advertising and Promotion

Title: Optimal Control of the Vidale-Wolfe Advertising Model

Author: Suresh P. Sethi

Source: Operations Research, Vol. 2, No. 4, July-August 1973, 998-1013

Summary: An optimal-control problem for the dynamics of the Vidale-Wolfe advertising model is considered, the optimal control being the rate of advertising expenditure to achieve a terminal market share within specified limits in a way that maximizes the present value of net profit streams over a finite horizon. First, the special polar cases of fixed and free end points are solved with and without an upper limit on advertising rate. The complete solution to the general problem is then constructed from these polar cases. The fixed-end-point case with no upper limit on the advertising rate is solved by using Green's theorem, while the other cases require additional use of switching-point analysis based on the maximum principle. The optimal control is characterized by a combination of bang-bang, impulse, and singular control, with the singular arc forming a turnpike.

Model: Vidale and Wolfe [#72] argue that changes in the rate of sales of a product depend on two effects: response to advertising that acts (via the response constant p) on the unsold portion of the market, and loss due to forgetting that acts (via the decay constant k) on the sold portion of the market.

$$\dot{x} = \rho u(1-x) - kx, \quad x(0) = x_o, \qquad (1)$$

where x is the market share (i.e., the rate of sales expressed as a fraction of the market potential or saturation level) and u is the rate of advertising expenditure (a control variable) at time t.

The objective to be maximized is the present value of the profit stream up to the horizon T. With the maximum sales-revenue potential π (this assumes a constant margin per unit product), the maximum allowable rate of advertising expenditures Q, and the discount rate i, the optimal control problem is

$$\max_{Q \geq u(t) \geq 0} \left\{ J = \int_o^T [\pi x(t) - u(t)] \exp[-it] \, dt \right\}, \qquad (2)$$

subject to (1) and the terminal constraint $x(T) \in [x_T^1, x_T^2]$.
The problem is known as a fixed-end-point problem if
$x_T^1 = x_T^2 = x_T$, and a free-end-point problem if $x_T^1 = 0$,
$x_T^2 = 1$. Solving for $Q = \infty$ (i.e., no upper limit on the
advertising rate), we substitute $udt = (dx + kxdt)/p(1-x)$
derived from (1) into the objective function (2) to obtain
the line integral along any curve Γ in (t,x) space:

$$J_\Gamma = \int_\Gamma \{ [\pi x - kx/\rho(1-x)] \exp[-it] dt - [1/\rho(1-x)] \exp[-it] dx \}.$$

For a simple closed curve Γ, we can use Green's theorem to
express this line integral over the area R bounded by Γ:

$$J_\Gamma = \iint_R \{ (\partial/\partial t) [-\exp[-it]/\rho(1-x)]$$

$$- (\partial/\partial x) [\pi x - kx/\rho(1-x)] \exp[-it] \} \, dt \, dx$$

$$\tag{3}$$

$$= (1/\rho) \iint_R [k/(1-x)^2 + i/(1-x) - \pi \rho] \exp[-it] \, dt \, dx.$$

To specify the optimal control, we equate the integrand
to zero, and solve the resulting quadratic equation in
$(1-x)^{-1}$ to provide us with

$$x = 1 - 2k/(\pm \sqrt{i^2 + 4\pi \rho k} - i).$$

Since we are only interested in the values of x between 0
and 1, we define

$$x^S = \max[1 - 2k/(\sqrt{i^2 + 4\pi \rho k} - i), \ 0],$$

and the corresponding control

$$u^S = kx^S/\rho(1-x^S) = (\sqrt{i^2 + 4\pi \rho k} - i - 2k)/2\rho,$$

which will maintain the state at x^S by using (1). The
superscript s refers to 'singular arc'. If we define
$\alpha = (\pi \rho)/(k+i)$, $x^S = 0$, $u^S = 0 \iff \alpha \leqq 1$.

Theorem 1. Let $P(a,b,t)$ denote the magnitude of an advertising pulse necessary to increase the market share $x(t) = a$ at time t to $x(t^+) = b$ at time t^+ (just after t); then

$$P(a,b,t) = (1/\rho)\ln[(1-a)/(1-b)] = P(a,b)$$

for all $0 \leqq a \leqq b \leqq 1$ and for all t. Furthermore, if $a \leqq c \leqq b$, then $P(a,b,t) = P(a,c,t) + P(c,b,t)$.

Theorem 2. With respect to the switching diagrams of Figure 1, $u*(I) = 0$, $u*(II) = u^S$, and $P*(x,\cdot,t) = \sup_{x \in III} P(x,y,t)$ for $(t,x) \in III$ is a unique optimal feedback policy.

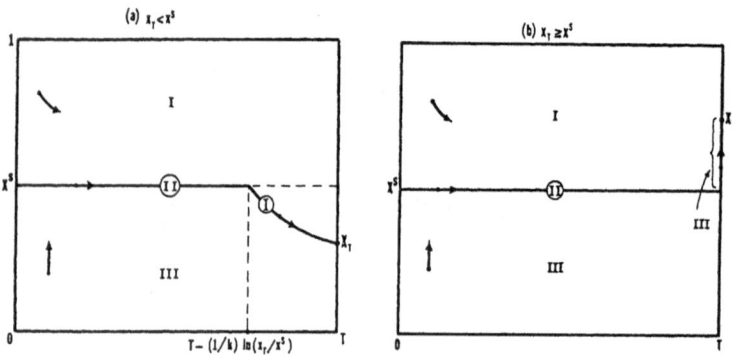

Figure 1. Switching Diagrams

We form the current-value Hamiltonian as $H = \pi x - u + v[\rho u(1-x) - kx]$, where the current-value adjoint variable v satisfies $\dot{v} = iv + v(\rho u + k) - \pi$, or the variable $\lambda = \rho v$ satisfies

$$\dot{\lambda} = i\lambda + \lambda(\rho u + k) - \pi\rho. \qquad (4)$$

The transversality condition on the variable λ for the general problem of Section 1 is

$$\lambda(T)[x - x(T)] \geq 0, \quad \forall x \in [x_T^1, x_T^2]; \tag{5}$$

Theorem 3 (maximum principle). There exists an optimal control. Furthermore, for u*(t) to be an optimal control with corresponding trajectory x*(t), it is necessary and sufficient that there exist a nonzero trajectory λ*(t) satisfying (4) and (5) and the Hamiltonian maximizing condition

$$H[x^*(t), u^*(t), \lambda^*(t)] \geq H[x^*(t), u, \lambda^*(t)], \quad u \in [0, Q] \tag{6}$$

for each $t \in [0, T]$.

Since the Hamiltonian H is linear in u, the condition (6) in the theorem can be replaced by

$$u^*(t) = \begin{cases} 0, & \text{if } W(t) = -1 + \lambda(t)[1 - x(t)] < 0, \\ [0, Q], & \text{if } W(t) = 0, \\ Q, & \text{if } W(t) > 0, \end{cases} \tag{7}$$

for each t.

This type of control is known as "bang-bang" control. However, interior control is possible on an arc along which $W(t) = 0$, known as a 'singular arc.' We define $u^{ss} = \min[u^s, Q]$, with $x^{ss} = \rho u^{ss}/(\rho u^{ss} + k)$ and $\lambda^{ss} = \pi \rho/(i + k + \rho u^{ss})$ being the associated state and adjoint variables. Furthermore, we define $x^a = 1 - (1/\lambda^{ss})$ and $\beta(Q) = \pi \rho k/(k + i + \rho Q)(\rho Q + k)$.

Theorem 4. (i) $\beta(Q) \leq 1 \Rightarrow u^{ss} = u^s$, $x^{ss} = x^s = x^a$, $\lambda^{ss} = \lambda^s$, and $\lambda^{ss}(1 - x^{ss}) = 1$.

(ii) $\beta(Q) > 1 \Rightarrow u^{ss} = Q < u^s$, $x^{ss} = \rho Q/(\rho Q + k) < x^s < x^a$, $\lambda^{ss} = \pi \rho/(i + k + \rho Q) > \lambda^s$, $\lambda^{ss}(1 - x^{ss}) = \beta(Q)$, and $\lambda^{ss}(1 - x^a) = 1$.

Define the reverse time $r = T - t$. Let $\dot{y} = dy/d\tau$; then

$\dot{y} = -\dot{y}$. The case $u^s > Q$ has three subcases:
(i) $x_T \leq x^{ss}$, (ii) $x^{ss} < x_T \leq x^s$, and (iii) $x_T > x^s$. We
treat subcase (i). To have $W(\underline{x}', \tau') = 0$ at τ' on the
switching manifold, we must initialize $\underline{\lambda}' = 1/(1-\underline{x}')$.

Define $W'(\tau) = -1 + \lambda'(\tau)[1 - x'(\tau)]$ to be the
reverse time-switching function starting from $(\underline{x}', \tau')$.
Theorem 5. $W'(\tau)$ is strictly concave; there exists a
unique $\tau(\underline{x}') \in (\tau', \infty)$ such that $W'[\tau(\underline{x}')] = 0$, $\forall \underline{x}' \in (x^{ss}, x^s)$.
Furthermore, $x^s < x'[\tau(\underline{x}')] < x^a$.

Theorem 6. Let $\underline{x}'' > \underline{x}'$ and $\tau'' > \tau'$. Then (i) $\tau(\underline{x}'') < \tau(\underline{x}')$,
and (ii) $x''[\tau(\underline{x}'')] < x'[\tau(\underline{x}')]$.
The analysis of

$$W(\tau, x_T) = -1 + [\pi\rho/(k+i)](1 - \exp[-(k+i)\tau])(1 - x_T\exp[k_T]).$$

yields the following theorem:

Theorem 7. For each $x_T \in \{0, x_T^s = x^s[\rho u^s/(k+i+\rho u^s)]^{k/(k+i)}\}$,
there exists a unique $\tau'(x_T)$ such that (i) $W(\tau'(x_T), x_T) = 0$
and (ii) $W(\tau, x_T) = 0 \Rightarrow \tau \geq \tau'(x_T)$. Furthermore, $\tau'(x_T)$
is a continuous strictly monotonically increasing function
of x_T with $\tau'(0) = -[1/(k+i)]\ln[(\pi\rho - k-i)/\pi\rho]$ and $\tau'(x_T^s) = -[1/(k+i)]\ln[\rho u^s/(k+i+\rho u^s)]$. Finally, $\tau'(x_T) < -(1/k)\ln x_T$.

Theorem 8. For $\beta(Q) > 1$, $x(t) < x^a \Rightarrow u^*(t) = Q$ and
$x(t) > x^a \Rightarrow u^*(t) = 0$. When $\beta(Q) \leq 1(x^a = x^s)$, we have
$x(t) = x^s \Rightarrow u^*(t) = u^s$ in addition. Furthermore, $\{x^{ss}, u^{ss}, \lambda^{ss}$
is the unique, optimal long-run stationary equilibrium.

Theorem 9. Let $[\overline{x}_T^1, \overline{x}_T^2]$ denote the largest feasible subset
of $[x_T^1, x_T^2]$ and let x_T^f be the terminal state in the optimal
solution of the free-end-point problem. Then the optimal
control $u^*(x,t)$ can be defined as:

(i) If $x_T^f \in [\bar{x}_T^1, \bar{x}_T^2]$, $u^*(x,t)$ is the solution of the free-end-point problem.

(ii) Otherwise, if $\bar{x}_T^2 < x_T^f$, set $x_T = \bar{x}_T^2$, and, if $\bar{x}_T^1 > x_T^f$, set $x_T = \bar{x}_T^1$; $u^*(x,t)$ is the solution of the fixed-end-point problem obtained in Theorem 7.

The optimal control for the Vidale-Wolfe advertising model has the turnpike property, the turnpike being the singular arc, in the cases where it is feasible to ride along the turnpike. The optimizing control is the feasible control that spends maximum time along the turnpike, i.e., it uses fastest entry and exit ramps to and from the turnpike.

References: #69 and 72.

Subject: Sales Response to Advertising and Promotion

Title: Advertising and Cigarettes

Author: Lester G. Telser

Source: Journal of Political Economy, Vol. 70, No. 5, October 1962, 471-99

Summary: Measurement of various aspects of competitive advertising to determine how advertising affects sales.

Model:

Studying the relation between market share and relative advertising, let

Q = sales of all brands in the market (per period)

Q_i = sales of all brands, excluding sales of brand i

q_i = sales of brand i

m_i = market share of brand i

x_i = advertising outlay for brand i

X = advertising outlay of whole industry

X_i = advertising outlay of all brands, excluding i

$a_i = x_i/X$

$s_i = x_i/X_i$

The differential of market share with respect to x_{it} is

$$Q \frac{dm_i}{dx_i} = (1 - m_i) \frac{dq_i}{dx_i}$$

$$- m_i \frac{dQ_i}{dx_i} = QL_{2i} \frac{ds_i}{dx_i} \tag{1}$$

where

$$\frac{dq_i}{dx_i} = \frac{\partial q_i}{\partial x_i} + \frac{\partial q_i}{\partial X_i} \frac{dX_i}{dx_i} \tag{2}$$

is the differential of the firm's sales with respect to its advertising, taking into account the effect on the advertising of its competitors induced by a change in its advertising. There is, therefore, a relation

$$X_i = h(x_i) \tag{3}$$

that indicates the competitors' reactions to the firm's advertising. Similarly,

$$\frac{dQ_i}{dx_i} = \frac{\partial Q_i}{\partial x_i} + \frac{\partial Q_i}{\partial X_i} \frac{dX_i}{dx_i} \tag{4}$$

$$\frac{ds_i}{dx_i} = \frac{1}{X_i} [1 - \frac{x_i}{X_i} \frac{dX_i}{dx_i}] = \frac{1}{X_i} \lambda_i, \tag{5}$$

where λ_i is one minus the elasticity of response of competitors' advertising to a change in the given firm's advertising. Equation (1) simplifies to

$$\gamma_{x_i} = (1 - m_i)(\theta_i - \psi_i) = \gamma_{x_i} \lambda_i, \tag{6}$$

where

$$\gamma_{x_i} = \frac{x_i}{m_i} (\frac{dm_i}{dx_i}) = \text{market share elasticity with respect to i's advertising}$$

$$\theta_i = \frac{x_i}{q_i} (\frac{dq_i}{dx_i}) = \text{absolute sales advertising elasticity}$$

$$\psi_i = \frac{x_i}{Q_i} (\frac{dQ_i}{dx_i}) = \text{competitive advertising elasticity}$$

$$\gamma_{s_i} = \frac{s_i}{m_i} (\frac{dm_i}{ds_i}) = \text{market share elasticity with respect to relative advertising.}$$

Subject: Sales Response to Advertising and Promotion
Title: Bayesian Decision Theory in Advertising
Author: Paul E. Green
Source: Journal of Advertising Research, Vol. 2, No. 4, December
            1962, 33-41
Summary: Use of Bayesian analysis in evaluating the decision
            implications of a point-of-purchase merchandising
            experiment.
Model:

The competitor of the small firm A specializing in
the sale of soft drinks vended by machine had been
increasing its market share, largely at A's expense.
Since both firms offer the same variety of drinks
and comparable product quality, A's management feels
that the competitor's gain stems from their more
appealing machines.

To cover the cost of machine conversion, monthly gross
sales would have to increase $200 per machine to
break even.
An experiment was carried out; the average increase
in sales of the converted machines amounted to $190 per
machine per month.

Table 1

A's Management's Prior Probabilities
for Vending Machine Problem

Probability Assignment	Mean Sales Difference; New vs. Old (Dollars Per Machine Per Month)	Expected Value
0.00	Under $100	$ 0.00
0.10	$100 - 150	12.50
0.30	150 - 200	52.50
0.50	200 - 250	112.50
0.10	250 - 300	27.50
0.00	Over $300	0.00
1.00		$205.00

Calculating the conditional probability of getting the sample difference, $190, under each possible state of nature:

Table 2

Conditional Probabilities
for Vending Machine Problem

Possible Differences	Midpoint	$Z = (\bar{x}-\mu)/\sigma(\bar{x})$	Ordinate	$P(\bar{x}/\mu)$
$100 - 150	$125	$(190-125)/15$	.00004	$.00004\Delta\bar{x}/15$
150 - 200	175	$(190-175)/15$	.24200	$.24200\Delta\bar{x}/15$
200 - 250	225	$(190-225)/15$	.02833	$.02833\Delta\bar{x}/15$
250 - 300	275	$(190-275)/15$	.00000	$.00000\Delta\bar{x}/15$

Combining Tables 1 and 2:

Table 3

Posterior Probabilities
for Vending Machine Problem

State of Nature	Prior Probabilities	Conditional Probabilities	Joint Probabilities	Posterior Probabilities
$125	.10	$.00004\Delta\bar{x}/15$	$.000004\Delta\bar{x}/15$	.00
175	.30	$.24200\Delta\bar{x}/15$	$.072600\Delta\bar{x}/15$	.84
225	.50	$.02833\Delta\bar{x}/15$	$.014165\Delta\bar{x}/15$	.16
275	.10	$.00000\Delta\bar{x}/15$	$.000000\Delta\bar{x}/15$	.00
			$.086769\Delta\bar{x}/15$	1.00

Applying the posterior probabilities to the admissible state of nature and deriving a new expected value based on the combined judgmental and sample evidence:

E.V. = .00(125) + .84(175) + .16(225) + .00(275) = $183.

Subject: Sales Response to Advertising and Promotion

Title: Differential Equation Approach to Marketing

Authors: Shiv K. Gupta and K.S. Krishnan

Source: Operations Research, Vol. 15, September–December 1967, 1030–39

Summary: The effect of promotional efforts in the previous periods on the sales in the present period is studied. A differential equation approach to study the relation between sales and promotional effort for multicompetitors and multiperiods is presented. Using these relations, the problem of finding optimum promotional effort has been formulated for multiperiods, multicompetitors, and multi-products. For the case of a single product, the optimal values have been obtained.

Model:

Single Period, Single Firm and Single Product

(a) Sales as a Function of Advertising Only

Let $S = a[1 - \exp(-\gamma A)]$,

where S = total sales,

A = advertising expenditure,

a = maximum market potential,

γ = constant,

then $dS/dA = a\gamma \exp(-\gamma A)$,

$$= \gamma(a - S).$$

The relation between sales and advertising is postulated by

$$dS/dA = \Phi_1(a - S), \tag{1}$$

where $\Phi_1(a - S)$ is some function of the untapped market potential $(a - S)$.

(b) Sales as a Function of Price Only

$$dS/dp = -\Phi_2(S), \tag{2}$$

where $\Phi_2(S)$ is some function of S.

(c) Sales as a Function of Both Price and Advertising

$$dS = (\partial S/\partial A)dA + (\partial S/\partial p)dp.$$

Using the models postulated in (a) and (b)

$$dS = \phi_1(a-S)\ dA - \phi_2(S)dp, \tag{3}$$

where a is the sales for a given value of p when $A \to \infty$.

(d) Sales as a Function of Price, Advertising, and Number of Outlets

It is assumed that the relation between sales and outlets (0) is of the same form as the relation between sales and advertising and can be postulated as

$$\partial S/\partial A = \phi_1(a-S),$$

$$\partial S/\partial p = -\phi_2(S),$$

$$\partial S/\partial 0 = \phi_3(b-S),$$

where a is the maximum market potential for a fixed p and 0 and b is the maximum market potential for a fixed p and A.

In this case

$$dS = \phi_1(a-S)dx_1 + \phi_2(S)dx_2 + \phi_3(b-S)dx_3,$$

where $x_1 = A$, $x_2 = -p$, and $x_3 = 0$,

a is a function of x_2 and x_3,

and b is a function of x_1 and x_2.

(e) Sales as a Function of w-Variables

$$dS = \sum_{j=1}^{j=w} \phi_j dx_j.$$

Multiperiods

(a) Single Firm, Single Product

Let A_t be the promotional expenditure in the ith period.

The effect of promotional effort is distributed over several periods and diminishes at a constant rate and E_t is the effective promotional expenditure up to time t. In terms of the actual promotional expenditure A_t, we have

$$E_t = \sum_{i=1}^{i=t} k_{t+i-1} A_i, \qquad (4)$$

where k_1, k_2, ..., k_t are constants. If S_t and S_{t-1} are sales in time t and $t-1$ respectively, the sales due to E_t are

$$1/(1-b)S_t - b/(1-b)S_{t-1},$$

where b is the retention rate.

Hence,

$$S_t = f_t(E_t) + bf_{t-1})(E_{t-1}) + \ldots + b^{t-1} f_1(E_1), \qquad (5)$$

where $f_t(E_t)$ is some unknown function of E_t.

$$dS_t = (\partial S_t/\partial E_1) \cdot dE_1 + (\partial S_t/\partial E_2) \cdot dE_2 + \ldots + (\partial S_t/\partial E_t)dE_t \qquad (6)$$

$$= b^{t-1}\phi_1(a_1 - S_1)dE_1 + b^{t-2}\phi_2(a_2 - S_2)dE_2 + \ldots$$

$$+ \phi_t(a_t - S_t)dE_t,$$

where a_i is the maximum potential in the ith period for given values of E_1, E_2, ..., E_{i-1}.

Let m_t be the unit profit in the ith period. Then net revenue, R_t, in the ith period is

$$R_t = m_t S_t - A_t.$$

Total revenue, R, in T periods is

$$R = \sum_{t=1}^{t=T} \alpha^{t-1} R_t,$$

where α is the discounting factor. The maximum value of R is given by

$$\partial R/\partial A_t = 0. \qquad (t = 1,2, \ldots, T) \qquad (7)$$

In general, if $\phi_t(a_t - S_t) = \lambda_t(a_t - S_t)$ for $t = 1,2, \ldots$, then

$$S_t = \gamma_t[1 - \exp(-\lambda_t E_t)]$$
$$+ \sum_{i=1}^{t-1} \gamma_i b^{t-i}[\exp(-\lambda_{i+1}E_{i+1}) - \exp(-\lambda_i E_i)], \tag{8}$$

where γ_i = maximum sales in the ith period; note $\gamma_1 = a_1$.

(b) Multicompetitors Single Product

Let A_{ti} = promotional expenditure of ith competitor in t^{th} period; $i = 1,2, \ldots, n$. E_{ti} = effective promotional expenditure of ith competitor in t^{th} period.

Total sales of ith competitor in t^{th} period is

$$S_{ti} = S_t E_{ti} / \sum_{i=1}^{i=n} E_{ti}.$$

Total revenue of the ith competition in T periods is

$$R_i = \sum_{t=1}^{t=T} \alpha^{t-1}[(S_t E_{ti} / \sum_{i=1}^{i=n} E_{ti}) m_{ti} - A_{ti}],$$

$$E_{ti} = \sum_{t=1}^{t=T} \alpha^{t-1} R_{ti} \quad \text{for} \quad i = 1,2, \ldots, n, \tag{9}$$

where α is the discounting factor.

(c) Multicompetitors Multiproducts

Let A_{tip} = promotional expenditure of the ith competitor for p^{th} product in t^{th} period.
Total sales of the ith competitor in the tth period for the pth product is

$$S_{tp} E_{tip} / \sum_{i=1}^{i=n} E_{tip}.$$

The total revenue of the ith competitor for periods $1,2, \ldots,$ T is then

$$R_i = \sum_{t=1}^{t=T} \sum_{p=1}^{p=m} \alpha^{t-1}[(S_{tp} E_{tip} / \sum_{i=1}^{i=n} E_{tip}) m_{tip} - A_{tip}]. \tag{10}$$

Subject: Sales Response to Advertising and Promotion

Title: Consumer Response to Promotions

Authors: Alfred A. Kuehn and Albert C. Rohloff

Source: Patrick J. Robinson (ed.), Promotional Decisions Using
          Mathematical Models, Boston: Allyn and Bacon, Inc.,
          1967, 43-85

Summary: The relative profitability of specific consumer-oriented
          promotions is compared by measuring the effect of the
          promotions on consumer trial and conversion. The model
          is applied to hypothetical data on sales volumes moved
          by a price-off label and by a coupon promotion.

Model: Gain equation: If brand i is purchased on the nth
          purchase,

$$p_{i,n+1} = (1-\lambda)p_{i,n} + \phi s_i + (\lambda-\phi) \qquad (1)$$

where

$p_{i,n}$ = the probability that brand i will be purchased
on the nth purchase

s_i = the equilibrium (projected) market share for
brand i, $\sum s_i = 1$

λ and ϕ are constants depending on the product class,
independent of a particular brand

$\lambda > \phi$

Loss equation: If brand i is not purchased on the nth
purchase,

$$p_{i,n+1} = (1-\lambda)p_{i,n} + \phi s_i. \qquad (2)$$

Letting $p_{i,n+1}$ denote the expected (average) value
of $p_{i,n+1}$:

$$\bar{p}_{i,n+1} = \bar{p}_{i,n}[(1-\lambda)\bar{p}_{i,n} + \phi s_i + \lambda - \phi] + (1 - \bar{p}_{i,n})[(1-\lambda)\bar{p}_{i,n}$$
$$+ \phi s_i] = (1-\phi)\bar{p}_{i,n} + \phi s_i. \qquad (3)$$

The rate at which $\bar{p}_{i,n}$ approaches s_i, the equilibrium brand
share, depends only on ϕ and is independent of λ.
Rewriting (3),

$$\bar{p}_{i,n+1} = (1 - \phi_1 - \phi_2)\bar{p}_{i,n} + \phi_1 \beta_{i,n} + \phi_2 s_i \qquad (4)$$

where $\beta_{i,n}$ = the effective share of merchandising activity.

The corresponding gain and loss equations for equation (4) are:

Gain: $\quad P_{i,n+1} = (1-\lambda)P_{i,n} + \phi_1 \beta_{i,n} + \phi_2 s_i + (\lambda - \phi_1 - \phi_2)$ (5)

Loss: $\quad P_{i,n+1} = (1-\lambda)P_{i,n} + \phi_1 \beta_{i,n} + \phi_2 s_i.$ (6)

Rewriting (5) to show expected change in the next purchase probability among those families who buy the brand at time n,

$$P_{i,n+1} - P_{i,n} = \lambda(1 - P_{i,n}) - \phi_1(1 - \beta_{i,n}) - \phi_2(1-s_i) \quad (7)$$

$$P_{i,T} - s_i = (P_{i,0} - s_i)e^{-T(a+\sigma/N)} = (P_{i,0} - s_i)e^{-aT - aA} \quad (8)$$

where $P_{i,T}$ = probability of purchasing brand i at time T, given that the brand was purchased at T = 0

$\quad P_{i,0}$ = probability of purchasing brand i at T = 0

$\quad T$ = time elapsed

$\quad N$ = number of purchases occurring in the interval (0,T)

$\quad A$ = average time between purchases

α and σ are parameters for the product class.

Equation (8) may be rewritten as

$$\log_e(P_{i,T} - s_i) = \log_e(P_{i,0} - s_i) - \alpha T - \sigma T \quad (9)$$

$(\lambda - \phi)$ depends on the time between successive purchases and is a measure of the importance of the present purchase decision:

$$(\lambda - \phi) = P_{i,n+1}/(\text{gain}) - P_{i,n+1}/(\text{loss}). \quad (10)$$

Estimating the _relative_ increase in volume for deal k:

$$W_k = \frac{\sum f_i V_i Q_{ik}}{\sum f_i} \quad (11)$$

where f_i = panel family projection factor, used to project results to U.S. totals;

V_i = 1965 volume of the family;

Q_{ik} = change in purchase probabilities (conversion) for family i, given a purchase of deal k;

W_k = relative increase in volume for deal k;

and the sums are taken over all families who purchased the deal.

Subject: Sales Response to Advertising and Promotion

Title: A Simultaneous Equation Regression Study of Advertising and Sales of Cigarettes

Author: Frank M. Bass

Source: Journal of Marketing Research, Vol. 6, August 1969, 291-300

Summary: Application of simultaneous equation regression methods in analyzing limited time series data for sales and advertising. In testing sales and advertising relationships for filter and nonfilter cigarette brands, a model in which the advertising elasticity for filter brands is substantially greater than that for nonfilter brands is not rejected.

Model:

DEMAND FOR FILTER BRANDS

For every year t, if the demand for filter brands is considered in isolation from the rest of the system,

$$y_{1t} = \beta_1 y_{3t} + \beta_2 y_{4t} + \gamma_1 x_{1t} + \gamma_2 x_{2t} + \gamma_3 + \mu_{1t}, \quad (1)$$

where

y_{1t} = logarithm of sales of filter cigarettes (number of cigarettes) divided by population over age 20

y_{3t} = logarithm of advertising dollars for filter cigarettes divided by population over age 20 divided by advertising price index

y_{4t} = logarithm of advertising dollars for nonfilter cigarettes divided by population over age 20 divided by advertising price index

x_{1t} = logarithm of disposable personal income divided by population over age 20 divided by consumer price index

x_{2t} = logarithm of price per package of nonfilter
cigarettes divided by consumer price index

$E(\mu_{1t}) = 0$

$\text{Var } (\mu_{1t}) = E(\mu_{1t}^2) = \omega_{\mu_1}^2$

$\beta_1 + \beta_2 = 0.$

DEMAND FOR NONFILTER BRANDS

$$Y_{2t} = \beta_3 Y_{3t} + \beta_4 Y_{4t} + \gamma_4 x_{1t} + \gamma_5 x_{2t} + \gamma_6 + \mu_{2t}, \quad (2)$$

where

Y_{2t} = logarithm of sales of nonfilter cigarettes
(number of cigarettes) divided by population over
age 20

$E(\mu_{2t}) = 0$

$\text{Var } (\mu_{2t}) = E(\mu_{2t}^2) = \omega_{\mu_2}^2$

$\beta_3 + \beta_4 = 0.$

ADVERTISING BEHAVIOR OF FILTER BRANDS

For every year t, if the advertising of filter
cigarettes is considered in isolation from the rest of
the system,

$$Y_{1t} = \beta_5 Y_{2t} + \beta_6 Y_{3t} + \gamma_7 + \mu_{3t} \quad (3)$$

where

$E(\mu_{3t}) = 0$

$\text{Var } (\mu_{3t}) = E(\mu_{3t}^2) = \omega_{\mu_3}^2$

$.6 \leq \beta_6 \leq .7$

$-1.0 \leq \beta_5 \leq -.9.$

ADVERTISING BEHAVIOR OF NONFILTER BRANDS

$$y_{2t} = \beta_7 y_{1t} + \beta_8 y_{4t} - \gamma_8 - \mu_{4t} \qquad (4)$$

where

$$E(\mu_{4t}) = 0$$
$$\text{Var } (\mu_{4t}) = E(\mu_{4t}^2) = \omega_{\mu_{4t}}^2$$
$$-1.0 \le \beta_7 \le -1.5$$
$$-3.0 \le \beta_8 \le -3.5.$$

MODEL OF SALES AND ADVERTISING OF FILTER AND NONFILTER CIGARETTES

For two competing products:

$$-y_{1t} + 0y_{2t} + \beta_1 y_{3t} + \beta_2 y_{4t} + \gamma_1 X_{1t}$$
$$+ \gamma_2 X_{2t} + \gamma_3 + \mu_{1t} = 0$$

$$0y_{1t} - y_{2t} + \beta_3 y_{3t} + \beta_4 y_{4t} + \gamma_4 X_{1t}$$
$$+ \gamma_5 X_{2t} + \gamma_6 + \mu_{2t} = 0$$

$$-y_{1t} + \beta_5 y_{2t} + \beta_6 y_{3t} + 0y_{4t} + 0X_{1t}$$
$$+ 0X_{2t} + \gamma_7 + \mu_{3t} = 0$$

$$\beta_7 y_{1t} - y_{2t} + 0y_{3t} + \beta_8 y_{4t} + 0X_{1t}$$
$$+ 0X_{2t} + \gamma_8 + \mu_{4t} = 0,$$

or

$$\boldsymbol{\beta} y_t + \mathbf{T} X_t + u_t = 0.$$

The reduced-form equations are:

$$Y_{1t} = \alpha_1 X_{1t} + \alpha_2 X_{2t} + \alpha_3 + \eta_{1t}$$

$$Y_{2t} = \alpha_4 X_{1t} + \alpha_5 X_{2t} + \alpha_6 + \eta_{2t}$$

$$Y_{3t} = \alpha_7 X_{1t} + \alpha_8 X_{2t} + \alpha_9 + \eta_{3t}$$

$$Y_{4t} = \alpha_{10} X_{1t} + \alpha_{11} X_{2t} + \alpha_{12} + \eta_{4t}.$$

Subject: Sales Response to Advertising and Promotion

Title: The Effect of Advertising on Liquor Brand Sales

Author: Julian L. Simon

Source: Journal of Marketing Research, Vol. 6, August 1969, 301-13

Summary: Estimation of the revenue and profit generated by increments of advertising expenditures for liquor brands. Various distributed-lag regressions are fitted to ten-year periods for 15 brands. Annual retention rates around .75, and a logarithmic response function are found.

Model:

The general model follows:

$$S_{j,t} = f(S_{j,t-1}, A_{j,t}, R_{k,t}, P_{j,t}, F_{j,t}, \ldots) \quad (1)$$

where

$S_{j,t}$ = sales of Brand j in Year t

$A_{j,t}$ = advertising expenditures for Brand j in Year t

$R_{k,t}$ = sales of all brands of the same type k as Brand j, e.g., blended whiskey, gin, etc.

$F_{j,t}$ = proof of Brand j in Year t

$P_{j,t}$ = price of Brand j in Year t.

To estimate the decay rate or retention rate the simplest model is

$$S_{j,t} = a_j + b_j S_{j,t-1} + c_j A_{j,t} + h_k R_{k,t} \quad (2)$$

where

$S_{j,t}$ = sales of Brand j in monopoly states in adjusted cases (1,000's) in Year t

$R_{k,t}$ = sales of product-type k in monopoly states in adjusted cases (1,000's) in Year t

$A_{j,t}$ = advertising ($1,000's) of Brand j in Year t

b_j = retention rate.

However, the matrix of observed simple correlation coefficients showed very high multicollinearity between S_{t-1} and R_t. And stepwise multiple regressions run with this form showed extreme instability in the coefficient b and h because of the multicollinearity.

Another possibility is

$$\frac{S_{j,t}}{R_{k,t}} = a_j + b_j \left(\frac{S_{j,t-1}}{R_{k,t-1}}\right) + c_j A_{j,t}. \tag{3}$$

Forcing (3) through the origin:

$$\frac{S_{j,t}}{R_{k,t}} = b_j \left(\frac{S_{j,t-1}}{R_{j,t-1}}\right) + c_j A_{j,t}. \tag{4}$$

Varying (4) by using the logarithm of advertising:

$$\frac{S_{j,t}}{R_{k,t}} = b_j \left(\frac{S_{j,t-1}}{R_{k,t-1}}\right) + c_j \log A_{j,t}. \tag{5}$$

(5) generally seemed best for estimating b_j.

The estimate used for the retention rate for a brand whose sales are stable from year to year will be .75. Substituting this estimate into equation (5)

$$\frac{S_t}{R_t} - .75 \left(\frac{S_{t-1}}{R_{t-1}}\right) = a + c_j \log A_t. \tag{6}$$

To determine the effects of advertising as a function of other variables we seek the marginal number of cash sales in the short run (MVH_j), which can be used in a model such as:

$$S_{j,t} = b_j S_{t-1} + c_j A_{j,t} \tag{7}$$

$$MVH_j = c_j Z, \tag{8}$$

where Z is a factor of conversion to all U.S. data.

For the equation standardized by product type we substitute $(R_t + R_{t-1})/2$ for each R_t and R_{t-1}, and obtain for Model (3)

$$MVH_j = c_j ZR_{k,t}. \tag{9}$$

For (4),

$$MVH_j = c_j z\overline{R}_k \ \frac{\log \overline{A}_j}{\overline{A}_j}, \tag{10}$$

where $\overline{A}$ and $\overline{R}$ are mean values.

Marginal Dollar Sales Effect, Long Run (MDL_j)

= price x $MVL_j \approx 4$ x price x MVH_j. (11)

Marginal Profit per Advertising Dollar (MNL_j)

= (price - all costs except advertising - 1)

x (MVL_j). (12)

A firm's last adjustment is for the interest discount on money.

Subject: Sales Response to Advertising and Promotion

Title: Estimating Dynamic Effects of Market Communications Expenditures

Authors: David B. Montgomery and Alvin J. Silk

Source: Management Science, Vol. 18, No. 10, June 1972, B-485-501

Summary: The problem is the measuring of market response to a "communications mix"--the various means which a firm employs to transmit sales messages to potential buyers. Distributed lag models are applied to time series data for an ethical drug to estimate the short-run, intermediate, and long-run effects on market share of expenditures made for journal advertising, direct mail advertising, and samples and literature. Important differences are found among the communications variables with respect to the magnitude and over-time pattern of effect each had on market share. The results indicate that the average historic allocation of resources to alternative communication vehicles has been in inverse relation to measured market response.

Model: The basic distributed lag model is:

$$LMS(t) = a_o + \sum_{i=0}^{I} a_{i+1} LJA(t-i) + \sum_{j=0}^{J} b_{j+1} LSL(t-j)$$
$$+ \sum_{k=0}^{K} c_{k+1} LDM(t-k) + e(t), \qquad (1)$$

where

L = the log of a variable

MS = market share

JA = journal advertising

SL = samples and literature

DM = direct mail advertising

$e(t)$ = residual.

Whenever a communication variable is zero for some period t, it is set equal to \$1 for that t to avoid a log value of minus infinity. In view of the problems of multi-collinearity and degrees of freedom we modify the model:

$$LMS(t) = a_o + a_1 LJA(t) + a_2 LJA(t-1) + a_3 LJA(t-2)$$
$$+ a_4 \sum_{i=0}^{\infty} \lambda^i LJA(t-3-i) + b_1 LSL(t) + b_2 LSL(t-1)$$
$$+ b_3 \sum_{i=0}^{\infty} \lambda^i LSL(t-2-i) + c_1 LDM(t) \qquad (2)$$
$$+ c_2 \sum_{i=0}^{\infty} \lambda^i LDM(t-1-i) + e(t),$$

where $0 < \lambda < 1$.

The following points should be noted:

(a) The geometric decay in the effect of the communication variables may set in at different points in time for each variable.

(b) Once the geometric decay sets in, the same rate of decay $(1-\lambda)$ is assumed to hold for all exogenous variables.

(c) The decay terms form an infinite series.

(d) Specific lags are included because there is reason to believe that certain of the variables may have a greater effect after one or two periods than they do in the period during which the expenditure was made. In addition, the specific lags allow each variable to exhibit an individual decay rate up to the period in which the geometric decay sets in.

The model in (2) may be transformed into a form which is readily estimated. First, write (2) for LMS(t-1), multiply both sides through by λ, and subtract it from (2). This yields:

$$
\begin{aligned}
\text{LMS}(t) = {}& a_0(1-\lambda) + a_1 \text{LJA}(t) + (a_2 - \lambda a_1)\text{LJA}(t-1) \\
& + (a_3 - \lambda a_2)\text{LJA}(t-2) + (a_4 - \lambda a_3)\text{LJA}(t-3) \\
& + b_1 \text{LSL}(t) + (b_2 - \lambda b_1)\text{LSL}(t-1) + (b_3 - \lambda b_2)\text{LSL}(t-2) \\
& + c_1 \text{LDM}(t) + (c_2 - \lambda c_1)\text{LDM}(t-1) \\
& + \lambda \text{LMS}(t-1) + e(t) - \lambda e(t-1) \qquad (3) \\
= {}& \alpha_0 + \alpha_1 \text{LJA}(t) + \alpha_2 \text{LJA}(t-1) + \alpha_3 \text{LJA}(t-2) \\
& + \alpha_4 \text{LJA}(t-3) + \beta_1 \text{LSL}(t) + \beta_2 \text{LSL}(t-1) \\
& + \beta_3 \text{LSL}(t-2) + \gamma_1 \text{LDM}(t) + \gamma_2 \text{LDM}(t-1) \\
& + \lambda \text{LMS}(t-1) + U(t),
\end{aligned}
$$

where $U(t) = e(t) - \lambda e(t-1)$ and

$$a_0 = \alpha_0/(1-\lambda), \qquad\qquad b_1 = \beta_1,$$

$$a_1 = \alpha_1, \qquad\qquad b_2 = \beta_2 + \lambda\beta_1,$$

$$a_2 = \alpha_2 + \lambda\alpha_1, \qquad\qquad b_3 = \beta_3 + \lambda\beta_2 + \lambda^2\beta_1, \quad (4)$$

$$a_3 = \alpha_3 + \lambda\alpha_2 + \lambda^2\alpha_1, \qquad\qquad c_1 = \gamma_1,$$

$$a_4 = \alpha_4 + \lambda\alpha_3 + \lambda^2\alpha_2 + \lambda^3\alpha_1, \qquad c_2 = \gamma_2 + \lambda\gamma_1.$$

The α's, β's, and γ's in (3) may be directly estimated using ordinary least squares. Estimates of the a's, b's, and c's which are the coefficients in the basic Koyck distributed lag form (2) may be derived from the estimates of the α's, β's, and γ's in (3) by the equations given in (4).

Reference: #148.

Subject: Sales Response to Advertising and Promotion

Title: Toward a Normative Model of Promotional Decision Making

Author: David A. Aaker

Source: Management Science, Vol. 19, No. 6, February 1973,
            593-603

Summary: A normative model of promotional decision making that
            emphasizes the long-run impact of promotions and draws
            upon stochastic buyer-behavior model technology.
            In particular, a stochastic model is used to predict
            the level of brand acceptance obtained from a group of
            new triers attracted by a promotion. This brand
            acceptance is made a function of the composition of
            the new-trier group. Attention is focused upon
            the probability distribution of those attracted by
            the promotion, conditional on the nature of the
            promotion. This distribution is used to develop
            expressions for the expected long-term worth of a
            new-trier group attracted by a specific promotion.

Model: The objective function $V(\theta)$ represents the net present

            value to the firm of the incremental sales stream

            generated by the promotion:

            select θ to max $V(\theta)$,   (1)

            subject to: $V(\theta) \geq 0$, Budget Constraints,

where:

            $\theta = (p,s,x,t, \ldots)$, the decision parameter vector,

            $V(\theta)$ = the net present value to a firm expected
                    from a promotion described by θ.

The promotion is largely an investment expected to return

future sales that would not materialize without it. Let

            $V(\theta) = S(\theta) - A - C(\theta) + L(\theta) + W(\theta),$ (2)

where:

            $S(\theta)$ = total gross margin contributed by the brand
                    during and just after the promotion,

              A = total gross margin that would have been
                    generated in the absence of the promotion
                    during a comparable time period,

            $C(\theta)$ = direct promotion cost, excluding the unit cost,

            $L(\theta)$ = the value of any increase in loyalty among
                    existing customers as a result of the promotion,

$W(\theta)$ = the long-run value of a group of new triers of a brand attracted during promotion.

Measuring $W(\theta)$:

$$W(\theta) = \sum_{n=0}^{\infty} (Nvm)g^{n}P - W_{o} = (Nvm)(P - W_{o})/(1-g)$$

$$= W_{1} - W_{o}, \tag{3}$$

where:

n = index of a time period of analysis (i.e., one year),

N = the number of new triers in the group,

v = average per capita product-class purchasing volume per time period of the new-trier group,

m = gross margin of the brand,

g = discount factor representing the cost of capital and the risks of the market,

P = factor representing the brand's long-run share of the group's purchasing volume--operationally, the asymptotic market-share prediction of a brand choice stochastic model,

W_{o} = long-run worth of new triers who would have been attracted in the absence of the deal,

W_{1} = $(Nvm)(1/1 - g)P$.

P is likely to depend upon promotion descriptors, θ, and upon the type of new trier attracted during the promotion, Y:

$$P = P(y, \theta). \tag{4}$$

For example, (4) can take the form of a linear regression model with four independent variables:

$$P^{*} = a + a_{2}v + a_{3}\ell + a_{4}\ell + a_{5}s, \tag{5}$$

where:

v = index of purchasing volume,

ℓ = index of brand loyalty,

d = dummy variable--coded as 1 if the first purchase was associated with a promotion; otherwise as 0, s = size of the promotion,

    P* = the asymptotic market share of a brand choice
          stochastic model conditional upon a given
          purchase sequence.

Thus, in this case:

$$y = (v, \ell) \quad \text{and} \quad \theta = (d, s).$$

It is now possible to develop an expression for $E[W(\theta)]$.
In (3), W_1 is presented as a function of five terms:

$$W_1 = W_1(N, v, m, g, P). \tag{6}$$

Substituting (4) into (6), we have:

$$W_1 = W_1(N, v, m, g, y, \theta). \tag{7}$$

In (7), the term, v, is now redundant. As (5) showed,
the term, y, logically can contain v as one descriptor
of the new-trier group. Dropping v leaves:

$$W_1 = W_1(N, m, g, y, \theta). \tag{8}$$

The terms m and g are determined internally by the
firm. The term θ is a vector of firm decision parameters.
The remaining two terms, N and y, are random variables
determined by the market in accordance with the probability
distribution $f(N, y(\theta)$.

    Taking the expectation of W_1, with respect to
$f(N, y|\theta)$, yields $E_f(W_1|0)$. W_0, in (3) is simply
$E_f[W_1|s = 0]$. It is the expected long-term value of the
new-trier group that would be obtained, on the average, in
the absence of a promotion. The expression, $s = 0$, here
symbolizes that no promotion is offered.

Subject: Sales Response to Advertising and Promotion

Title: On-Line and Adaptive Optimum Advertising Control by a Diffusion Approximation

Author: Charles S. Tapiero

Source: Operations Research, Vol. 23, No. 5, September-October 1975, 890-907

Summary: A diffusion approximation to a stochastic advertising model of the Vidale-Wolfe type [#72] is formulated. This formulation allows solution of problems of optimum and on-line sales forecasting, parameter identification, and advertising control under uncertainty. For practical solutions, approximations are suggested, and simulation to forecast the probabilistic response of sales to an advertising program are used. Examples contrasting the stochastic process and the diffusion approximations approaches are included.

Model: Assume that advertising expenditures affect the probability of sales and that in a small time interval Δt, the probability that sales will increase by one unit is a function of this advertising rate. Consider a set of nonnegative integers $\{x = 0,1,2,3, \ldots, M\}$, where x represents a level of sales and M is the total market potential. Denote by $P(x,t)$ the probability of selling x at time t. At time $t+\Delta t$, the probability of selling x is given by

$$P(x,t+\Delta t) = P(x+1,t)\ m(x+1)\Delta t + P(x,t)\ [1 - m(x)\Delta t]$$
$$[1 - q(M- x,\ a(t))\Delta t] + P(x-1,t)\ q[M- x,\ a(t)]\Delta t, \qquad (1)$$

where $m(x)\Delta t$ is the probability that a unit of sales is lost by forgetting. Assume that $m(x) = mx$ and

$$q[M - x,\ a(t)] = qa(t)(M - x). \qquad (2)$$

Inserting (2) into (1) and taking the limit, we obtain

$$dP(x,t)/dt = m(x+1)\ P(x+1,t) - [mx + qa(t)(M-x)]\ P(x,t)$$
$$+ qa(t)(M- x+1)\ P(x - 1,t), \qquad (3)$$

$$x = 1,2,\ \ldots,\ M-1,$$

$$dP(0,t)/dt = mP(1,t) - Mqa(t)\ P(0,t),$$

$$dP(M,t)/dt = -mMP(M,t) + qa(t)\ P(M-1,t)$$

which describes the probabilistic evolution of sales at time t as a function of the advertising rate $a(t)$ and the forgetting m.

If $s(t)$ and $v(t)$ are the evolution of the mean sales and variance at time t, then (dropping the time subscripts)

$$ds/dt = -ms + qa(M - s), \quad s(0) = s^O,$$

and $\hspace{10cm}$ (4)

$$dv/dt = qaM + (m-qa)s - 2v(m+qa), \quad v(0) = 0.$$

To transform (3) into a stochastic differential equation, we first replace $P(x+1,t)$ and $P(x-1,t)$ by the first two terms of their Taylor's series expansion. This yields a Fokker-Planck diffusion equation whose solution is a stochastic differential equation in the sense of the Itô Calculus. This equation is given by

$$dx = [-mx + qa(M-x)]dt + [mx + qa(M-x)]^{1/2}dw, \quad x(0) = s^O \quad (5)$$

where dw is a Wiener process; that is,

$$E(dw) = 0, \quad E[dw(t)dw(r)] = \delta(t-r) \hspace{3cm} (6)$$

with $\delta(t-\tau)$, the Dirac-delta function. (5) can be reduced, using the Itô differential rule to being additive in the noise source dw. Specifically, letting

$$y = 2/(m-qa)[mx + qa(M-x)]^{1/2}, \hspace{3cm} (7)$$

we obtain

$$dy - f(y,t)dt = dw, \hspace{4cm} (8)$$

$$f(y,t) = qa\dot{y} + 2qa[4Mm - y^2(m-qa)]/4y(m-qa)^3$$

$$+ y(m-qa)(1 - m/2)/2 - Mmqa/y + 1/2y$$

with $\dot{a} = da/dt$.

$w(t+\Delta t) - w(t)$ is a zero-mean random variable with variance given by Δt. Replacing the differentials 'd' in (8) by the difference 'Δ' yields a discrete time formulation for the advertising model. If we assume $\Delta t = 1$, (8) can be rewritten as

$$y(t+1) - y(t) - f(y,t) = \varepsilon(t), \qquad (9)$$

where $\varepsilon(t)$ is a zero-mean, unit variance random variable.

Equations (8) and (9) are two nonlinear equations with additive error terms. The variable y no longer describes sales but $2/(m-qa)$ times the sales standard deviation. For continuous sales information, we use (8) or (5), while for discrete time sales information, we use (9).

Optimum Sales Estimation and Forecasting

Assume that a sales time series is given by $z(t)$

$$z = x + \eta,$$

$$E(\eta) = 0, \quad E[\eta(t)\eta(r)] = \delta(t-r)\theta, \qquad (10)$$

where θ is a measurement error variance. Define by $z(t)$ the continuous time series

$$Z(t) = [z(\sigma)|\sigma \leq t], \qquad (11)$$

and let X be the conditional mean sales estimate at time t, based on the time series Z; that is,

$$X = E(x|Z). \qquad (12)$$

The conditional error variance is

$$V = E[(x - X)^2|Z]. \qquad (13)$$

Supposing first-order approximation to (5),

$$[mx + qa(M-x)]^{1/2} \simeq [mX + qa(M-X)]^{1/2}. \qquad (14)$$

The corresponding optimum filter when m, q, M, and a are given is

$$dX/dt = -mX + qa[M-X] + V[z-X]/\theta, \quad X(0) = s^{o},$$
$$\qquad\qquad\qquad\qquad\qquad\qquad\qquad\qquad (15)$$
$$dV/dt = -2(m+qa)V + mX + qa(M-X) - V^2/\theta, \quad V(0) = 0,$$

where V/θ is the Kalman gain and $y - X$ is the on-line observation of the error forecast. Thus, given M, m, q, and the

open-loop advertising program a, optimum sales estimates
are given by integration of (15). An extrapolation of (15)
at time t with the filter estimate X(t) as an initial
condition would provide an optimum estimate of mean sales
response to an advertising strategy.

In practice, the measurement model we have considered
may not always be used. The error variance θ is usually
unknown and there are measurement delays. A common
measurement model would be

$$z(t) = x(t-\tau).\tag{16}$$

We have

$$dz(t)/dt = dy(t-\tau)/dt,$$

and   (17)

$$dz(t)/dt = f(y,t-\tau) + w(t-\tau).$$

Define

$$r(t) = dz(t)/dt,$$

and   (18)

$$\eta(t) = w(t-\tau).$$

(17) together with (8) may then be written as

$$dy(t)/dt = f(y,t) + w(t)$$

and   (19)

$$r(t) = f(y,t-\tau) + \eta(t),$$

where $R(t) = \{r(\sigma)|\sigma \leq t\}$ is now a time series, and w(t)
and $\eta(t)$ are two identically and independently distributed
random variables with zero means and variance given by
$\delta(t-\tau)$. Application of approximate filtering techniques to
(19) is now straightforward. For example,

$$dY/dt = f(Y,t) + W\partial f(Y,t-\tau)/\partial Y[r(t) - f(Y,t-\tau)]\tag{20}$$

and

$$dW/dt = 2W\partial f(Y,t)/\partial Y + 2W\partial g(Y,t)/\partial Y + 1,$$

where $g(Y,t) = \partial f / \partial Y (Y,t-\tau)[r(t) - f(Y,t-\tau)]$, $t > \tau$, and f
is defined in (8). An approximate expression for the sales
estimate (in this case, forecast) at time t is then given by

$$(m - qa)Y^2/4 - qaM/(m - qa). \tag{21}$$

Computations similar to these conducted for the continuous
time model (8) can be repeated for the discrete time model (9).

Best-Fit Advertising Effectiveness and Forgetting Rate Parameters

Consider the measurement model (16) and assume parameters
m, q, and M to be constants. Define the squared error

$$\varepsilon^2(t) = [y(t-\tau) - Y(t-\tau)]^2 \tag{22}$$

whose expected value is the variance $W(t-\tau)$ defined in (20).
Replacing y by (7) wherever appropriate, we can determine
an optimum control problem with parameters m, q, M unknown.
A simple formulation of this problem:

$$\text{Min}_{m,q,M} \, J(T),$$

subject to

$$J(T) = \int_{\tau}^{T-\tau} W(t-\tau)dt \quad \text{and} \quad (20), \tag{23}$$

with $W(t)$, the advertising strategy $a(t)$ and sales $x(t)$ known.
Necessary conditions for optimum parameters of (23) can be
determined theoretically by calculus. An analytical solution,
however, is complicated, requiring solution of a system of
three nonlinear equations in three unknowns.

Optimum Advertising Under Uncertainty

For operational purposes it is necessary to use
approximations. If we assume an open-loop advertising
strategy, the filter equations (15) or (20) (without the
innovation term) can be used as 'deterministic equivalent'
models in determining an optimum control strategy. For
example, let t be the present time, and let T be the
planning time. Assume a revenue $\pi(x,t)$ induced by an

advertising strategy a(t). The expected value of this revenue is assumed to involve the first two probability moments; that is, $E\{\pi(x,t)\} = \Pi(X, V, t)$, where X and V are the expected sales and sales variance at time t. If we further assume a constant discount rate δ, the optimum advertising control problem is stated as

   max $J(t,T)\,|a(\tau),\ \tau\,\varepsilon\,[t,T]$,

subject to:

$$J(t,T) = \int_{t}^{t+T} e^{-\delta\tau}[\Pi(X,V,\tau) - a]d\tau, \qquad (24)$$

where A is a constraint set on the advertising expenditure rate (15) and $a(\tau)\ \varepsilon A,\ \forall[t,T]$, and without the innovation random sequence. An alternative formulation would express the revenue $\pi(x,t)$ as a function of y defined in (7) and determine an expected value, $\Pi(Y,W,t)$. The optimum advertising policy would then be found by minimizing the discounted expected returns in [t,T] subject to (20).

Reference: #72.

Subject: Sales Response to Advertising and Promotion

Title: The Product Life Cycle and Time-Varying Advertising Elasticities

Author: Leonard J. Parsons

Source: Journal of Marketing Research, Vol. 12, November 1975, 476-80

Summary: A model of sales response to advertising using advertising elasticities that change over the product life cycle. The model is applied to the case history of a household cleanser.

Model: Assume the sales response function

$$\ln S_t = \alpha_1 + \alpha_2 \ln (A_t/P_t) + \alpha_3 \ln S_{t-1} \qquad (1)$$

where:

S_t = units sales in one-half gross at time t,

A_t = advertising expenditures at time t,

P_t = general price index at time t.

Traditionally, the coefficients are assumed to be constant so that they might be estimated by ordinary least squares regression.

The coefficients of the explanatory variables in the model are elasticities. This can be shown by differentiating the sales response function by each of the explanatory variables, i.e.,

$$\frac{d \ln S}{d \ln (A/P)} = \alpha_2 = \frac{dS/S}{d(A/P)/(A/P)} = \varepsilon_{A/P}.$$

Consequently, the elasticities are constant.

Product life-cycle theory indicates that these elasticities should vary over time. Therefore, a sales response function is proposed in which the advertising elasticity is expressed as exponential functions of time, while the elasticity of sales in the previous period is postulated to be a modified exponential function:

$$\ln S_t = (\beta_1 e^{-\beta_2 t} + \beta_3) \ln (A_t/P_t)$$

$$+ [\beta_4 (1 - e^{-\beta_5 t}) + \beta_6] \ln S_{t-1}. \qquad (2)$$

Thus, at $t = 0$, $\varepsilon_{A/P} = \beta_1 + \beta_3$ and $\varepsilon_{S_{t-1}}$ β_6. As $t \to \infty$, $\varepsilon_{A/P} \to \beta_3$ and $\varepsilon_{S_{t-1}} \to \beta_4 + \beta_6$. All parameters are conjectured to lie within the zero-one interval. The sums $\beta_1 + \beta_3$ and $\beta_4 + \beta_6$ are also conjectured to lie within the zero-one interval near its upper boundary. Furthermore, β_3 is conjectured to be zero, while β_5 may or may not be zero depending, perhaps, on the level of time aggregation.

An important feature of this model is that the constant elasticity model is a special case in which $\beta_2 = 0$ and $\beta_5 = \infty$. Thus the model does not force the elasticities to be time varying.

Subject: Advertising and Sales Promotion

Title: A Study in Promotional Competition

Author: Harland D. Mills

Source: Research Paper No. 101-103, December 1959, Mathematica, Princeton, New Jersey. Reprinted in Bass, et. al. (eds.), Mathematical Models and Methods in Marketing, Homewood, Illinois: Richard D. Irwin, Inc., 1961, 271-301.

Summary: A game-theoretic approach to promotional competition to define the equilibrium conditions in three models: (a) Competition between two brands in a fixed market, (b) competition between any number of brands in a fixed market, (c) general promotional competition among any number of brands.

Model:

COMPETITION BETWEEN TWO BRANDS IN A FIXED MARKET

If the competition is purely promotional the two profit models are:

$$P = mv - x - f$$

$$\overline{P} = \overline{mv} - \overline{x} - \overline{f}$$

where

P = profit

m = unit logistic margin ($= p - c$)

p = unit price

c = unit variable producing and distributing cost

v = unit volume

x = promotional outlay in creative selling

f = allocated and/or fixed costs.

$$v = \frac{(\alpha x)^{e} V}{(\alpha x)^{e} + (\overline{\alpha x})^{e}}, \quad \overline{v} = \frac{(\overline{\alpha x})^{e} V}{(\overline{\alpha x})^{e} + (\alpha x)^{e}}$$

where

α = coefficient of relative brand promotion effectiveness

e = exponent of market promotion effectiveness.

Necessary conditions for a competitive equilibrium point $(x, \overline{x}) > 0$ are

$$\frac{\partial P}{\partial x} = 0, \qquad \frac{\partial \overline{P}}{\partial \overline{x}} = 0,$$

which can be rewritten as

$$\frac{Vme(\alpha\overline{\alpha}x\overline{x})^e}{x} = [(\alpha x)^e + (\overline{\alpha x})^e]^2 = \frac{V\overline{m}(\overline{\alpha}\alpha\overline{x}x)^e}{\overline{x}},$$

and has the solution, which can be verified as an equilibrium point,

$x° = mK, \ \overline{x}° = \overline{m}K,$ where $K =$ set of brands

$$K = \frac{Ve(m\alpha)^e(\overline{m\alpha})^e}{[(m\alpha)^e + (\overline{m\alpha})^e]^2}$$

with the equilibrium payoffs

$$P° = [(\frac{\alpha m}{\alpha m})^e + 1 - e]\frac{x°}{e} - f, \ \overline{P}° = [(\frac{\overline{\alpha m}}{\alpha m})^e + 1 - e]\frac{\overline{x}°}{e} - \overline{f}.$$

SIMPLE PROMOTIONAL COMPETITION AMONG n BRANDS

Consider a fixed market which is shared in proportion to competing promotional outlays. We simplify our notation to the game defined by payoff functions

$$P_i(x) = \frac{Vm_i x_i}{\sum_j x_j} - x_i - f_i, \qquad i = 1,2, \ldots, n, \quad (1)$$

Results:

    1. There exists a unique competitive equilibrium point for each game defined by (1).

    2. A finite algorithm exists for determining this point.

Consider the derivatives

$$\frac{\partial P_i(x)}{\partial x_i} = \frac{Vm_i \sum_{j \neq i} x_j}{(\sum_j x_j)^2} - 1, \quad i = 1,2, \ldots, n \qquad (2)$$

$$\frac{\partial^2 P_i(x)}{\partial x_j^2} = - \frac{2Vm_i \sum\limits_{j \neq i} x_j}{(\sum\limits_{j} x_j)^3}, \quad i = 1, 2, \ldots, n. \quad (3)$$

When $x_i = 0$, (2) reduces to

$$\frac{\partial P_i(x | x_i = 0)}{\partial x_i} = \frac{Vm_i}{\sum\limits_{j \neq i} x_j} - 1, \quad i = 1, 2, \ldots, n. \quad (4)$$

$$\sum_{j} x_j = \frac{(n-1)V}{\sum\limits_{j} \frac{1}{m_j}} = V(\frac{n-1}{n}) M_n, \quad (5)$$

where M_n is the harmonic mean of $m_1, m_2, \ldots, m_n$,

$$x_i = V(\frac{n-1}{n}) M_n [1 - (\frac{n-1}{n}) \frac{M_n}{m_i}] = V(\frac{n-1}{n}) M_n b_i \quad (6)$$

where

$$b_i = 1 - (\frac{n-1}{n}) \frac{M_n}{m_i} \quad (7)$$

$$P_i = Vm_i b_i^2 - f_i. \quad (8)$$

If the x_i are all positive in this solution, they constitute an equilibrium point.

Algorithm

1. Relabel players, if necessary, so

   $m_1 \geq m_2 \geq \ldots \geq m_n$.

2. Set an integer parameter $t = 2$ (t will correspond to the number of players i with $x_i > 0$ at each stage of the algorithm).

3. Find an equilibrium point $(x_1^0, x_2^0, \ldots, x_t^0)$ for the game containing players $1, 2, \ldots, t$ by using (6).

4. If $t < n$, test whether

$$m_{t+1} > \left(\frac{t-1}{t}\right) M_t.$$

If yes, go to step 2 and replace t by $t+1$.

If no, then $(x_1^0, x_2^0, \ldots, x_t^0, 0, \ldots, 0)$ is the unique equilibrium point for the game with all players $1, 2, \ldots, n$.

Theorem. Given a game defined by (1), there exists a unique competitive equilibrium point, and the Algorithm just above determines this point in a finite number of steps.

GENERAL PROMOTIONAL COMPETITION AMONG n BRANDS

The results for Simple Promotional Competition can be extended to include varying degrees of promotional effectiveness and market expansion. We consider a game defined by payoff functions

$$P_i(x) = \frac{(V + \beta \sum_j \alpha_j x_j) m_i \alpha_i x_i}{\sum_j \alpha_j x_j} - x_i - f_i \tag{9}$$

and note

$$\frac{\partial P_i(x)}{\partial x_i} = \frac{V m_i \alpha_i \sum_{j \neq i} \alpha_j x_j}{(\sum_j \alpha_j x_j)^2} - (1 - \beta m_i \alpha_i) \tag{10}$$

$$\frac{\partial^2 P_i(x)}{\partial x_i^2} = - \frac{2 V m_i \alpha_i \sum_{j \neq i} \alpha_j x_j}{(\sum_j \alpha_j x_i)^3}. \tag{11}$$

When (10) = zero,

$$\sum_j \alpha_j x_j = \frac{V(n-1)}{\sum_j \frac{1}{\alpha_i m_i} - n\beta} \tag{12}$$

$$\alpha_i x_i = b_i \sum_j \alpha_j x_j \tag{13}$$

where

$$b_i = 1 - \frac{(n-1)\left(\frac{1}{\alpha_i m_i} - \beta\right)}{\sum_j \frac{1}{\alpha_j m_j} - n\beta} \tag{14}$$

$$P_i = V m_i b_i^2 - f_i. \tag{15}$$

If $b_i > 0$, all i, then $(x_1, x_2, \ldots, x_n)$ constitutes an equilibrium point. Otherwise, with one additional restriction, given below, the Algorithm of the preceding section may be extended to this case as follows.

Algorithm (Extension)

1. Relabel players, if necessary, so

   $$\alpha_1 m_1 \geq \alpha_2 m_2 \geq \ldots \geq \alpha_n m_n.$$

2. Set an integer parameter $t = 2$.

3. Find an equilibrium point $(x_1^0, x_2^0, \ldots, x_t^0)$ for the game containing players 1, 2, ..., t by using (13).

4. If $t < n$, test whether

   $$(t-1)\left(\frac{1}{\alpha_{t+1} m_{t+1}} - \beta\right) < \sum_{j=1}^{t} \frac{1}{\alpha_j m_j} - n\beta.$$

If yes, go to step 2 and replace t by $t+1$.

If no, then $(x_1^0, x_2^0, \ldots, x_n^0, 0, \ldots, 0)$ is the unique equilibrium point for the game with all players 1, 2, ..., n.

Theorem. Given a game defined by (9) for which

$$\beta < \frac{1}{\alpha_i m_i}, \quad i = 1, 2, \ldots, n$$

there exists a unique competitive equilibrium point, and the Algorithm just above determines this point in a finite number of steps.

References: #140 and 141.

Subject: Advertising and Sales Promotion

Title: Imperfect Markets Through Lack of Knowledge

Author: S.A. Ozga

Source: Quarterly Journal of Economics, February 1960, 29-52

Summary: Some general principles of the diffusion of knowledge are described. The optimum advertising appropriation and optimum output are determined as a function of the proportion of the individuals in the market who have been informed of the firm's offering.

Model:

Let

N = number of people (or firms) among whom information is being passed

K_t = number of people possessing relevant information at time t

c = contact coefficient, number of contacts made by every member of the relevant group during a given interval

$cK_t\,dt$ = number of contacts during which information can be passed in a short interval

a = advertising coefficient, proportion of people who see advertisement and are <u>informed</u> thereby in their future purchasing decision

g = rate of growth of relevant group of people

r = rate of removal of members of relevant group (such as death or change of residence)

$N_t = N_0 e^{gt}$ = size of the relevant group of people at time t, where N_0 represents the size of the group at time zero.

The basic equation of the model is

$$\frac{dK_t}{dt} = (c + a)K_t\left(1 - \frac{K_t}{N_0 e^{gt}}\right) - rK_t \qquad (1)$$

with the solution

$$\frac{K_t}{N_t} = \frac{(c+a) - (g+r)}{(c+a) + [(\frac{N_o}{K_o} - 1)(c+a) - \frac{N_o}{K_o}(g+r)]e^{-[(c+a)-(g+r)]t}}$$

which gives us the total number of people informed, in relation to the total size of the group, as a function of t.

This principle is applied to the theory of the firm. Assuming identical individual buyers, constant marginal production costs, and constant advertising coefficient, for a single firm, the total cost of advertising is

$$A_t = paB_t \tag{2}$$

where

    p = price of making one buyer see firm's advertisement

    B = number of buyers who know of the firm's offer.

The maximum profit output in relation to the total size of the market is given by

$$\frac{Q_t}{kN_L} = 1 - \sqrt{\frac{g+r}{c+m}} \tag{3}$$

and the optimum advertising coefficient by

$$a = \sqrt{(g+r)(c+m)} - c \tag{4}$$

where

$$m = \frac{k(P-M)}{p}$$

    P = firm's maximum profit price

    M = marginal costs.

The elasticity of demand and the marginal revenue product of advertising are equal to one another:

$$- \frac{P}{Q_t} \frac{dQ_t}{dP} = \frac{d(PQ_t)}{dA_t} = \frac{kP}{pm} \tag{5}$$

The increase in the number of customers of the ith firm on the market during a short interval dt is equal to the number of potential buyers who either have been contacted during that interval by that firm's customers or have seen its advertisements (and have not so far been customers of any other firm);

$$\frac{dB_t^i}{dt} = (c+a^i)B_t^i(1 - \frac{B_t}{N_t}) - rB_t^i. \tag{6}$$

For the jth firm, we have thus

$$(\frac{B_t^j}{B_o^j})(c+a^i) = (\frac{B_t^i}{B_o^i})(c+a^j) \; e^{r(a^j-a^i)t}.$$

References: #55 and 86.

Subject: Advertising and Sales Promotion

Title: The Economics of Information

Author: George J. Stigler

Source: The Journal of Political Economy, Vol. 69, No. 3, June 1961, 213-25

Summary: The identification of sellers and the discovery of their prices is given as an example of the role of the search for information in economic life.

Model: If sellers' asking prices (p) are uniformly distributed between zero and one, it can be shown that the distribution of minimum prices with n searches is

$$n(1-p)^{n-1}. \tag{1}$$

Whatever the distribution of prices, increased search will yield diminishing returns as measured by the expected reduction in the minimum asking price.

For any buyer the expected savings from an additional unit of search will be approximately the quantity (q) he wishes to purchase times the expected reduction in price as a result of the search, or

$$q \left| \frac{\partial P_{min}}{\partial n} \right|. \tag{2}$$

The expected saving from given search will be greater, the greater the dispersion of prices.

Assume each dealer sets a selling price, p, and makes sales to all buyers for whom this is the minimum price. With a uniform distribution of asking prices by dealers, the number of buyers of a total of N_b possible buyers who will purchase from him is

$$N_i = KN_b n(1-p)^{n-1}, \tag{3}$$

where K is a constant. The number of buys from a dealer increases as his price is reduced, and at an increasing rate. Moreover, with the uniform distribution of asking

prices, the number of buyers increases with increased
search if the price is below the reciprocal of the amount
of search. The condition for optimum search (with
perfect correlation of successive prices) would be:

$$q \left| \frac{\partial p}{\partial n} \right| = i \times \text{marginal cost of search},$$

where i is the interest rate.

Suppose that a given advertisement of size a will
inform c per cent of the potential buyers in a given
period, so c = g(a). This contact function will presumably
show diminishing returns, at least beyond a certain size
of advertisement. A certain fraction, b, of potential
customers will be "born" (and "die") in a stable population,
where "death" includes not only departure from the market
but forgetting the seller. The value of b will vary with
the nature of the commodity. In a first period of
advertising (at a given rate) the number of potential
customers reached will be cN, if N is the total number of
potential customers. In the second period cN(1-b) of
these potential customers will still be informed, cbN new
potential customers will be informed, and
c[(1-b)n - cN(1-b)] old potential customers will be reached
for the first time, or a total of cN[1 + (1-b)(1-c)].
This generalizes, for k periods, to

$$cN[1 + (1-b)(1-c)$$

$$+ \ldots + (1-b)^{k-1} (1-c)^{k-1}],$$

and, if k is large, this approaches

$$\frac{cN}{1 - (1-c)(1-b)} = \lambda N. \tag{4}$$

If each of r sellers advertises the same amount, λ is
the probability that any one seller will inform any buyer.
The distribution of N potential buyers by the number of contacts
achieved by r sellers is given by the binomial distribution:

$$N(\lambda + [1 - \lambda])^r,$$

with, for example,

$$\frac{Nr!}{m!(r-m)!} \quad \lambda^m (1 - \lambda)^{r-m}$$

buyers being informed of exactly m sellers' identifies. The number of sellers known to a buyer ranges from zero to r, with an average of $r\lambda$ sellers and a variance of $r\lambda(1-\lambda)$.

A monopolist will advertise (and price the product) so as to maximize his profits,

$$\pi = Npq\lambda - \phi(N\lambda q) - ap_a,$$

where $p = f(q)$ is the demand curve of the individual buyer, $\phi(Nq\lambda)$ is production costs other than advertising, and ap_a is advertising expenditures. The maximum profit conditions are

$$\frac{\partial \pi}{\partial q} = N\lambda \ (p + q \ \frac{\partial p}{\partial q}) - \phi'N\lambda = 0 \tag{5}$$

and

$$\frac{\partial \pi}{\partial a} = Npq \ \frac{\partial \lambda}{\partial a} - \phi'Nq \ \frac{\partial \lambda}{\partial a} - p_a = 0. \tag{6}$$

On the assumption that all firms are identical and that all buyers have identical demand curves and search equal amounts, we obtain the maximum-profit equation for the competitive firm:

$$\text{Production cost} = p(1 + \frac{1}{\eta_{qp} + \eta_{Kp}}), \tag{7}$$

where η_{qp} is the elasticity of a buyer's demand curve and η_{Kp} is the elasticity of the fraction of buyers purchasing from the seller with respect to his price. The latter elasticity will be of the order of magnitude of the number of searches made by a buyer. With a uniform distribution of asking prices,

increased search will lead to increased advertising by
low-price sellers and reduced advertising by high-price
sellers. The amount of advertising by a firm decreases
as the number of firms increases. If prices are advertised
by a large portion of the sellers, the price differences
diminish sharply.

References: #53, 55, 85.

Subject: Advertising and Sales Promotion

Title: Can Advertising Differentiate the Product?

Author: Lester G. Telser

Source: Journal of Political Economy, Vol. 72, No. 6, December 1964,
          "Advertising and Competition, Appendix I," 559-62

Summary: The problem whether advertising can change the price elasticity
          of demand is approached directly by estimating the demand
          function and indirectly by drawing inferences about the price
          elasticity from a study of the effects of changes of production
          and advertising costs on the price of the product and the
          number of advertising messages transmitted. Both methods
          proved to be inconclusive.

Model:

In the direct method the demand equation is

$$x = D(p,y,v) \tag{1}$$

where

    x = rate of sales
    p = deflated price
    y = real income
    v = quantity of advertising messages.

If the price elasticity is independent of the level of
advertising, the demand function can be written as

$$x = F(p,y)G(v). \tag{2}$$

The four types of demand equations estimated (linear in
all of the variables; linear in quantity and linear in
the logs of all other variables; linear in the logs of
all variables; linear in the logs of quantity and linear
in all other variables) proved inconclusive in the case
of cigarettes.

In the indirect method the problem is to discover
from the signs of the coefficients of the reduced-form

equations whether advertising can reduce the price elasticity.

Consider the cost function

$$c = H(x,v) + bv + mx \qquad (3)$$

where bv and mx represent, respectively, that part of total cost that is linear in output and advertising messages; $H(x,v)$ is the nonlinear part of total cost.

Net revenue, R, is defined by

$$R = pD(p,v) - c. \qquad (4)$$

To maximize R with respect to p and v it is necessary that

$$
\begin{aligned}
R_p &= x + (p - H_x)D_p - mD_p = 0, \\
R_v &= -H_v + (p - H_x)D_v - b \\
&\quad - mD_v = 0.
\end{aligned} \qquad (5)
$$

It is sufficient for maximum R that the Hessian B is negative definite where B is defined as follows:

$$B = \begin{pmatrix} R_{pp} & R_{vp} \\ R_{vp} & R_{vv} \end{pmatrix}. \qquad (6)$$

Since the rate of sales x is a function of p and v, the two equations in (5) implicitly define two reduced-form equations for p and v. The slopes of the reduced-form equations with respect to b satisfy

$$
B \begin{bmatrix} \frac{\partial p}{\partial b} \\ \\ \frac{\partial v}{\partial b} \end{bmatrix} = - \begin{bmatrix} R_{bp} \\ \\ R_{bv} \end{bmatrix} = \begin{bmatrix} 0 \\ \\ 1 \end{bmatrix}. \tag{7}
$$

Similarly, the slopes with respect to m satisfy

$$
B \begin{bmatrix} \frac{\partial p}{\partial m} \\ \\ \frac{\partial v}{\partial m} \end{bmatrix} = - \begin{bmatrix} R_{mp} \\ \\ R_{mv} \end{bmatrix} = \begin{bmatrix} D_p \\ \\ D_v \end{bmatrix}. \tag{8}
$$

Therefore,

$$
\frac{\partial p}{\partial b} = - \frac{R_{vb}}{\det B}, \quad \frac{\partial v}{\partial b} = \frac{R_{pp}}{\det B} < 0. \tag{9}
$$

$$
\frac{\partial p}{\partial m} = \frac{D_p R_{vv} - D_v R_{vp}}{\det B},
$$
$$
\frac{\partial v}{\partial m} = \frac{D_v R_{pp} - D_p R_{vp}}{\det B}. \tag{10}
$$

By hypothesis the signs of the partial derivatives are
as follows:

$$
D_p < 0, \qquad D_v > 0,
$$
$$
R_{vv} < 0, \quad \text{and} \quad R_{pp} < 0. \tag{11}
$$

On the basis of (A10) it follows that

$$
R_{vp} \leq 0
$$

implies

$$
\frac{\partial p}{\partial b} \geq 0, \quad \frac{\partial p}{\partial m} > 0, \quad \frac{\partial v}{\partial m} < 0. \tag{12}
$$

If the slopes of p and v, have signs given by (12), then it must be true that R_{vp} is nonpositive.

R_{vp} can be positive but it cannot be too large.

From (5)

$$R_{vp} = -D_p (H_{xx} D_v + H_{xv}) \qquad (13)$$
$$+ D_v + (p - H_x - m) D_{vp}.$$

Since by definition the price elasticity of demand η is $(p/x) D_p$,

$$\frac{\partial \eta}{\partial v} = \frac{p}{x} (D_{pv} - \frac{D_p D_v}{x}). \qquad (14)$$

Using (5) equation (13) reduces to

$$R_{vp} = -D_p [H_{xx} D_v + H_{xv}] + D_v - x \frac{D_{vp}}{D_p}$$

$$= -D_p [H_{xx} D_v + H_{xv}]$$

$$- \frac{x}{D_p} [D_{vp} - \frac{D_v D_p}{x}].$$

Finally from (14),

$$R_{vp} = -D_p (H_{xx} D_v + H_{xv}) - \frac{x}{\eta} \frac{\partial \eta}{\partial v}. \qquad (15)$$

Assume that the two reduced-form equations are available and that the slope of p with respect to b is positive. As given by (A8), R_{vp} must be negative. Assume in addition that $H_{xx} > 0$, $H_{xv} > 0$ and $D_v > 0$. These imply

$$(H_{xx} D_v + H_{xv}) > 0. \qquad (16)$$

Since sales vary inversely with price, (16) plus the assumption of a negative R_{vp} means that the price elasticity increases in size as v increases.

Now what can be learned from finding that $\partial p/\partial b < 0$? In this case R_{vp} is positive, which might result entirely from the cost conditions as shown in (16). Thus an inverse relation between the price of the product and the average cost of advertising messages is consistent with either a positive or a negative sign of $\partial \eta/\partial v$. Since a rise in the level of b reduces v, even though the product is less advertised and the price falls, it is incorrect to conclude that the reduction in the advertising raised the price elasticity in magnitude.

Knowledge of the effects of changes in m on v and p give no assistance. Even if the signs of both $\partial p/\partial m$ and $\partial v/\partial m$ were known, one could not be sure of how much of the effect on price to attribute to advertising's effect on the price elasticity.

Subject: Advertising and Sales Promotion

Title: Zur Frage optimaler Diffusionspunkte in einem Modell der Mund-zu-Mund-Propaganda

Author: Edgar Topritzhofer

Source: Zeitschrift für betriebswirtschaftliche Forschung, Vol. 25, No. 7, July 1973, 694-98

Summary: Graph theory model of the "Two Step Flow of Communications" hypothesis: the advertising message reaches a primary group whose members are treated as diffusion points in a communication network.

Model: ## Graphentheoretisches Modell

Gegeben ist der Kommunikationsdigraph $N = (C,D,T)$ bestehend aus

--der Menge der Umworbenen, repräsentiert durch die Knotenmenge $C = \{c_1, \ldots, c_i, c_j, \ldots, c_n\}$,

--der Menge der zwischen den Umworbenen bestehenden Kommunikationswege, repräsentiert durch die Menge der gerichteten Kanten D,

--und einer Funktion $T: D \to C \times C$.

Zwischen zwei beliebigen Knoten c_i and c_j besteht daher der Kommunikationsweg $d \in D$, wenn gilt $T(d) = (c_i c_j)$.

Es bestehen zwei Möglichkeiten, daß ein Umworbener c in den Besitz einer Werbebotschaft gelangt: entweder er erhält sie im Mund-zu-Mund-Weg von einem anderen Umworbenen, von dem aus er erreichbar ist, oder die Werbebotschaft kommt an ihn von außen heran. Jeder Umworbene, der in den Besitz der Botschaft gelangt, gibt diese an sämtliche anderen Umworbenen, die er erreichen kann, weiter.

$E(c)$ = Menge aller Umworbenen, die von c erreichbar sind

$E(\overline{c})$ = Menge aller Umworbenen, die von jedem der Umworbenen $c \in \overline{C}$ erreichbar ist.

Die gestellte Aufgabe lautet, jene Knotenmenge S zu bestimmen, für die folgende zwei Bedingungen erfüllt sein müssen:

(I) $E(S) = C$ $\wedge$ (II) $\not\exists$ $\overline{S} \subset S: E(\overline{S}) = C$

Reduziert man den Digraphen auf seine streng zusammenhängenden Komponenten, so erhält man einen Digraphen $\tilde{N}$, dessen Knoten die streng zusammenhängenden Komponenten z von N sind und der überall dort eine Kante $(z_i z_j)$ aufweist, wo in N für $c_i \in z_i$ und $c_j \in z_j$ $(\underset{\sim}{c_i c_j}) \in D$ ist. Der von N reduzierte Kommunikationsdigraph $\tilde{N}$ besitzt eine einzige Erreichbarkeitsmenge $\tilde{S}$, welche den Bedingungen (I) and (II) genügt: ihre Elemente umfassen sämtliche Knoten z von $\tilde{N}$ für die gilt $\delta - (z) = 0$.

Numerische Lösung mit Hilfe des Matrizenkalküls

Gegeben sei eine Menge von $n = 6$ Umworbenen. Die dem Kommunikationsdigraphen N entsprechende Adjazenzmatrix $N_{6x6} = (n_{ij})$ habe die Gestalt

$$N = \begin{bmatrix} 1 & & 1 & & & \\ & 1 & & & & \\ & & & 1 & & \\ & & & 1 & 1 & \\ 1 & 1 & & 1 & & 1 \\ & & & & & 1 \end{bmatrix} \qquad \text{wobei } n_{ij} = \begin{cases} 1, \text{ wenn } (c_i c_j) \in D \\ 0, \text{ wenn } (c_i c_j) \notin D \end{cases}$$

Die Erreichbarkeitsmatrix $E_{6x6} = (e_{ij})$ erhält man durch schrittweises Zusammensetzen aus den Adjazenzmatrizen für die Pfadlängen 1 (= N), 2(= N^2), ... bis $n-1(=N^{n-1})$. Da für die Erreichbarkeitsmatrix gilt

$$e_{ij} = \begin{cases} 1, \text{ wenn in N mindestens ein Pfad von } c_i \text{ nach } c_j \text{ existiert} \\ 0, \text{ wenn in N kein Pfad von } c_i \text{ nach } c_j \text{ existiert} \end{cases}$$

erfolgt die schrittweise Zusammensetzung durch Boolesche Addition

$$E = I \boxplus N \boxplus N^2 \boxplus \ldots \boxplus N^{n-1}$$

wobei I die Einheitsmatrix darstellt und

$$N^X = N \boxdot N \boxdot \ldots \boxdot N \text{ (x mal)}$$

Im Beispiel hat die Erreichbarkeitsmatrix die Gestalt

$$\mathbf{E} = \begin{bmatrix} 1 & 1 & 1 & 1 & 1 & 1 \\ & 1 & 1 & & & 1 \\ & & 1 & & & 1 \\ 1 & 1 & 1 & 1 & 1 & 1 \\ 1 & 1 & 1 & 1 & 1 & 1 \\ & & & & & 1 \end{bmatrix}$$

Die elementweise Produktbildung der Erreichbarkeitsmatrix $\mathbf{E}$ mit ihrer Transponierten $\mathbf{E}^T$

$$\mathbf{E} \times \mathbf{E}^T = (e_{ij} \times e_{ji})$$

wobei

$$e_{ij} \times e_{ji} = \begin{cases} 1, \text{ wenn } c_i \text{ and } c_j \text{ voneinander wechselweise erreichbar sind} \\ 0, \text{ wenn das nicht der Fall ist} \end{cases}$$

$$\mathbf{E} \times \mathbf{E}^T = \begin{bmatrix} 1 & & & 1 & 1 & \\ & 1 & & & & \\ & & 1 & & & \\ 1 & & & 1 & 1 & \\ 1 & & & 1 & 1 & \\ & & & & & 1 \end{bmatrix}$$

Durch Umordnung erhält man unmittelbar die streng zusammenhängenden Komponenten z_1, z_2, z_3 und z_4.

Die Matrix des reduzierten Digraphen $\tilde{\mathbf{N}} = (\tilde{n}_{ij})$ hat daher die Gestalt:

$$\tilde{N} = \begin{bmatrix} 0 & 1 & 1 & 1 \\ 0 & 0 & 1 & 0 \\ 0 & 0 & 0 & 1 \\ 0 & 0 & 0 & 0 \end{bmatrix}$$

wobei

$$\tilde{n}_{ij} = \begin{cases} 1, & \text{wenn für ein } c_i \in z_i \text{ und } c_j \in z_j \ (i \neq j) n_{ij} = 1 \text{ ist} \\ 0, & \text{wenn für kein } c_i \in z_i \text{ und } c_j \in z_j \ (i \neq j) n_{ij} = 1 \text{ ist.} \end{cases}$$

Da nur $\delta-(z_1) = 0$ ist, besteht die Erreichbarkeitsmenge S des reduzierten Graphen aus dem einzigen Element z_1. Für die optimale Versorgung des Kommunikationsnetzes genügt es daher, die Information bei c_1, c_4 oder c_5 von außen in das Netz einzuführen, wobei alle drei Lösungen vom Standpunkt der Kommunikationsweitergabe gleichwertig sind.

Subject: Pricing

Title: Short-Term Price and Dealing Effects in Selected Market Segments

Authors: William F. Massy and Ronald E. Frank

Source: Journal of Marketing Research, Vol. 2, May 1965, 171-85

Summary: Changes in relative price and dealing activity are likely to affect different segments of the market in different ways. A distributed lag model is developed for predicting a firm's market share over a period of weeks from knowledge of the changes in these variables. It is tested on aggregate data for a metropolitan market, and applied to data for three classes of underlying segments. The bases for segmenting the market are by family purchasing characteristic, package size, and channel of distribution.

Model: The relation is a distributed lag equation in several variables.

We assume a hyperbolic demand relation, but if all variables are understood to have been transformed to logarithms before estimation it can be written as a classical linear regression.

$$S_t = a_0 + a_1 P_t + a_2 P_{t-1} + a_3 P_{t-2} + a_4 D_t$$
$$+ a_5 D_{t-1} + a_6 D_{t-2} + a_7 PD_t + a_8 ES_t \qquad (1)$$
$$+ a_9 ES_{t-1} + \lambda S_{t-1} + v_t.$$

Since the variables are logarithms it is possible to interpret the parameters a_1 through a_9 as elasticities. After the adjustment to be noted shortly, each of them represents the percentage change in the market share of Brand M that, on the average, has been associated with a one per cent change in the particular explanatory variable, other things remaining the same. The time subscripts refer to the week at which the variable is evaluated, relative to the week for which market share is to be predicted; e.g., P_{t-1} means "price in the week just prior to the one for which market share is to be predicted."

Equation (1) is derived from an underlying model that assumes the effect of each variable dies away to zero according to an exponential decay law with parameter λ, for times in excess of the maximum lag that is explicitly considered in the model. The existence of S_{t-1} on the right hand side of the equation gives the model this property.

The derivation can be demonstrated in terms of a reduced model involving only the variables for market share and price. The distributed lag equation is derived from an underlying structural equation of the form:

$$S_t = b_o + p_1 P_t + b_2 P_{t-1} + b_3 \sum_{j=0}^{\infty} \lambda^j P_{t-j-2} + u_j. \quad (2)$$

Lagging this equation by one week, multiplying through by λ, and subtracting from the original yields the following linear regression of current and lagged price and lagged share on current share.

$$S_t = (1-\lambda)b_o + b_1 P_t + (b_2 - \lambda b_1) P_{t-1}$$
$$+ (b_3 - \lambda b_2) P_{t-2} + (u_t - \lambda u_{t-1}). \quad (3)$$

In order for the parameters in (1) to be interpretable as elasticities, certain adjustments to their raw values are necessary. The following formulas are derived from the underlying exponential model.

$$b_o = a_o/(1-\lambda) \qquad b_4 = a_4$$
$$b_1 = a_1 \qquad\qquad b_5 = a_5 + \lambda b_4 \qquad (4)$$
$$b_2 = a_2 + \lambda b_1 \qquad \text{etc.}$$
$$b_3 = a_3 + \lambda b_2$$

The b's are elasticities of market share with respect to the price and dealing variables. The a's are the raw values from (1). The cumulative or "long run" effects for price (P) and dealing (D) are as follows:

$$\text{Long-run price effect} = b_1 + b_2 + b_3/(1-\lambda)$$
$$\text{Long-run deal effect } = b_4 + b_5 + b_6/(1-\lambda). \tag{5}$$

The variables included in (1) are defined as:

S = log of market share for Brand M,

P = log of relative price index,

D = log of relative deal magnitude index,

PD = log of proportion of sales on deal,

ES = log of expected market share.

Expected share at time t is based upon the average probability of buying Brand M and the total consumption of the product, as measured over the whole sample period, for each family in the panel that made a purchase during the week in question. Define W_i as the total number of ounces of the product purchased by the ith family and P_{iM} as the proportion of ounces devoted to Brand M by the family. Then,

$$ES_t = \frac{\sum_i W_i P_{iM}}{\sum_i W_i},$$

where the summation is taken over only those families that purchased the product in week t. Thus ES is the statistical expectation for M's market share for those families that were in the market, conditional on the constancy of their relative purchase volumes and propensity to buy the brand.

Subject: Pricing
Title: The Use of Models in Marketing Timing Decisions
Author: Sidney W. Hess
Source: Operations Research, Vol. 15, July-August 1967, 17-34
Summary: A model to aid pricing of obsolescent products.

Model: At time t = 0, a lower priced product (modern) is introduced
         nationally to replace the existing, but similar, product
         (ancient). Retail sales of ancient continue but at a decreasing
         rate as customers switch to modern. Excess ancient is
         gradually returned to the manufacturer for credit. It is
         assumed that the manufacturer, after t = 0, no longer
         sells ancient.

         Two mutually exclusive courses of action are available to
         the manufacturer to prevent the returns:

         1. At time T decrease the price of ancient to the price of
            modern making up the wholesaler's and retailer's losses
            with equivalent quantities of free modern.
         2. At time θ call in all ancient from the field making up
            wholesaler's and retailer's losses with equivalent
            quantities of free modern.

         A third course of action is, of course, to do nothing and
         accept the returns. Let

             $S = S(t)$ = unit sales (or returns) of ancient per month
             $S' = S'(t)$ = unit sales (or returns) of modern per month
             $I = I(t)$ = inventory of ancient in the field.

         All of these variables are functions of time, t. To indicate
         the direction and/or location of the movement or inventory
         we will use the following subscripts:

             m = manufacturer
             w = wholesaler
             r = retailer
             c = customer.

         Unit prices and costs are given by

             p = ancient price to wholesaler
             p' = modern price to wholesaler
             c_v = variable manufacturing cost of ancient
             c_v' = variable manufacturing cost of modern

c_s = salvage value (before tax basis) for returned ancient

c_f = freight on ancient

c_f' = freight on modern.

We will construct a mathematical model of the manufacturer's variable profit as a function of action times, T or θ, and then determine the values of T or θ maximizing this profit.

Course of Action No. 1--Price Reduction

At any time before the ancient price cut, the manufacturer's variable profit per month will be the difference between marginal profits from modern and the cost of returned ancient. The total profit up to the price cut is the integral of this profit per month. Hence,

Profit from time 0 to T =

$$= \int_0^T (p' - c_v' - c_f') S_{mw'} dt - \int_0^T (p + c_f - c_s) S_{wm} dt. \quad (1)$$

At the time of the price cut, T, an amount of free modern proportional to the ancient inventory in the field at that time will be introduced at a cost of $c_v' + c_f'$ per unit. Letting α = units of free modern per unit of ancient in the field then

Cost of free goods = $(c_v' + c_f') \; \alpha [I_w(T) + I_r(T)]. \quad (2)$

Let β = the fraction of free goods that "steal" normal manufacturer sales. Then,

Profit from time T to ∞

$$= (p' - c_v' - c_f') \; \{\int_T^\infty S_{mw'} dt - \beta \alpha [I_w(T) + I_r(T)] \}. \quad (3)$$

Subtracting (2) from the sum of (1) and (3), total profit as a function of T, is given by

$$Z(T) = \int_0^\infty (p' - c_v' - c_f') S_{mw'} dt - \int_0^T (p + c_f - c_s) S_{wm} dt$$

$$- \alpha [\beta p' + (1 - \beta)(c_v' + c_f')] [I_w(T) + I_r(T)]. \quad (4)$$

Our maximization problem reduces to minimizing the cost of returns plus the cost of free goods (both out-of-pocket and the cost of "stolen" modern sales). Denoting this cost as $L(T)$,

$$L(T) = (p + c_f - c_s) \int_0^T S_{wm} dt$$
$$+ \alpha[\beta p' + (1 - \beta)(c_v' + c_f')][I_w(T) + I_r(T)]. \quad (5)$$

Conditions for minimum loss are

$$\frac{dL(T)}{dT} = 0, \qquad \text{and} \qquad (6)$$

$$\frac{d^2 L(T)}{dT^2} > 0. \qquad (7)$$

But

$$\frac{dL(T)}{dT} = (p + c_f - c_s) S_{wm}(T)$$
$$+ \alpha[\beta p' + (1 - \beta)(c_v' + c_f')] \{\frac{d}{dt}[I_w(T) + I_r(T)]\}, \quad (8)$$

and the rate of change of inventory is given by a material balance around the wholesaler and retailer, or

$$\frac{d}{dt}[I_w(T) + I_r(T)] = -S_{rc}(T) - S_{wm}(T). \qquad (9)$$

Substituting (9) into (8) and equating to zero

$$(p + c_f - c_s) S_{wm}(T) = \alpha[\beta p' + (1 - \beta)(c_v' + c_f')][S_{rc}(T) + S_{wm}(T)]. \quad (10)$$

If we combine constants by letting

$$k + 1 = \frac{(p + c_f - c_s)}{\alpha[\beta p' + (1 - \beta)(c_v' + c_f')]}, \qquad (11)$$

(10) will reduce to

$$k S_{wm}(T) = S_{rc}(T). \qquad (12)$$

By (7), (12) will be a minimum if

$$k \frac{dS_{wm}(T)}{dT} > \frac{dS_{rc}(T)}{dT}. \qquad (13)$$

Let

$$a = p + c_f - c_s.$$

Then by (5) and (11)

$$L(\infty) = a \int_0^\infty S_{wm} dt, \tag{14}$$

and

$$L(T_1) = a \int_0^{T_1} S_{wm} dt + \frac{a}{k+1} \left[\int_{T_1}^\infty S_{wm} dt + \int_{T_1}^\infty S_{rc} dt \right], \tag{15}$$

$$L(\infty) < L(T_1) \iff a \int_{T_1}^\infty S_{wm} dt < \frac{a}{k+1} \int_{T_1}^\infty S_{wm} dt + \frac{a}{k+1} \int_{T_1}^\infty S_{rc} dt$$

$$\iff k \int_{T_1}^\infty S_{wm} dt < \int_{T_1}^\infty S_{rc} dt. \tag{16}$$

(16) states that it is cheaper not to cut the price if k times the total returns from T_1 on is less than the total retail sales from T_1 on.

Courses of Action No. 2--Ancient Recall

This model differs from the previous one only to the extent that at some time θ we call back all outstanding ancient and have correspondingly higher free goods. The total cost to be minimized, $G(\theta)$, is similar to (5) and given by

$$G(\theta) = (p + c_f - c_s) \int_0^\theta S_{wm} dt + \{\delta [\beta p' + (1 - \beta)(c_v' + c_f')]$$
$$+ (c_f' - c_s)\} [I_w(\theta) + I_r(\theta)], \tag{17}$$

where

δ = units of free modern per unit of ancient in the field.

Letting

$$h + 1 = \frac{(p + c_f - c_s)}{\delta [\beta p' + (1 - \beta)(c_v' + c_f')] + c_f' - c_s}, \tag{18}$$

then

$$G(\theta) = a \int_{o}^{\theta} S_{wm}dt + \frac{a}{h+1} [I_w(\theta) + I_r(\theta)]. \qquad (19)$$

This is the same equation form as (8) and, therefore, has a similar solution, namely that θ such that

$$hS_{wm}(\theta) = S_{rc}(\theta), \qquad (20)$$

and

$$\frac{hdS_{wm}(\theta)}{d\theta} > \frac{dS_{rc}(\theta)}{d\theta} . \qquad (21)$$

For the particular numbers for this case (11) and (18) show $k > h$. This means that the first minimum for policy No. 1 occurs before the first minimum for policy No. 2, i.e., $T_1 < \theta_1$.

Subject: Pricing

Title: Determining Optimum Price Promotion Quantities

Authors: David A. Goodman and Kavin W. Moody

Source: Journal of Marketing, Vol. 34, October 1970, 31-39

Summary: A method for the determination of an optimal selling quantity for price-promoted items is described.

Model: Let

$$S = f(X_1, X_2, \ldots X_N),$$

where S = sell-through quantity (which actually reaches the consumer) and $X_1, X_2, \ldots X_N$ = independent factors such as price, advertising expenditures, length of promotion, normal sales volume, selling effort. Assume the functional form

$$S = aX_1 + bX_2 + cX_3 + d.$$

The values for a, b, c, and d can be obtained using a multiple regression routine. Let C_S = marginal opportunity cost of shortage (which is incurred if additional units could have been sold had more been shipped to distributors) and C_I = marginal opportunity cost of inventory (if the last unit shipped is left in inventory).

Define P to be the probability of selling through one additional unit. In order to minimize total expected opportunity costs, the size of promotional shipments should be increased up to the point where the expected marginal opportunity cost from shipping one more unit, if it sells through, is just equal to the expected marginal opportunity cost from shipping that unit, if it remains in inventory. The quantity to ship in order to minimize total expected opportunity costs is, therefore,

$$P \cdot C_S = (1-P) \cdot C_I. \tag{1}$$

Let $P^* = P$ for the minimizing equation. Solving (1) for P^* gives

$$P^* = \frac{C_I}{C_s + C_I} \qquad (2)$$

P^* represents the minimum probability of selling at least one additional unit. Therefore, additional units should be shipped so long as the probability of selling through these units is greater than P^*.

The cost of shortage is the opportunity loss of not selling the item at the special price. The opportunity cost of inventory is the difference in revenue between selling the item at its regular price and selling it at the special or reduced price. Let

C_v = the variable cost of production

P_s = the special or promotional price

P_r = the regular price.

Then

$C_s = P_s - C_v$

$C_I = P_r - P_s$.

Substituting these expressions into (2) gives

$$P^* = \frac{P_r - P_s}{P_r - C_v}. \qquad (3)$$

Let s = standard error of the estimate of S, which is assumed to have a normal distribution, and $P(S)$ = probability of S. The estimate $\bar{S}$ is such that 50% of the area under the curve lies on each side of it. P^* is a "break-even" probability corresponding to an area under the standard curve. The objective is to find S^*, the optimal selling quantity which should be shipped, which is the S corresponding to P^*. Let Z = unit normal transformation of $\bar{S}$, and k_σ^* correspond to S^*. Then k^* represents the point on the

standard normal curve corresponding to the point
$\frac{S^* - \overline{S}}{s}$ on the normal curve of interest; hence, the
desired conversion yielding the optimal solution is
given by

$$S^* = \overline{S} + k^*s \qquad \text{or} \qquad (4)$$

$$\int_{S^*}^{\infty} N(\overline{S}, s^2) dS = P^*.$$

Subject: Pricing

Title: A Quasi-Game Theory Approach to Pricing

Authors: Ambar G. Rao and Melvin F. Shakun

Source: Management Science, Vol. 18, No. 5, January (Part 2) 1972, P-110-123

Summary: A quasi-game theory approach to market-entry pricing is taken for a product class where price is the only indicator of quality. The model is based on certain hypotheses regarding individual customer behavior, which is then aggregated to find demand for a product given its price. Using game theoretic thinking and assumptions about brand behavior, optimal pricing policies are derived.

Model: Consider a market where product quality is judged as a function of price (examples: beer, gasoline, soap, razor blades). We use the following hypotheses:

1. A homogeneous group of customers is considered.

2. The minimum price that a member of the group will pay is a log normally distributed random variable.

3. For each person there is a fixed interval of constant length   a   on the log (price) scale, the range of acceptable prices.

Suppose there are   n   brands that are priced in such a way as to be acceptable to a given individual.

For   n = 2, we assume that:

1. A proportion λ of the customers purchase the higher priced brand in the interval and are called "quality conscious."

2. A proportion $(1-\lambda)$ of the customers purchase the lower priced product in the range and are called "price conscious."

For   n = 3,   we assume that:
3. A proportion λ_1 of the price conscious customers will continue to purchase the lowest priced brand when a middle price is available to them. The remainder of the price conscious customers will buy the middle priced brand.

4. A proportion λ_2 of the quality conscious customers continue to purchase the highest priced brand when a middle priced brand is available to them. The remainder of the quality conscious group will buy the middle priced brand.

The probability that a customer picks the middle priced brand, given that all three brands are within his acceptable price range, is

$$(1-\lambda)(1-\lambda_1) + \lambda(1-\lambda_2).$$

The quantity λ is easily obtained by inquiring from customers which brand (the higher or lower priced) they would buy if two are in the acceptable region. Estimates for λ_1 and λ_2 can be obtained through similar questioning. We can also estimate these parameters through more formal experimentation of the laboratory variety.

Let P_i = Price of the ith brand,

$p_i = \ln P_i$,

ρ_i = Probability that a randomly chosen customer buys brand i,

$\phi(x)$ = cumulative distribution function of the standard normal random variable.

Suppose there is only one brand in the market priced at P_1. The probability that a randomly chosen customer buys Brand 1 is

$$\rho_1 = \phi\left(\frac{p_1 - \mu}{\sigma}\right) - \phi\left(\frac{p_1 - \mu - a}{\sigma}\right). \tag{1}$$

Now suppose there are two brands in the market priced at P_1 and P_2, $P_1 < P_2$.

If $p_2 - p_1 < a$, the probability that the demand of a randomly chosen individual will not be satisfied is

$$\left\{1 - \phi\left(\frac{p_2 - \mu}{\sigma}\right) + \phi\left(\frac{p_1 - a - \mu}{\sigma}\right)\right\}. \tag{2}$$

If $p_2 - p_1 \geq a$, the probability is

$$\{1 - \phi(\frac{p_2 - \mu}{\sigma}) + \phi(\frac{p_1 - a - \mu}{\sigma}) + \phi(\frac{p_2 - a - \mu}{\sigma}) - \phi(\frac{p_1 - \mu}{\sigma})\}. \quad (3)$$

Multiplying the probabilities above by the total number of customers in the group and their average purchase quantity say N and q, we can find the unsatisfied demand.

The probability that a randomly chosen customer will choose Brand 1 is

$$\rho_1 = \{\phi(\frac{p_2 - a - \mu}{\sigma}) - \phi(\frac{p_1 - a - \mu}{\sigma})$$

$$+ (1-\lambda)(\phi(\frac{p_1 - \mu}{\sigma}) - \phi(\frac{p_2 - a - \mu}{\sigma}))\} \quad (4)$$

and that he will choose Brand 2 is

$$\rho_2 = \{\phi(\frac{p_2 - \mu}{\sigma}) - \phi(\frac{p_1 - \mu}{\sigma}) + \lambda(\phi(\frac{p_1 - \mu}{\sigma}) - \phi(\frac{p_2 - a - \mu}{\sigma}))\} \quad (5)$$

if $p_2 - p_1 < a$. If $p_2 - p_1 \geq a$, then the terms multiplied by λ and $(1-\lambda)$ will be evaluated at $\lambda = 1$.

BRAND ENTRY

We assume that the goal of each brand relates to its sales rather than profits.

One Brand Case

If a new product enters a market where there is no existing competition, the price that maximizes sales is

$$p_1^* = \mu + \frac{a}{2}. \quad (6)$$

Two Brand Case

If Brand 1 is already established in the market and Brand 2 wishes to enter, the loss to Brand 1 is proportional to

$$\lambda [\phi (\frac{p_1 - \mu}{\sigma}) - \phi (\frac{p_2 - \mu - a}{\sigma})] \equiv L(p_2) \qquad (7)$$

with $P_2 > P_1$.

If $P_2 = P_1$ the loss to Brand 1 will be

$$\frac{1}{2} \{\phi (\frac{p_1 - \mu}{\sigma}) - \phi (\frac{p_1 - a - \mu}{\sigma})\} \equiv L(p_1). \qquad (8)$$

Comparing this to (7) for $L(p_2)$ we find that

$$\phi (\frac{p_2 - \mu - a}{\sigma}) \gtreqqless \phi (\frac{p_1 - \mu - a}{\sigma})$$

so long as $p_2 \geqq p_1$. Thus the loss to Brand 1 increases the closer Brand 2 is priced to it.

Now consider this problem in the context of game theory.

1. Suppose noncooperative equilibrium type of behavior for both brands. We obtain

$$p_1^* = \frac{\sigma^2}{a} \ln (1-\lambda) + \mu + \frac{a}{2} \qquad (9)$$

and

$$p_2^* = - \frac{\sigma^2}{a} \ln (\lambda) + \mu + \frac{a}{2}. \qquad (10)$$

2. Suppose both brands wish to cooperate, i.e., they wish to maximize joint sales. We obtain

$$p_1^* = \mu \quad \text{and} \quad p_2^* = \mu + a. \qquad (11)$$

3. Suppose Brand 1 wishes to maximize its own sales, while Brand 2 wishes to maximize industry sales.
We consider $\rho_1 + \rho_2$, in the two cases when $p_2 + p_1 \leqq a$ and $p_2 - p_1 > a$.
No matter what the value of p_1, the optimal value of p_2

is $p_2^* = p_1 + a$. Once Brand 2 has entered the market at $p_1 + a$, Brand 1 recomputes its price by maximizing ρ_1. This yields

$$p_1^* = \mu + \frac{a}{2} + \frac{\sigma^2}{a} \ln (1-\lambda),$$

which is the noncooperative equilibrium solution. Once Brand 1 has adjusted its price, Brand 2 follows and sets its price to

$$p_2^* = \mu + \frac{3a}{2} + \frac{\sigma^2}{a} \ln (1-\lambda).$$

4. Suppose P_2 is selected to maximize S_2 while P_1 is selected so as to minimize ($\max_{P_2} S_2$). Then

$$p_2^* = - \frac{\sigma^2}{a} \ln (\lambda) + \mu + a/2.$$

Using this value in (5) gives

$$\rho_2 = \phi(\frac{a/2 - \sigma^2/a \ln \lambda}{\sigma}) - \lambda\phi(\frac{\sigma^2/a \ln \lambda - a/2}{\sigma})$$

$$- (1-\lambda)\phi(\frac{p_1 - \mu}{\sigma}).$$

(12)

To minimize ρ_2 we would like to pick p_1 as large as feasible. Since if two brands have equal price the probability that a randomly chosen customer buys each is equal, p_1^* should be either equal to p_2^* or slightly below it, depending on whether λ is greater than or less than $\frac{1}{2}$.

Three Brand Case

First, assume that all three brands wish to maximize their own sales. Determining the noncooperative equilibrium prices for a high, medium, and a low priced brand H, M, and L say, we obtain

$$p_L^* = \mu + \frac{a}{2} + \frac{\sigma^2}{a} \ln \lambda_1 (1-\lambda),$$

$$p_M^* = \mu + \frac{a}{2} - \frac{\sigma^2}{a} \ln \frac{\lambda}{1-\lambda}, \qquad (13)$$

$$p_H^* = \mu + \frac{a}{2} - \frac{\sigma^2}{a} \ln \lambda_2 \lambda.$$

The following inequalities hold:

$$p_L^* < p_1^* \quad \text{and} \quad p_H^* > p_2^*, \qquad (14)$$

where p_1^* and p_2^* are the noncooperative equilibrium prices in the two brand case. Brand 3 can pick one of the three prices p_L^*, p_M^* and p_H^* to enter the market. The decision will depend on which of ρ_L^*, ρ_M^*, and ρ_H^* is the greatest. Because of (14) we can assert that:

> (i) If Brand 3 elects to pick p_M^*, Brands 1 and 2 will reduce and increase their prices respectively to p_L^* and p_H^*.

> (ii) If Brand 3 elects to pick p_H^*, Brands 1 and 2 will change their prices to p_L^* and p_M^* respectively.

> (iii) If Brand 3 elects to pick p_L^*, Brands 1 and 2 change their prices to p_M^* and p_H^*.

Unfortunately, the decision on which price Brand 3 should pick depends on the values of λ, λ_1, and λ_2 and must be evaluated for each combination. It can be shown that if $\lambda = \lambda_1 = \lambda_2 = \frac{1}{2}$ then Brand 3 should pick p_M^*.

Finally, consider the case where Brands 1 and 2 are at the noncooperative equilibrium prices and Brand 3 through its price selection behavior signals a desire to cooperate with Brand 1. If Brand 1 responds to Brand 3, and adjusts its price P_1, and Brand 2 wishes to maximize its own sales, we obtain

$$p_1^* = \mu + \frac{a}{2} + \frac{\sigma^2}{a} \ln [(1-\lambda)\lambda_1 - \lambda(1-\lambda_2)],$$

$$p_2^* = \mu + \frac{a}{2} + \frac{\sigma^2}{a} \ln [\frac{\lambda}{1-\lambda}],$$

$$p_3^* = \mu + \frac{a}{2} - \frac{\sigma^2}{a} \ln [\lambda\lambda_2 - (1-\lambda)(1-\lambda_1)].$$

Subject: Pricing

Title: Adaptive Pricing by a Retailer

Authors: Leonard J. Parsons and W. Bailey Price

Source: Journal of Marketing Research, Vol. 9, No. 2, May 1972,
          127-33

Summary: A retailer in an environment in which a competitor is
          the price leader must determine how to adapt his prices
          to those of this competitor. The sequential decision
          problem of the manager is formulated as a Markov process
          with rewards.

Model: Assume three states of store sales performance:

1. 4% or more above quota for the period

2. 4% or more below quota

3. within $\pm$ 4% of quota.

The three pricing alternatives for a list of most highly
identifiable items are:

1. 3% above competitor's prices

2. at the same price

3. 3% below competitor's price.

The manager must decide which alternative to follow:
According to Howard's policy determination method (Howard,
R.A., Dynamic Programming and Markov Processes, Cambridge:
The M.I.T. Press, 1960), if an N-state Markov process
earns r_{ij} dollars when it makes a transition from state i
to state j, then r_{ij} is the reward associated with the
transition, and the set of rewards for the process may be
described by a reward Matrix R with elements r_{ij}.
Further, q_i may be defined as the reward expected in the
next transition out of state i, designated as the
expected immediate reward for state i. Therefore:

$$q_i = \sum_{j=1}^{N} p_{ij} r_{ij}$$

where p_{ij} is the conditional probability that a system
which occupies state i will occupy state j after its
next transition.

A superscript k may be used to indicate the alternatives available in a particular state; each alternative has its own associated reward and probability distributions for transitions out of the state. The quantity q_i^k is defined as the expected reward from a single transition from state i under alternative k. Thus:

$$q_i^k = \sum_{j=1}^{N} p_{ij}^k r_{ij}^k.$$

Also, $v_i(n)$ may be defined as the total expected return in n stages starting from state i—if an optimal policy is followed. Therefore:

$$v_i(n+1) = \max_k \sum_{j=1}^{N} p_{ij}^k [r_{ij}^k + v_j(n)], \quad n = 0,1,2, \ldots$$

or, in terms of expected immediate rewards from each alternative:

$$v_i(n+1) = \max_k [q_i^k + \sum_{j=1}^{N} p_{ij}^k v_j(n)].$$

Howard focusses upon the average earnings of the process per unit of time and defines an optimal policy as one that maximizes the gain. His iteration cycle is as follows:

Value determination operation

Use p_{ij} and q_i for a given policy to solve:

$$g + v_i = q_i + \sum_{j=1}^{N} p_{ij} v_j, \quad i = 1, \ldots, N$$

for all relative values v_i and g by setting $v_N = 0$.

Policy improvement routine

For each state i, find the alternative k' that maximizes:

$$q_i^k + \sum_{j=1}^{N} p_{ij}^k v_j$$

using the relative values v_i of the previous policy. Then k' becomes the new decision in the ith state, $q_i^{k'}$ becomes q_i and $p_{ij}^{k'}$ becomes p_{ij}.

The gain, g, is maximized when policies on two successive iterations are identical.

The probability and reward matrices were subjectively estimated by the manager because historical data were not available. In determining the reward values associated with a transition from one state to another under three possible pricing policies, three considerations were involved:

1. Reward values should vary directly with the resultant (state j) level of sales performance.

2. Rewards should vary directly with the state i level of sales performance; e.g., a higher reward should be associated with sales transitioning from near quota to above quote performance than from below quota to quota performance.

3. Rewards should vary directly with the prices of items because the higher the prices charged, the larger the store's gross margin.

In calculating the r_{ij}^k associated with each transition the formula used was:

r_{ij}^k = Percentage gross margin x expected sales
   + change in number of customers expected x 50 cents
   per customer.

The value of 50¢ is near the gross margin per customer. Result: Decision policy (3,3,3), i.e., 3% below competitor's price.

The optimal solution to the sequential decision problem can be found by linear programming as well as by Howard's policy iteration method. Minimize g subject to:

$$g + v_i \geq q_i^k + \sum_{j=1}^{N} p_{ij}^k v_j$$

for each i and k. The corresponding dual is:

maximize: $\sum_{i=1}^{N} \sum_{k=1}^{K_i} q_i^k x_i^k$ (expected revenue per period)

subject to: $\sum_{i=1}^{N} \sum_{k=1}^{K_i} x_i^k = 1$ (normalization restriction)

$\sum_{k=1}^{k_j} x_j^k - \sum_{i=1}^{N} \sum_{k=1}^{K_i} p_{ij}^k x_i^k = 0$ for each j (conservation of probability)

all $x_i^k \geq 0$

where N is the number of states in the system and K_i is the number of alternative decisions. The x_i^k can be interpreted as the joint probability that the system is at state i and decision k is made. For each state i, there exists at most only one x_i^k which has a nonzero value.

The dual is applicable to the manager's problem.

Subject: Pricing

Title: Eine Preisabsatzfunktion zur optimalen Preis- und
Qualitätspolitik bei heterogenen Gütern

Author: Klaus P. Kaas

Source: Zeitschrift für betriebswirtschaftliche Forschung,
Vol. 25, 1973, 604-23

Summary: Consumer preferences for two substitute goods are described
as a probability process. The demand function has the
following parameters: the price of the competing product,
product heterogeneity, and the quality difference expressed
in money units.

Model: <u>Berechnung der Kaufwahrscheinlichkeiten</u>

Die Wahrscheinlichkeit W_1, dass ein beliebiger Konsument das
Produkt 1 kauft, ist gleich der Wahrscheinlichkeit, daß die
Zufallsvariable V einen beliebigen Wert kleiner als $-\pi$ annimmt:

$$W_1 = \int_{-\infty}^{-\pi} f(v)\,dv, \text{ mit } f(v) = \text{Dichtefunktion von v.} \quad (1)$$

Entsprechend ist die Kaufwahrscheinlichkeit W_2 gleich:

$$W_2 = \int_{-\pi}^{\infty} f(v)\,dv. \quad (2)$$

Als Approximation für die Normalverteilung verwenden wir die
Ableitung der logistischen Funktion:

$$F(v) = \frac{1}{1 + e^{a-bv}}, \quad (3)$$

$$\frac{dF(v)}{dv} = f(v) = \frac{be^{a-bv}}{(1 + e^{a-bv})^2}. \quad (4)$$

Als Dichtefunktion muß $f(v)$ folgenden Bedingungen genügen:

$$\int_{-\infty}^{+\infty} f(v)\,dv = \lim_{v \to +\infty} F(v) - \lim_{v \to +\infty} F(v) = 1.$$

$$f(v) \leq 1 \quad \text{für} \quad -\infty < v < +\infty$$

$f(v)$ hat bei $v = \frac{a}{b}$ ein Maximum mit dem Ordinatenwert
$f(v) = \frac{b}{4}$. Daraus folgt eine Beschränkung des Parameters b
auf Werte $b \leq 4$.

Um als Approximation der Normalverteilung dienen zu können, muss $f(v)$ symmetrisch zur Senkrechten in $v = \frac{a}{b}$ sein.

Infolgedessen müssen die Ordinatenwerte für $v = \frac{a}{b} + c$ und $\frac{a}{b} - c$ für $-\infty < c < +\infty$ einander gleich sein:

$$\frac{be^{-bc}}{(1 + e^{-bc})^2} = \frac{be^{bc}}{(1 + e^{bc})^2} \; \Big| \Big| \cdot \frac{e^{bc}}{b} \; .$$

Daraus folgt:

$$\frac{1}{(1 + e^{-bc})^2} = \frac{1}{e^{-2bc}(1 + e^{bc})^2}$$

und

$$\frac{1}{(1 + e^{-bc})^2} = \frac{1}{(e^{-bc} + 1)^2} .$$

$f(v)$ ist also symmetrisch zur Senkrechten in $\frac{a}{b}$. Daraus folgt, daß der Mittelwert der Dichtefunktion $f(v)$ gleich dem Abszissenwert der Symmetrieachse ist, nämlich $\bar{v} = \frac{a}{b}$. Das bedeutet, daß die Dichtefunktion für $\frac{a}{b} > 0$ in Richtung der positiven, für $\frac{a}{b} < 0$ in Richtung der negativen v-Achse verschoben ist. Wir können somit Gleichung (4) als Approximation der Normalverteilung in (1) und (2) einsetzen und die Kaufwahrscheinlichkeiten berechnen:

$$W_1 = \int_{-\infty}^{-\pi} \frac{be^{a-bv}}{(1 + e^{a-bv})^2} \, dv = \left[\frac{1}{1 + e^{a-bv}} \right]_{-\infty}^{-\pi} = \frac{1}{1 + e^{a+b\pi}} \quad (5)$$

$$W_2 = \int_{-\pi}^{\infty} \frac{be^{a-bv}}{(1 + e^{a-bv})^2} \, dv = \left[\frac{1}{1 + e^{a-bv}} \right]_{-\pi}^{\infty} = \frac{1}{1 + e^{-a-b\pi}} \quad (6)$$

Diese Ausdrücke bezeichnen die Wahrscheinlichkeit, mit der sich ein zufällig ausgewählter Konsument bei seinem Kauf für das eine oder andere der beiden Substitutionsgüter entscheidet, wenn der Preisunterschied zwischen ihnen $P_1 - P_2 = \pi$ beträgt.

Berechnung der Preisabsatzfunktion

Gewichtet man die gesamte auf die Produktgruppe entfallende
Nachfrage X mit den Kaufwahrscheinlichkeiten, so erhält man
die Absatzmengen der beiden Produkte, x_1 und x_2:

$$x_1 = \overline{X}W_1 = \frac{\overline{X}}{1 + e^{a+b\pi}}$$

$$x_2 = \overline{X}W_2 = \frac{\overline{X}}{1 + e^{-a-b\pi}}$$

Die Auflösung dieser Gleichungen nach π ergibt

$$\pi = -\frac{a}{b} + \frac{1}{b} \ln \frac{\overline{X} - x_1}{x_1}. \tag{7}$$

Ersetzt man π durch $P_1 - P_2$, so erhält man die
Preisabsatzfunktion in der üblichen, nach P_1 aufgelösten
Form:

$$P_1 = P_2 - \frac{a}{b} + \frac{1}{b} \ln \frac{\overline{X} - x_1}{x_1}. \tag{8}$$

Subject: Pricing

Title: A Mathematical Model for Price Promotion

Author: Yoram Kinberg, Ambar G. Rao, Melvin F. Shakun.

Source: Management Science, Vol. 20, No. 6, February 1974 ,
           948-59

Summary: A market where two groups of brands, premium (higher
           priced) and private label (lower priced) are sold is
           considered. It is assumed price is the only indicator of
           quality. Using hypotheses about consumer behavior in
           such markets, a model for changes in market share that
           would result from temporary price reductions by one of
           the premium brands is constructed. The model is used to
           develop promotional strategies for one of the premium
           brands, given various assumptions about competitive
           behavior. Methods for use of the model are suggested.

Model:

Assume the following:

Customers who are price conscious and usually purchase
competitive premium brands will switch to the promoted
brand due to its temporarily reduced price. If they
usually buy private label brands, they will not switch
to the promoted brand, so long as the promotional price
is above the normal private label price.

Customers who are quality conscious and usually buy a
private label brand because the premium brands are normally
priced above their acceptable range will switch to the
promoted brand if its price enters their acceptable range.
If they usually buy other premium brands, they will
not switch to the promoted brand.

Let

P_1, P_0 = prices of premium and private label brands,
respectively; $P_0 < P_1$.

ΔP_1 = discount offered by the promoting brand, $\Delta P_1 < P_1 - P_0$.

$P_1, P_0, \Delta P_1$ = prices on a log scale.

$\Phi(x)$ = cumulative distribution function of the standard
normal random variable.

δ = share of premium market share normally enjoyed by
the promoting brand.

P_1, P_2 = probabilities that a randomly chosen customer buys
a premium brand and a private label brand, respectively, $P_1 + P_2 \leq 1$.

α = proportion of customers who are quality conscious.

$p, p+a$ = minimum, maximum log prices a randomly chosen
customer is prepared to pay.

When $P_1 - P_0 < a$, then

$$\rho_1 = \Phi(\frac{P_1-\mu}{\sigma}) - \Phi(\frac{P_0-\mu}{\sigma}) + \alpha[\Phi(\frac{P_0-\mu}{\sigma}) - \Phi(\frac{P_1-a-\mu}{\sigma})], \quad (1)$$

$$\rho_0 = \Phi(\frac{P_1-a-\mu}{\sigma}) - \Phi(\frac{P_0-a-\mu}{\sigma}) + (1-\alpha)[\Phi(\frac{P_0-\mu}{\sigma}) - \Phi(\frac{P_1-a-\mu}{\sigma})]. \quad (2)$$

The probability that a customer of a competing premium brand will switch to the promoted brand is

$$\theta_1 = (1-\delta)(1-\alpha)[\Phi(\frac{P_1-\mu}{\sigma}) - \Phi(\frac{P_0-\mu}{\sigma})]. \quad (3)$$

The probability that a randomly chosen customer will switch from private label brands is

$$\theta_0 = \alpha[\Phi(\frac{P_1-\mu-a}{\sigma}) - \Phi(\frac{P_1-\Delta P_1-\mu-a}{\sigma})]. \quad (4)$$

Therefore, the probability that a randomly chosen customer will switch to the promoted brand is

$$\theta(\Delta p_1) = \theta_1 + \theta_0. \quad (5)$$

The incremental revenue (sales) per customer during the promotion is

$$\Delta R = q_0[\theta(\Delta p_1) \cdot (P_1 - \Delta P_1) - \delta\rho_1\Delta P_1], \quad (6)$$

where q_0 is the average quantity purchased on each occasion. In the analysis below we set $q_0 = 1$.

Substituting (3) and (4) into (6) and maximizing ΔR with respect to ΔP_1 we can determine the optimal price off ΔP_1^*. Thus,

$$\partial\Delta R/\partial\Delta P_1 = (P_1 - \Delta P_1)\partial\theta/\partial\Delta P_1 - \theta - \delta\rho_1$$
$$= 0 \text{ for a maximum.} \quad (7)$$

The optimizing value of ΔP_1 can now be obtained numerically from (7).

Subject: New Product

Title: Product Search and Evaluation

Authors: Paul Stillson and E. Leonard Arnoff

Source: Journal of Marketing, Vol. 22, No. 1, July 1957,
           33-39

Summary: Simple model to determine the minimum sales required of
           a product to cover different costs.

Model:

The total cost of the production process for all sizes of a product can be expressed by:

$$\text{Total cost} = A + \sum_{i=1}^{n} \{FP_i + IP_i +$$

$$(R + D + S + C)(xP_i) + xc_iP_i\} \tag{1}$$

where

A = fixed costs for new equipment, including amortization

FP_i = fixed costs allocated to each size

IP_i = regulated costs allocated to each size

R = royalties as per cent of total gross sales

D = distributor discount

S = shipping costs

C = sales commission

xP_i = gross sales for each size

xc_iP_i = total prime cost for each size.

Since the sum of all of the percentages of gross sales attributable to each size equals unity, that is, $\sum_{1=i}^{n} P_i = 1$, it follows that:

Total cost =

$$A + F + I + x(R+D+S+C) + x\sum_{i=1}^{n} c_iP_i \tag{2}$$

where

x = total gross sales in dollars

F = total fixed costs for other than new equipment

I = total regulated costs.

The net dollar profit for the product is

$$Zx = x - F - A - I - x(R + D + S + C) - x \sum_{i=1}^{n} c_i P_i. \quad (3)$$

For a determination of the expected dollar return for any level (x) of gross sales:

$$x = \frac{F + A + I}{1 - (R + D + S + C) - Z - \sum_{i=1}^{n} c_i P_i}. \quad (4)$$

Subject: New Product

Title: Early Prediction of Market Success for New Grocery Products

Authors: Louis A. Fourt and Joseph W. Woodlock

Source: Journal of Marketing, Vol. 25, No. 2, October 1960, 31-38

Summary: Consumer panel statistics are linked to a mathematical model of penetration to predict the success of new grocery products.

Model: 1. Experience with a large number of earlier new products is used to predetermine the general functional form or shape of the cumulative penetration as a function of time. By penetration is meant the proportion of households that make an initial purchase of an item. Observations of penetration for the particular new product are then used to determine its unique constants. Having estimated these constants, we can then extend the penetration as far into the future (for the same market) as desired.

2. The first repeat ratio (= fraction of initial buyers who make a second purchase) is applied to this extended penetration curve to derive a cumulative first repeat purchase curve.

3. Subsequent repeat ratios as needed are similarly applied in turn. These are the ratio of third purchases to second, fourth to third, etc.   The actual number of such ratios used depends on the frequency of purchase. The values used for these ratios are not merely those achieved to date, but are estimates for later periods, allowing each group of buyers an opportunity to make a repurchase.

4. The time intervals between purchases and the average size of transactions are observed, the latter separately for new and repeat buyers, and applied to obtain volume estimates.

Observation of numerous annual cumulative penetration curves shows that (1) successive increments in these curves decline, and that (2) the cumulative curve seems to approach a limiting penetration less than 100 per cent of households-- frequently far less.

A simple model with these properties states that the increments in penetration for equal time periods are pro- portional to the remaining distance to the limiting "ceiling" penetration. In other words, in each period the ceiling is approached by a constant fraction of the remaining distance.

Let x = ceiling and r = proportional increment that occurs in each period. Then

Time period	Formula
1	$r(x-0) = rx$
2	$r(x-rx) = rx(1-r)$
3	$rx(1-r)^2$
i	$rx(1-r)^{i-1}$

That is, each increment is simply $1-r$ times the preceding increment:

$$\Delta_i = rx(1-r)^{i-1}.$$

Reference: #110.

Subject: New Product

Title: A Theory of Market Behavior After Innovation

Author: George H. Haines, Jr.

Source: Management Science, Vol. 10, No. 4, July 1964,
           634-58

Summary: A study of innovation of consumer nondurable products;
           a simple hypothesis about consumer behavior in short run,
           non-equilibrium situations is presented. Following
           classical economic methodology, the behavior of individuals
           is then aggregated. This aggregate equation form is
           tested and shown to be not rejected by the data.

Model:

Suppose two events--purchasing and not purchasing a given
new nondurable product--are considered. Then the vector $T_i p$
may be written:

$$T_i p = \alpha_i p + (1 - \alpha_i)\lambda_i.$$

The first element of $T_i p$ can be denoted $Q_1 p$; the second $Q_2 p$.
Then:

$$Q_1 p = \alpha_1 p + (1 - \alpha_1)\lambda_1$$

$$Q_2 p = \alpha_2 p + (1 - \alpha_2)\lambda_2. \tag{1}$$

Take $Q_1 p$ as referring to the probability of purchase; hence
$Q_2 p$ refers to the probability of not purchasing.

Now focus on a specific model:

$$Q_1 p = \alpha_1 p + (1 - \alpha_1)\lambda_1$$

$$Q_2 p = p. \tag{2}$$

The event E_2 associated with Q_2 has no influence on the
response probability.

Take (2) as a stochastic model of choice. This model contains three parameters: λ, which represents the long-run probability of purchase, α_1, which measures the speed of approach to equilibrium, and p_0, the initial probability of purchase. To complete the model, one simple extension is made. It is asserted that there is a set of variables, denoted θ_k which affect λ and another set of variables, denoted z_j, which affect α_1.

Suppose α_1 is taken as constant, or estimated independently by other methods, and that attention is focused upon examination of specific factors (θ_k) affecting the equilibrium position. Then:

$$Q_1 p = \alpha_1 p + (1 - \alpha_1) \lambda [\theta_k] \qquad k = 1, \ldots, b \qquad (3)$$

$$Y = \exp (- (1 - \alpha_1) n) p_0 + (1 - \exp (- (1 - \alpha_1) n)) \lambda \quad (3a)$$

where y is probability of purchase.

$$dY_n/dn = Y_n (1 - \alpha_1)(\lambda - Y_n)$$

$$\qquad\qquad\qquad (4)$$

$$= (1 - \alpha_1) \lambda Y_n - (1 - \alpha_1) Y_n^2$$

which may be solved by separation of variables:

$$\frac{dY_n}{(1 - \alpha_1) \lambda Y_n - (1 - \alpha_1) Y_n^2} = dn.$$

Integrating,

$$\ln \left[\frac{-Y_n}{-Y_n + \lambda}\right] = (1 - \alpha_1) \lambda n + c^1.$$

Take antilogarithms:

$$\frac{-Y_n}{\lambda - Y_n} = \exp\ (a_1 + (1 - \alpha_1)\lambda n)$$

$$Y_n - Y_n \exp\ ((1-\alpha_1)\lambda n + a_1) = \lambda \exp\ ((1-\alpha_1)\lambda n + a_1)$$

$$Y_n = \frac{-\lambda \exp\ ((1-\alpha_1)\lambda n + a_1)}{1 - \exp\ ((1 - \alpha_1)\lambda n + a_1)} \tag{5}$$

$$Y_n = \frac{\lambda}{1 - \exp\ (-a_1 - (1-\alpha_1)\lambda n)}$$

Assume that trials can be roughly approximated by the dimension time, since trials are not always directly observable. Hence:

$$Y_t = \frac{\lambda}{1 - \exp\ (-a_1 + [(\alpha_1 - 1)\lambda]t)} \tag{5a}$$

In some applications, λ is fixed or known. If a maximum number of responses (say, N) can be defined

$$m_t = \frac{N\lambda}{1 + \exp\ (a_1 + [(\alpha_1 - 1)\lambda]t)} \tag{6}$$

If

$$\lambda = 1, \ \alpha_1 - 1 = f(z_j)$$

$$M_t = \frac{N}{1 + \exp\ (a_1 + f(z_j)t)} \tag{7}$$

The z_j are interpretable as factors which affect the rate of organizational learning.

Subject: New Product

Title: Dynamics of New Product Campaigns

Author: Harlan D. Mills

Source: Journal of Marketing, Vol. 28, October 1964, 60-63

Summary: Model for introducing a new consumer product on a nation-wide basis.

Model: Let

$$T = \frac{Nx}{A + x} \qquad (1)$$

where

    T = the number of consumers trying,

    N = the total number of consumers in the market,

    x = the dollars of promotion used,

    A = new product "market resistance."

A is a number that varies from one campaign to the next, and is adjusted to fit the statistical data.

Let

    R = average rate of purchases per year, generated by a trial purchase,

    H = time horizon,

    p = unit price,

    c = unit cost (without cost of introductory campaign).

Then the volume of the product over its profit horizon will be

$$V = TRH. \qquad (2)$$

The profit of the product over this horizon will be

$$P = V(p - c) - x. \qquad (3)$$

Substituting (1) and (2) into (3),

$$P = (\frac{Nx}{A + x}) \; RH(p - c) - x, \qquad (4)$$

where N, p, and c are constants.

We can simplify (4) by arbitrarily introducing a new statistic B, in place of R, by the definition

$$B = NH(p - c)R, \qquad (5)$$

which we interpret as

B = new product "profit potential."

Then, the profit of the new product over its horizon P becomes

$$P = \frac{Bx}{A + x} - x. \tag{6}$$

Maximizing P in (6) when A and B are assumed to be known (or estimated),

$$x = \begin{cases} \sqrt{AB} - A & \text{if } B > A \\ \\ 0 & \text{if } B \le A. \end{cases} \tag{7}$$

This decision rule says: initiate a new product campaign only if, first, "profit potential" B exceeds "market resistance" A, and; second, it gives the level of effort in terms of A and B.

On the basis of initial estimates, suppose that a partial amount s of x has already been invested in a new product campaign, and that now new estimates A* and B* are available for A and B. What should the revision x* of x be?

Let x* = s + y*; that is, y* will be the additional (incremental) level of effort, beyond s, to be put into the campaign. The total profit over the horizon, for any y*, can be determined as

$$P = \frac{B^* (s + y^*)}{A^* + (s + y^*)} - (s + y^*). \tag{8}$$

Differentiating and checking for a maximum, we find

$$y^* = \begin{cases} \sqrt{A^* B^*} - A^* - s, & \text{if } B^* > A^* + 2s + \frac{s^2}{A^*}; \\ \\ 0, & \text{if } B^* \le A^* + 2s + \frac{s^2}{A^*}. \end{cases} \tag{9}$$

Subject: New Product

Title: Competitive Strategies for New Product Marketing over
          the Life Cycle

Author: Philip Kotler

Source: Management Science, Vol. 12, No. 4, December 1965,
          B-104-119

Summary: Formulation of a long-run competitive marketing strategy
          for a new product introduced into a market with classic
          growth, seasonal, and merchandising characteristics.
          The first part describes the market model as well as the
          accounting model used by the firm to compute its profits.
          The second part discusses nine conceptually different
          classes of marketing strategies. The third part reports
          the results of a duopoly confrontation involving various
          pairs of competitive strategies. The last part suggests
          additional variations in the market model and in the
          strategies which would increase the significance of
          the findings.

Model:

I. Sales and Profit Models
Total Market Sales Sub-Model
Growth Factor

A Gompertz equation with the following parameters is
used:

$$G_t = 4,000 \ (.2)^{.9^t} \qquad (1)$$

where G_t = level of industry sales at time t due to
the growth factor, t = an index for the particular month.
Because of the given parameters, industry sales are 800
units a month at t = 0. As t→∞, industry sales grow
asymptotically in S-curve fashion toward 4,000 units a
month, most of this occurring by t = 60.

Seasonal Factor

$$V_t = 1 + .1 \sin (30t + 180) \qquad (2)$$

where V_t = index for the level of industry sales at time t due to the seasonal factor. Equation (2) produces a yearly seasonal index with an amplitude of .90 to 1.10.

Merchandising Factor

$$M_t = [\sum_{i=1}^{n} P_{i,t}^{-2} A_{i,t}^{1/8} D_{i,t}^{1/4}/n(20^{-2}2500^{1/8}2500^{1/4})]^2(1.05)^{-t}$$

(3)

where M_t = index for the level of industry sales due to the merchandising factor

$P_{i,t}$ = the price charged by firm i at time t

$A_{i,t}$ = advertising expenditures of firm i at time t

$D_{i,t}$ = distribution expenditures of firm i at time t

n = the number of competitors.

Combining the three individual components for the determination of industry sales:

$$I_t = G_t \cdot V_t \cdot M_t$$

$$I_t = [4,000(.20)^{.9^t}][1 + .1(\sin(30t+180))] \quad (4)$$

$$\cdot [\sum_{i=1}^{n} P_{i,t}^{-2} A_{i,t}^{1/8} D_{i,t}^{1/4}/n(20^{-2}2500^{1/8}2500^{1/4})]^2(1.05)^{-t}$$

Company Sales Sub-Model

$$m_{i,t} = P_{i,t}^{-2} A_{i,t}^{1/8} D_{i,t}^{1/4} / \sum_{i=1}^{n} P_{i,t}^{-2} A_{i,t}^{1/8} D_{i,t}^{1/4} \quad (5)$$

where $m_{i,t}$ = market share of firm i at time t.

Company's actual sales:

$$S_{i,t} = m_{i,t} I_t \quad (6)$$

where $S_{i,t}$ = the sales of firm i at time t.

Company Cost and Profit Sub-Model

Assuming constant variable costs at $10 a unit and the following fixed costs on a monthly basis:

Depreciation	$1,000	(on a straight-line basis)
Overhead	2,167	
Advertising	A	(initially $2,500 a month)
Distribution	S	(initially $2,500 a month)

$3,167 + A + S

the total cost function is:

$$C_{i,t} = 10S_{i,t} + 3,167 + A_{i,t} + D_{i,t} \qquad (7)$$

where $C_{i,t}$ = total cost to firm i at time t for selling $S_{i,t}$. The total company profits are given by:

$$\pi_{i,t} = R_{i,t} - C_{i,t} \qquad \text{or}$$

$$\pi_{i,t} = (P_{i,t} - 10)S_{i,t} - 3,167 - A_{i,t} - D_{i,t}. \qquad (8)$$

II. Major Classes of Competitive Marketing Strategies

Non-Adaptive Strategy

A strategy where the initial marketing mix is held constant throughout the product's life cycle:

$$\begin{aligned} P_{i,t} &= P_{i,t-1} \\ A_{i,t} &= A_{i,t-1} \\ D_{i,t} &= D_{i,t-1} \end{aligned} \qquad (1)$$

Time-Dependent Strategy

Any strategy which provides for automatic marketing mix adjustments to take place through time:

$$\begin{aligned} P_{i,t} &= f(t) \\ A_{i,t} &= f(t) \\ D_{i,t} &= f(t) \end{aligned} \qquad (2)$$

Competitively Adaptive Strategy

Any strategy where firm i adjusts its marketing mix because of marketing mix changes made by firm j in previous periods:

$$P_{i,t} = f(P_{j,t-1}, P_{j,t-2}, \cdots P_{j,t-k})$$

$$A_{i,t} = f(A_{j,t-1}, P_{j,t-2}, \cdots P_{j,t-k}) \qquad (3)$$

$$D_{i,t} = f(D_{j,t-1}, P_{j,t-2}, \cdots P_{j,t-k})$$

Sales-Responsive Strategy

Any strategy which leads a company to adjust its marketing mix on the basis of its past sales results:

$$P_{i,t} = f(S_{i,t-1}, S_{i,t-2}, \cdots S_{i,t-k})$$

$$A_{i,t} = f(S_{i,t-1}, S_{i,t-2}, \cdots S_{i,t-k}) \qquad (4)$$

$$D_{i,t} = f(S_{i,t-1}, S_{i,t-2}, \cdots S_{i,t-k})$$

Profit-Responsive Strategy

A strategy where the marketing mix is adjusted in response to "significant" interperiod changes in company profits. An example of a profit-responsive strategy is shown below:

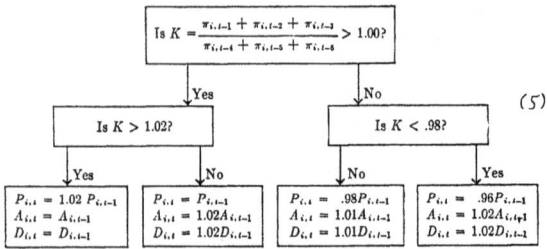

The cumulative, compounded net profits of firm i using strategy s for 60 months is given by:

$$\pi_{i,s} = \sum_{t=1}^{60} \pi_{i,t}(1.06)^{60-t}$$

$\pi_{i,s}$ serving as the index of the strategy's success.

Subject: New Product

Title: DEMON Mark II: Extremal Equations Solution and Approximation

Authors: A. Charnes, W.W. Cooper, J.K. DeVoe, and D.B. Learner

Source: Management Science, Vol. 14, No. 11, July 1968, 682-91

Summary: The DEMON model (Decision Mapping Via Optimum GO-NO Networks), a model for marketing new products, is formulated in terms of an extremal equation. The latter can be reduced to solution of a separated system of simpler equations which, for discrete distributions, can be solved by linear programming methods. The reduction also permits general characteristics of the solutions to be inferred. Methods of approximation and bounding are developed and interpreted for the general case.

Model: We characterize our knowledge or the "state" of our system by the pair (x,d) where

> x = amount of funds remaining from an originally specified study budget,
>
> d = the chance variable of demand quantity, as currently known,
>
> c_j = cost of making the study (or combination of studies) designated by j,
>
> $\tilde{d}_j$ = the demand quantity observed if study j is performed,

$f_j(\tilde{d}|d)$ = conditional density of $\tilde{d}$ given d,

c_o = cost of a "GO" decision.

Then, we obtain the extremal equation for $\phi(x,d)$--which corresponds to MEMP (Maximum Expected Maximum Profit)-- starting from the state (x,d) as

$$\phi(x,d) = \begin{cases} E_d\ \pi(d) - c_o + x, & \text{if GO,} \\ \max_j \int_{-\infty}^{\infty} \phi(x - c_j,\ \tilde{d}) f_j(\tilde{d}|d) d(\tilde{d}), & \text{if ON,} \quad (2) \\ x, & \text{if NO,} \end{cases}$$

where $E_d\ \pi(d)$ is the expected total profit and the symbols $d(\tilde{d})$, $\int_{-\infty}^{\infty}$ are short-hand expressions for the multivariate volume element and integration, respectively. We rewrite (2) as

$$\phi(x,d) = \max_j \{ \int_{-\infty}^{\infty} \delta(d - \tilde{d}) [H_G(\tilde{d})(E_{\tilde{d}} \pi(\tilde{d}) - c_0 + x)$$

$$+ H_N(\tilde{d})x]d(\tilde{d}) + \int_{-\infty}^{\infty} \phi(x - c_j, \tilde{d})[1 - H_G(\tilde{d}) \quad (3)$$

$$- H_N(\tilde{d})] \; H(x - c_j)f_j(\tilde{d}|d)d(\tilde{d}) \},$$

where δ is the Dirac delta function and H_G is the characteristic function of the set G for which d is GO. We obtain

$$\phi(x,d) = \Psi(d) + \eta(d)x. \quad (4)$$

Substituting (4) into (3),

$$\Psi(d) + \eta(d)x = \max_j \{G_j(d) + R_j(d)x \quad (5)$$

$$+ \int_{-\infty}^{\infty} [\Psi(\tilde{d}) + \eta(\tilde{d})(x - c_j)]O_j(\tilde{d},d)d(\tilde{d}) \}$$

$$\Psi(d) = \max_j \{G_j(d) + \int_{-\infty}^{\infty} [\Psi(\tilde{d}) - c_j\eta(\tilde{d})]O_j(\tilde{d},d)d(\tilde{d}) \} \quad (6)$$

$$\eta(d) = \max_j \{R_j(d) + \int_{-\infty}^{\infty} \eta(\tilde{d})O_j(\tilde{d},d)d(\tilde{d}) \}.$$

(6) is reduced to solution of an extremal equation of the form

$$\eta(u) = \max_j \{R_j(u) + \int_{-\infty}^{\infty} \eta(\tilde{u})O_j(\tilde{u},u)d\tilde{u} \}. \quad (7)$$

If the random variables have finite discrete distributions the system (7), and thereby (6) as well, may be attacked as ordinary linear programming problems. Further developments are undertaken with reference to log-normal distributions and then illustrated by reference to the single-parameter case.

Subject: New Product

Title: SPRINTER Mod III: A Model for the Analysis of New Frequently Purchased Consumer Products

Author: Glen L. Urban

Source: Operations Research, Vol. 18, No. 5, September–October 1970, 805-54

Summary: A model-based information system is presented which is designed to analyze test-market results, to assist decision-making for a new frequently purchased consumer product, and to serve as an adaptive control mechanism during national introduction. The model, called SPRINTER (Specifications of PRofits from INTER-dependencies) is based on the behavioral process of the diffusion of innovation and can be used normatively in an interactive search mode to find the best marketing strategy for a new product. The input is obtained from test-market data analyzed by statistics and combined with subjective judgments. A GO, ON, or NO decision is made on the basis of the estimated profit and risk produced by the best marketing strategy. An application of this model to a real product is reported, this application using an on-line computer program that allows man-system communication.

Model: In the first period the number of people in the trial class is the current number in the target groups for the product. In succeeding periods, it is the target groups less the number in the preference and loyalty classes.

$$TRIAL_t = TGTGR_t - NPREF_{t-1} \qquad (1)$$
$$- NLOYL1_{t-1} - NLOYL2_{t-1} - NNLOYL_{t-1},$$

where

$TRIAL_t$ = number of people in trial class in period t,

$TGTGR_t$ = number of people in target group of the product in period t,

$NPREF_{t-1}$ = number of people in preference class in period t-1,

$NLOYL1_{t-1}$ = number of people in the loyalty I class in period t-1,

$NLOYL2_{t-1}$ = number of people in the loyalty II class in period t-1,

$NNLOYL_{t-1}$ = number of people in the nonloyalty class in period t-1.

The number of people in the target group for the product can be influenced by advertising expenditures or by the total

number of samples of the new product sent by all the firms
in the industry. Then,

$$TGTGR_t = FTGTGR_t \quad RADIND(ADIND_t/FADIND_t)$$
$$+ (SMIND_t - FSMIND_t) \tag{2}$$
$$\cdot (1 - FTGTGR_t/PWORLD_t) \cdot SAMPUS,$$

where

$FTGTGR_t$ = forecast reference number of people in the
target group,

$RADIND$ = advertising response function,

$ADIND_t$ = actual industry advertising expenditure,

$FADIND_t$ = forecast industry advertising expenditure,

$SMIND_t$ = total number of samples sent out by firms in
the industry,

$FSMIND_t$ = forecast reference number of samples to be sent
out by firms in the industry,

$PWORLD_t$ = potential number of people who could possibly
be users of this product,

$SAMPUS$ = per cent of people who receive samples who
use them and are pleased with the product.

The number in the trial class after our firm's sampling is

$$NTRIAL_t = TRIAL_t - SMFIRM_t \cdot (TRIAL_t/PWORLD_t) \cdot SAMPUS, \tag{3}$$

where

$TRIAL_t$ = number of people in the trial class before
sampling,

$SMFIRM_t$ = number of samples sent by our firm,

$SAMPUS$ = per cent of people who use the sample and
experience a pseudotrial.

The people who experience a trial because of sampling are moved
on to the preference model.

The awareness section of the trial class describes the
effects of advertising in creating flows of people into
awareness states:

$$DADAWT_t = (NTRIAL_t - TADAWT_{t-1}) \cdot RADAWT(ADFIRM_t/FADFRM_t), \tag{4}$$

where

DADAWT_t = number of people newly aware resulting from our advertising level (ADFIRM_t),

TADAWT_{t-1} = number of people remaining aware at the end of the last period,

RADAWT = response function representing the fraction of people aware of our brand, ads, or appeals at our advertising expenditure level (ADFIRM_t) compared to the reference level (FADFRM_t).

$$\begin{aligned}
\text{NAWT}_{t,J} = {} & \text{NAWTFW}_{t-1,J} + \text{DADAWT}_t \\
& \cdot \text{RADAPT}(\text{ADFIRM}_t/\text{FADFRM}_t) \qquad (5) \\
& \cdot \text{RAPSPT}(\text{ADFIRM}_t/\text{FADFRM}_{t,J}),
\end{aligned}$$

where

$\text{NAWT}_{t,J}$ = number of people in appeal-awareness state J,

$\text{NAWTFW}_{t-1,J}$ = number of people in appeal-awareness state J at the end of the last period after forgetting and word-of-mouth transfer [to be defined in equation (11)],

RADAPT = response function representing the proportion of people newly aware who become aware of some appeal at our advertising level ADFIRM_t relative to our reference level FADFRM_t,

RAPSPT = response function representing per cent of people who are aware of some appeal who become aware of specific appeal J.

$$\begin{aligned}
\text{NTRY}_t = {} & \sum_J \text{NAWT}_{t,J} \cdot \text{TRATE}_{t,J} \qquad (6) \\
& \cdot \text{RACOMT}[(\text{TCPTA}_t/\text{ADIND}_t)/(\text{FTCPTA}_t/\text{FADIND}_t)],
\end{aligned}$$

where

NTRY_t = number of people with intent to try in period t,

$\text{TRATE}_{t,J}$ = per cent of people in awareness state J who intend to try in month t,

RACOMT = response function describing the effects of total competitive advertising (defined as TCPTA_t) as a per cent of total industry advertising (i.e., ADIND_t) relative to the expected proportion of competitive advertising that is the reference competitive advertising FTCPTA_t divided by the reference total industry advertising FADIND_t.

$$AVAIL_{t,S} = AVAIL_{t-1,S} + SLCAL_{t,S}$$

$$\cdot \; (1 - AVAIL_{t-1,S}/NSTOR_{t,S}) \qquad\qquad (7)$$

$$\cdot \; RDEAL(DEAL/ADEAL) - DROP_{t,S},$$

where

$AVAIL_{t,S}$ = number of stores of type S that stock our product,

$SLCAL_{t,S}$ = number of sales calls on store type S,

$NSTOR_{t,S}$ = total number of stores of type S,

$RDEAL$ = response function representing the per cent of stores who stock our product at a specific middleman deal (DEAL) relative to the average competitive deal (ADEAL),

$DROP_{t,S}$ = number of stores who drop our product when its sales are below their expectations.

$$TFIND_t = \sum_S PSHOP_S \cdot NTRY_t \cdot AVLPCT_{t,S}$$

$$+ \sum_S \sum_{SS} PSHOP_S \cdot (1 - AVLPCT_{t,S}) \qquad\qquad (8)$$

$$\cdot \; NTRY_t \cdot PSWST_{S,SS} \cdot AVLPCT_{t,SS},$$

where

$TFIND_t$ = number of people who have intent to try to find the product,

$PSHOP_S$ = proportion of people who deem store S as their favored retailer for this type of product,

$AVLPCT_{t,S}$ = per cent of stores of type S carrying the product,

$PSWST_{S,SS}$ = proportion of people who do not find the brand at their first-choice store who will switch to store SS.

The number actually purchasing is

$$NTBUY_{t,S} = TFIND_{t,S} \cdot RPDIFT[(PR_{t,1} - SPR_t)/SPR_t]$$

$$\cdot \; RPOP(SD_t/FSD_t), \qquad\qquad (9)$$

where RPDIFT is a response function representing the per cent of people who will exercise their intent when presented with our specific price $PR_{t,1}$ relative to the price standard SPR_t.

RPOP is a response function representing the point-of-purchase effects of our special displays SD_t relative to the expected level of our display activity FSD_t. The people with a coupon perceive a lower shelf price and are described by an equation similar to (9) with the price equal to the shelf price less the coupon 'price off' amount

$$NAWTF_{t,J} = NAWTA_{t,J} + \sum_{\substack{K \\ K>J}} NAWTA_{t,K} \cdot RFRGT_{K,J}$$

$$- \sum_{\substack{K \\ K>J}} NAWTA_{t,J} \cdot RFRGT_{J,K'} \tag{10}$$

where

$NAWTF_{t,J}$ = number of people in awareness state J after forgetting in the trial class,

$NAWTA_{t,J}$ = number of people in awareness state J after trial purchasers have been moved to the preference class,

$RFRGT_{J,K}$ = per cent of people who forget from awareness state J to awareness state K in the trial class.

The awareness-state population after word-of-mouth is

$$NAWTFW_{t,J} = NAWTF_{t,J} - \sum_{\substack{K \\ K>J}} WOM_{t,K} \cdot (NAWTF_{t,J}/$$

$$TGTGR_t) + \sum_{\substack{K \\ K>J}} WOM_{t,J} \cdot (NAWTF_{t,K}/TGTGR_t), \tag{11}$$

where

$WOM_{t,K}$ = total number of word-of-mouth exchanges about appeal K in the pool,

$TGTGR_t$ = total number of people in the target group in period t.

This number of people in each awareness state is an input to the next period.

$$HLDP_{t,H} = HLDP_{t-1,H+1} + [TBUY_{t-1} + SMFIRM_t$$

$$\cdot SAMPUS(NTRIAL_t/PWORLD_t)] \cdot FREPR_H, \tag{12}$$

where

$HLDP_{t,H}$ = number of people who will be ready to purchase in H periods, H = 1, ..., h,

$FREPR_H$ = frequency of purchase defined by the per cent of consumers repeat purchasing every H+1 months.

$$NP1P_t = \sum_J NAWP_{t,J} \cdot P1RATE_{t,J}, \qquad (13)$$

where

$NP1P_t$ = number with a first preference for brand,

$P1RATE_{t,J}$ = per cent of people in awareness state J who have a first preference for the product,

$NAWP_{t,J}$ = number aware of appeal J in period t in preference class and ready to buy.

Similarly, the number with a second preference for the brand is

$$NP2P_t = \sum_J NAWP_{t,J} \cdot P2RATE_{t,J}. \qquad (14)$$

The number of people in the preference model who intend to repeat purchase and who will purchase some product in period t is

$$RPTSHP_t = (MP1P_t \cdot BRP1P + NP2P_t \cdot BRP2P)$$

$$\cdot AREL[(ADFIRM_t/TCPTA_t)/(FADFRM_t/FTCPTA_t)], \qquad (15)$$

where

BRP1P = per cent of people with first preference who are expected to convert that preference into intent to repurchase,

BRP2P = per cent of people with second preference who are expected to convert that preference into intent to repurchase,

AREL = response function representing the effects of competitive advertising by the proportionate reduction in the number intending to repurchase at our level of advertising $ADFIRM_t$ relative to total competitive advertising $TCPTA_t$ compared to the forecast ratio $FADFRM_t/FTCPTA_t$.

The number of switchers with intent to buy our brand is

$$SWSHP_t = (NPREF_t - NP1P_t - NP2P_t) \cdot SWRFK$$

$$\cdot ARELK[(TCPTA_t/ADIND_t)/(FTCPTA_t/FADIND_t)], \qquad (16)$$

where

$NPREF_t$ = number of people in preference class ready to buy in period t, but with no intent to redeem a coupon,

$SWRFK$ = per cent of people with no preference for our brand who develop an intent to buy our brand at reference competitive advertising,

$ARELK$ = response function reflecting proportionate change in switching rate as total competitive advertising $TCPTA_t$ as a per cent of industry $ADIND_t$ varies from the predicted reference ratio $FTCPTA_t/FADIND_t$.

The number of actual purchases in the preference model by those with intent and no coupon is

$$NPBUY_{t,S} = TSHOP_{t,S} \cdot K \cdot [(PR_{t,1,S}^{SPRIS} \, FA_{t,1,S}^{SFAIS} \, SD_{t,1,S}^{SSDIS} /$$

$$\sum_i PR_{t,i,S}^{SPRiS} \, FA_{t,i,S}^{SPAiS} \, SD_{t,i,S}^{SSDiS})^{EI}], \qquad (17)$$

where

$TSHOP_{t,S}$ = number of people entering store of type S carrying our brand with intent to purchase our brand (but with no coupon),

$PR_{t,i,S}$ = price of brand of firm i in store S in period t,

$FA_{t,i,S}$ = number of package facings exposed on the shelf of brand of firm i in store S,

$SD_{t,i,S}$ = per cent of stores of type S that have special displays for firm i's brand,

K = scale constant,

$SPRiS$ = sensitivity of price for firm i's brand in store S,

$SFAiS$ = sensitivity of facings for firm i's brand in store S,

$SSDiS$ = sensitivity of special displays for firm i's brand in store S,

EI = elasticity of in-store environment for consumers with intent to buy our brand.

The number of loyalty I buyers is

$$BUYL1_t = (NLOYL1_t - \sum_H HLDL1_H) \cdot REPT1$$

$$\cdot AREL1[ADFIRM_t/(ADIND_t/QFIRM_t)] \qquad (18)$$

$$\cdot PREL1[PR_{t,1}/(\sum_i PR_{t,i}/QFIRM_t)],$$

where

REPT1 = per cent of loyalty I consumers who intend to repeat purchase our brand at reference price and advertising levels,

$NLOYL1_t$ = number of people in loyalty I class in period t,

$HLDL1_H$ = number of people who will be ready to purchase in H periods,

AREL1 = response function representing the effects of our advertising $ADFIRM_t$ relative to the average level of advertising per firm $ADIND_t/QFIRM_t$, where $QFIRM_t$ = number of firms in industry in period t, by the proportionate reduction in the intent rate in the loyalty I model,

PREL1 = response function representing the effects of our price $PR_{t,1}$ relative to the average price by the proportionate reduction in our repeater loyalty I model buyers in the store.

In order to find the best strategy for this model, iterative techniques must be utilized.

Subject: New Product

Title: Stochastic Models for Monitoring New-Product Introductions

Author: William F. Massy

Source: F.M. Bass, C.W. King, E.A. Pessimier (eds.), Applications
          of the Sciences in Marketing Management, Wiley & Sons:
          New York-London-Sydney, 1968, 85-111

Summary: A mathematical model that can be used to describe the
          adoption process for new frequently purchased products
          and to make forecasts of long-run sales volumes. The
          model is "objective" in the sense that it depends only on
          sales statistics for the new product. The sales statistics
          to be utilized must be obtained from consumer panels.

Model:

Primary model for interpurchase times:

$$f\{t_{ki}|\mu_{ki}(t_{ki})\} = e^{-m_{ki}(t_{ki})} \mu_{ki}(t_{ki}),\qquad(1)$$

where

$$m_{ki}(t_{ki}) = \int_{1}^{t_{ik}} \mu_{ki}(t)\,\alpha t.$$

Secondary model for effect of time since conversion:

$$\mu_{ki}(t) = \mu t_{ki}^{\lambda},\qquad(2)$$

where

$$t_{ki} = T - \tau_{ki} - \tau_{k*} + 2 \geq 1.$$

Secondary model for mixed populations:

$$f(\mu|\alpha(\tau_{ki}),\beta) = \frac{\beta e^{-\beta\mu}(\beta\mu)^{\alpha(\tau_{ki})-1}}{\Gamma(\alpha(\tau_{ki}))}.\qquad(3)$$

Secondary model for effect of conversion time:

$$\alpha(\tau_{ki}) = \alpha\tau_{ki}^{\gamma},\qquad(4)$$

where

$$\tau_{ki} = T - \tau_{k*} + 1 \geq 1.$$

The variables used in the models are defined as follows:

i = An index used to designate families. Each family in the panel is assigned a unique value of i.

k = An index used to designate depth of repeat classes.

T = An index used to designate time. It may be defined as the total number of weeks (months, and so on) since the product was introduced.

t_{ki} = The length of time since the ith family made its kth purchase (entered the kth depth of repeat class), incremented by one.

τ_{ki} = The time period at which the ith family made its kth purchase, taken relative to the time origin of kth depth of repeat class, incremented by one.

τ_{k*} = The "time origin" of the kth depth of repeat class.

The model contains four free parameters, which may be interpreted as follows:

λ = The intensity with which a family's expected purchase rate changes with respect to time since the last purchase. The model requires $-1 < \lambda$. For $\lambda < 0$ the expected purchase rate declines with time, as would normally be expected. We require $\lambda < 1$ to avoid explosive behavior.

β = The scale factor in the distribution of initial expected purchase rates. The model requires $\beta > 0$.

α = The number of degrees of freedom in the distribution of expected purchase rates for the first few families to enter a given depth of repeat class. The parameter is proportional to the mean of the prior distribution. The model requires $\alpha > 0$.

γ The intensity with which the mean of the prior distribution of expected rates changes with respect to the period of time since the first family entered the given depth of repeat class. We should expect that $-1 < \gamma < +1$ in order to avoid explosive behavior.

Conditional density and distribution functions:

$$f(t|\lambda,\mu) = \mu t^{\lambda} e^{-[\mu/(\lambda+1)](t^{\lambda+1}-1)}, \quad t \geq 1, \qquad (5)$$

and

$$F(t|\lambda,\mu) = \frac{\mu e^{\mu/(\lambda+1)}}{(\lambda+1)} \int_{1}^{t^{\lambda+1}} e^{-[\mu/(\lambda+1)]x} \, dx = 1 - e^{-[\mu/(\lambda+1)](t^{\lambda+1}-1)}. \qquad (6)$$

Unconditional distribution of t:

$$f(t|\alpha,\beta,\gamma,\lambda;\tau) = \alpha\tau^{\gamma}\beta\alpha\tau^{\gamma}(\lambda+1)\alpha\tau^{\gamma} \frac{t^{\lambda}}{[t^{\lambda+1} + \beta(\lambda+1) - 1]\alpha\tau^{\gamma+1}}. \qquad (7)$$

$$F(t|\alpha,\beta,\gamma,\lambda;\tau) = 1 - [\frac{\beta(\lambda+1)}{t^{\lambda+1} + \beta(\lambda+1) - 1}]^{\alpha\tau^{\gamma}}, \quad t \geq 1. \quad (4.11)$$

The unconditional distribution of interpurchase times gives our best assessment of the probabilities that a certain consumer will make his next purchase in a particular time interval, provided that (1) we do not know that consumer's value of μ and (2) we do know the time τ of his last purchase and we know or have estimates of the remaining parameters of the model.

Subject: New Product

Title: New-Product Profit Evaluation Models

Author: Philip Kotler

Source: P. Kotler: Computer Simulation in the Analysis of New-Product Decisions, in: F.M. Bass, C.W. King, E.A. Pessemier (eds.), Applications of the Sciences in Marketing Management, New York: John Wiley & Sons, Inc., 1968, 283-325

Summary: Breakeven, cash-flow, and simple marketing-mix models are described.

Model: Breakeven Models

Breakeven volume equation:

$$Q_B = \frac{F}{P-V}$$

where

Q_B = breakeven volume

F = total fixed costs

P = price

V = unit variable costs.

Payback period equation:

$$m = \frac{F}{\sum\limits_{i=1}^{m} (P-V)Q_i} = 1$$

where

m = payback period

Q_i = expected sales in year i.

Cashflow Models

Present value equation:

$$W = \sum_{i=1}^{n} \frac{R_i - C_i}{(1+c)^i}$$

where

W = present value of investment

R_i = expected total revenue in year i

C_i = expected total cost in year i

c = company opportunity cost of capital

n = planning horizon.

Rate-of-return equation:

$$1 = \sum_{i=1}^{n} \frac{R_i - C_i}{(1+r)^i}$$

where

I = total investment

r = internal rate of return.

Simple Marketing Mix Models

Breakeven equations:

$$Z = (P-V)(Q - Q_B) \qquad \text{(profit equation)}$$

$$Q_B = \frac{F + A + D}{P-V} \qquad \text{(breakeven volume equation showing marketing factors)}$$

$$Q = F(P, A, D) \qquad \text{(demand function)}$$

where

Z = total profits

Q = expected sales

A = total advertising expenditures

D = total distribution expenditures.

Cash-flow equations:

$$I = \sum_{i=1}^{h} \frac{P_i [F(P_i, A_i, D_i)] - \overline{C}_i - A_i - D_i}{(1+r)^i}$$

(rate-of-return equation showing marketing factors)

$$M = \begin{pmatrix} P_1 P_2 & \cdots & P_i & \cdots & P_n \\ A_1 A_2 & \cdots & A_i & \cdots & A_n \\ D_1 D_2 & \cdots & D_i & \cdots & D_n \end{pmatrix} \text{ (marketing program matrix)}$$

$$Q = (Q_1 Q_2 \cdots Q_i \cdots Q_n) \text{ (sales time series vector)}$$

where

$\overline{C}_i$ = total nonmarketing costs

M = marketing program matrix

Q = sales time series vector.

Subject: New Product

Title: A New Product Analysis and Decision Model

Author: Glen L. Urban

Source: Management Science, Applications, Vol. 14, No. 8, April 1968, B-490-517

Summary: The factors surrounding the decision to add, or to reject, or to investigate more fully a new product proposal can be mathematically considered by four submodels in the areas of demand, cost, profit, and uncertainty. The demand model is structured to consider life cycle, industry, competitive and product interdependency effects, and will admit nonlinear and discontinuous functions. A cost minimization model is joined to the demand model to formulate a constrained profit maximization problem. The optimization is accomplished by the use of dynamic programming. The final decision is based on the business-man's criterion in combining uncertainty and the rate of return on investment.

Model: Demand for a New Product

The estimated quantity of a new product to be sold in each year is called the life cycle of the product. The life cycle estimate, supplied with a complete marketing program of price level, advertising expenditure, and distribution effort, is called the reference life cycle.

If the reference price level of the new product were changed, the estimate of the quantity to be sold would change. These changes might be noted by the term

$$X_{It} = k\overline{X}_{1t}P_{1t}^{EP}, \tag{1}$$

where

X_{It} = industry sales of product one in year t,

$\overline{X}_{1t}$ = reference industry life cycle sales estimate for product one in year t,

P_{1t} = average price of product one in year t,

EP = price elasticity,

k = scale constant.

The quantity sold in any year is

$$X_{It} = \overline{X}_{1t} PR_{1t} \tag{2}$$

where

   PR_{1t} = price response function for product 1 in
            year t.

The formulation can be extended to include advertising
and distribution responses. For example:

$$X_{It} = \overline{X}_{1t} PR_{1t} AR_{1t} DR_{1t}, \tag{3}$$

where

   X_{1t} = reference quantity of product one in year t,
   PR_{1t} = price response function for product one in
            year t,
   AR_{1t} = advertising response function for product
            one in year t,
   DR_{1t} = distribution response function for product
            one in year t,

   The aggregate industry demand for the new product
can be described as:

$$X_{It} = \overline{X}_{1t} PR_{1t} LPR_{1t} LLPR_{1t} AR_{1t} LAR_{1t} LLAR_{1t} DR_{1t} LDR_{1t} LLDR_{1t}, \tag{4}$$

where

   PR_{1t} = industry price response function for product one
            in year t,
   LPR_{1t} = one year lagged price response function for product
            one in year t,
   $LLPR_{1t}$ = two year lagged price response function for product
            one in year t,
   AR_{1t} = advertising response function for product one
            in year t,
   LAR_{1t} = one year lagged advertising response function for
            product one in year t,
   $LLAR_{1t}$ = two year lagged advertising response function for
            product one in year t,

DR_{1t} = distribution response function for product one in year t,

LDR_{1t} = one year lagged distribution response function for product one in year t,

$LLDR_{1t}$ = two year lagged distribution response function for product one in year t.

If all firms entered at the same time, the market share for firm one is:

$$MS_{11t} = \frac{PR_{11t}AR_{11t}DR_{11t}}{\sum_{i=1}^{m} PR_{i1t}AR_{i1t}DR_{i1t}}, \qquad (5)$$

where

MS_{11t} = market share for firm one in product market one in year t,

PR_{i1t} = price response function for firm "i" and product one in year t,

AR_{i1t} = advertising response function for firm "i" and product one in year t,

DR_{i1t} = distribution response function for firm "i" and product one in year t,

m = number of firms in the industry.

The complete equation for the new product is:

$$X_{ijt} = \overline{X}_{jt}[PR_{jt}LPR_{jt}LLPR_{jt}AR_{jt}LAR_{jt}LLAR_{jt}DR_{jt}LDR_{jt}LLDR_{jt}]$$

$$\left[\frac{\sum_{T=t-c}^{t} e_{ijT}PR_{ijT}AR_{ijT}DR_{ijT}}{\sum_{T=t-c}^{t} \sum_{i=1}^{m} e_{ijT}PR_{ijT}AR_{ijT}DR_{ijT}} \right] \cdot$$

$$\prod_{\substack{k=1 \\ k \neq j}}^{N} CPR_{ijkt}CAR_{ijkt}CDR_{ijkt} \qquad (6)$$

where

X_{ijt} = quantity of good j sold by firm i in period t,

$\overline{X}_{jt}$ = reference level of industry sales for product j in year t,

PR_{jt} = industry price response function for product j in year t,

LPR_{jt} = one year lagged price response function for product j in year t,

$LLPR_{jt}$ = two year lagged price response function for product j in year t,

AR_{jt} = industry advertising response function for product j in year t,

LAR_{jt} = one year lagged advertising response function for project j in year t,

$LLAR_{jt}$ = two year lagged advertising response function for project j in year t,

DR_{jt} = industry distribution response function for product j in year t,

LDR_{jt} = one year lagged distribution response function for product j in year t,

$LLDR_{jt}$ = two year lagged distribution response function for product j in year t,

PR_{ijt} = price response function for firm i on good j at time t,

DR_{ijt} = distribution response function for firm i on good j at time t,

e_{ijt} = efficiency of firm i's marketing program for product j in year t,

AR_{ijt} = advertising response function for firm i on good j at time t,

CPR_{ijkt} = cross price response of product k's price on product j in firm i in period t,

CDR_{ijkt} = cross distribution response of product k's price on product j in firm i in period t,

CAR_{ijkt} = cross advertising response of product k's advertising on product j in firm i in period t,

N = number of interdependent products.

Similar equations could be specified for the other products
in the firm's product line. When the optimum levels of the
demand variables are determined, the maximum total profit
generated by these products can be calculated and the new
line profit is specified. If the profits of the product
line without the new product are estimated and deducted from
the new line profits, the change in total line profits is
generated. This change is called the "differential profit"
and it is a measure of the profits generated by adding the
new product when demand interdependencies are considered.

Cost Structure for a New Product

Given production requirements for each product in
the line, the problem is to minimize:

$$\sum_{j=1}^{n} c_j I_j \tag{7}$$

subject to

$$\sum_{j=1}^{n} a_{ij} I_j \geqq X_i \quad \text{and} \quad \sum_{j=1}^{n} \tilde{a}_{kj} I_j \leqq q_k \quad \text{and} \quad I_j \geqq 0$$

where

c_j = cost per unit of input factor j,

I_j = amount of input factor j utilized,

X_i = minimum quantity of good i to be produced,

a_{ij} = technical production relationships,

$\tilde{a}_{kj} = 0$ if $k \neq j$,

$\tilde{a}_{kj} = 1$ if $k = j$,

q_k = constraint or input factor availability,

n = number of input factors.

Profit for the New Product

The differential profit in year t is expressed by:

$$DP_t = f(P_t, A_t, D_t, S_t) = f(m_t, S_t) \tag{8}$$

where m_t are combinations of the marketing variables in year t and where S_t are combinations of past marketing variables, $S_t = G(m_{t-1}, S_{t-1})$. Dynamic programming is suited to the analysis of the optimization. The recursion relationship for the maximization of the total discounted differential profit is:

$$TDDP_t(S_t) = \text{Max}_{m_i \epsilon M_t} \{ f(m_t, S_t) + TDDP_{t+1}[G(m_t, S_t)] \} \quad (9)$$

and

$$TDDP_{pp}(S_{pp}) = \text{Max}_{m_{pp} \epsilon M_{pp}} [f(m_{pp}, S_{pp})]$$

where

pp = last year in the planning period,

$TDDP_1(S_1)$ = maximum total discounted differential profit over the planning period,

$TDDP_t(S_t)$ = total discounted differential profit earned in year t and following years in the planning period,

$TDDP_{pp}(S_{pp})$ = discounted differential profit accrued in the last year of the planning period,

M_t = set of marketing variables to be considered in year t.

This deterministic process can be solved by the upstream algorithm of dynamic programming.

Uncertainty Associated with a New Product

The maximum differential profit of new products must be balanced against the uncertainty associated with the product proposal. The differential uncertainty is the change in the total line uncertainty. Using the variances of the new and old line profits as surrogates for uncertainty, the differential uncertainty could be measured by the standard deviation of the differential profit distribution.

$$DU^2 = V' + V - 2[COV(Pr, Pr')], \quad (10)$$

where

DU = differential uncertainty,

V' = variance of new line profits,

V = variance of old line profits,

$COV(Pr, Pr')$ = covariance of new and old line profits,

$$= E\{[Pr - E(Pr)] \cdot [Pr' - E(Pr')]\},$$

$$Pr = \text{old line profits},$$

$$Pr' = \text{new line profits},$$

$$E = \text{expected value operator}.$$

The total variance of the total profit of n products is:

$$V = \sum_{i=1}^{n} \sum_{j=1}^{n} \sigma_{ij} \tag{11}$$

$$\sigma_{ij} = \text{covariance of } i \text{ and } j = E\{[y_i - E(y_i)] \cdot [y_i - E(y_i)]\}.$$

If each product's profit is normally distributed, the variance can be expressed as:

$$V = \sum_{i=1}^{n} \sigma_i^2 + \sum_{i=1}^{n} \sum_{j=1}^{n} \sigma_{ij}$$

The variance of the joint profit distribution is:

$$\sigma_{Profit}^2 = \sigma_{Px}^2 + \sigma_{xC}^2 - 2 \, COV(Px, xC), \tag{12}$$

where

$$\sigma_{Px}^2 = \text{variance of distribution of price times quantity},$$

$$\sigma_{xC}^2 = \text{variance of distribution of cost times quantity},$$

$$COV(Px, xC) = \text{covariance of the two distributions of price times quantity and cost times quantity},$$

$$COV(Px, xC) = E\{[Px - E(Px)] \cdot [xC - E(xC)]\}.$$

The variance of the joint cost distribution for independence of unit cost and quantity is:

$$\sigma_{xC}^2 = \sigma_x^2 \sigma_C^2 + [E(x)]^2 \sigma_C^2 + [E(C)]^2 \sigma_x^2. \tag{13}$$

Decision for the New Product

The differential profit and differential uncertainty must be combined to indicate whether the new product should be introduced (GO decision), should be rejected (NO decision), or should be investigated more fully (ON decision). The risk and return plane must be divided into GO, ON, and NO areas. The GO, ON, and NO areas can be defined by two methods:

1. Define the total risk-return utility preference map and then by specifying a minimum utility for GO and maximum for NO divide the map into three areas.
2. Define constraints on the decision process that can be represented on the risk-return plane to divide the areas. These constraints need not be in terms of utility, but some other measure (e.g., profits).

The first approach is very difficult to carry out in practice, since determining a utility map for an individual is difficult and almost impossible for a corporation. There could be a question as to whether a corporation utility function actually exists.

The second approach is chosen; the constraints to divide GO, ON, and NO areas are:

1. For a GO decision the probability of obtaining a target discounted rate of return must be greater than a specified level.
2. For a NO decision the probability of obtaining a target discounted rate of return must be less than a specified level.

These constraints can be expressed in terms of the differential profit and differential uncertainty. For the GO decision the constraint is

$$P\left(\frac{TDDP}{1} \geqq 1\right) \geq A_G \qquad (14)$$

and for the NO decision

$$P\left(\frac{TDDP}{1} \geqq 1\right) \leqq A_N \qquad (15)$$

where

A_N = maximum probability for a NO decision,
A_G = minimum probability for a GO decision,
P = probability operator,

    I = total investment in new product (assumed to be
        known),

    TDDP = total discounted differential profit, discounted
        at the target rate of return. This return is
        achieved when TDDP/I = 1.

Equation (14) can be expressed as

$$P(TDDP \geqq 1) \geqq A_G$$

or

$$P(\frac{TDDP - E(TDDP)}{DU} \geqq \frac{I - E(TDDP)}{DU}) \geqq A_G,$$

where

    DU = differential uncertainty and DU > 0.

Assuming TDDP is normally distributed, $(TDDP - E(TDDP))/DU$
is normally distributed with a mean of zero and a variance
of one. The equation can be restated in an equivalent
form as $[(I - E(TDDP))/DU] \leqq t_{GO}$, where t_{GO} is the fractile
of $(TDDP - E(TDDP))/DU$ associated with A_G.

If $A_G > .5$, then $t_{GO} < 0$, so let $t_{GO} = -|t_{GO}|$, then

$$E(TDDP) \geqq |t_{GO}|DU + I \tag{16}$$

is the equation for the GO constraint level of probability
of achieving the specified rate of return.

    A decision is specified when the total discounted
differential profit generated by the dynamic programming
routine (see equation (9)) and the differential
uncertainty (see equation (10) are plotted on the certainty
equivalence plane. This decision format assumes that the
project has a single measure of uncertainty.
When more than one point lies in the GO area, the "optimum"
point has to be selected by a preference approach.

Subject: New Product

Title: The Theory of First Purchase of New Products

Authors: Frank M. Bass and Charles W. King

Source: A New Measure of Responsibility for Marketing, June Conference Proceedings of the American Marketing Association, Chicago, 1968, 263-72

Summary: Model of the timing of initial purchase of new products based upon behavioral response reflecting interdependent consumer utility functions.

Model:

Let

$$P(T) = p + q/m \, Y(T)$$

where

$Y(T)$ = number of previous buyers at T

$P(T)$ = probability that an initial purchase will be made at T

m = initial purchases of the product

$\left. \begin{array}{c} p \\ q/m \end{array} \right\}$ = constants

The likelihood of purchase at time T given that no purchase has yet been made is

$$\frac{f(T)}{1 - F(T)} = p + qF(T)$$

where $f(T)$ = likelihood of purchase at T,

$F(T) = \int_0^T f(t)dt$,

$F(0) = 0$. Therefore, initial purchases at T are

$$S(T) = mf(T) = [p + q \int_0^T \frac{S(t)dt}{m}] [m - \int_0^T S(t)dt].$$

Since $f(T) = [p + qF(T)][1 - F(T)] = p + (q-p)F(T) - q[F(T)]^2$,

$$F = \frac{(q - pe^{-(T+C)(p+q)}}{q(1 + e^{-(T+C)(p+q)})}$$

$$-C = \frac{1}{p+q} \operatorname{Ln}(q/p)$$

$$F(T) = \frac{(1 - e^{-(p+q)T})}{(q/pe^{-(p+q)T} + 1)}$$

$$f(T) = \frac{(p+q)^2}{p} \frac{e^{-(p+q)T}}{(q/pe^{-(p+q)T} + 1)^2}$$

To find the time at which the sales rate reaches its peak, we differentiate S and thus obtain

$$T^* = \frac{1}{p+q} \operatorname{Ln}(q/p)$$

and if an interior maximum exists, $q > p$.

Subject: New Product

Title: A New Product Growth Model for Consumer Durables

Author: Frank M. Bass

Source: Management Science, Vol. 15, No. 5, 1969, 215-27

Summary: A growth model for the timing of initial purchase of new products. The basic assumption of the model is that the timing of a consumer's initial purchase is related to the number of previous buyers. A behavioral rationale for the model is offered in terms of innovative and imitative behavior. The model yields good predictions of the sales peak and the timing of the peak when applied to historical data. A long-range forecast is developed for the sales of color television sets.

Model:

Let

m = initial purchases of the product over the life of the product.

The likelihood of purchase at time T given that no purchase has yet been made is

$$[f(T)]/[1 - F(T)] = P(T) = p + q/m \, Y(T) = p + q \, F(T),$$

where $f(T)$ is the likelihood of purchase at T and

$$F(T) = \int_{0}^{T} F(t) \, dt, \qquad\qquad F(0) = 0.$$

Since $f(T)$ is the likelihood of purchase at T and m is the total number purchasing during the period for which the density function was constructed,

$$Y(T) = \int_{0}^{T} S(t)dt = m \int_{0}^{T} f(t)dt = m \, F(T)$$

is the total number purchasing in the $(0,T)$ interval. Therefore, sales at T =

$$S(T) = mf(T) = P(T)[m - Y(T)] = [p + q \int_{0}^{T} S(t)dt/m][m - \int_{0}^{T} S(t)dt].$$

Expanding this product we have

$$S(T) = p\,m + (q - p)\,Y(T) - q/m\,[Y(T)]^2.$$

The behavioral rationale for these assumptions are summarized:

    (a) Initial purchases of the product are made by <u>both</u> "innovators" and "imitators," the important distinction between an innovator and an imitator being the buying influence. Innovators are not influenced in the timing of their initial purchase by the number of people who have already bought the product, while imitators are influenced by the number of previous buyers. Imitators "learn," in some sense, from those who have already bought.

    (b) The importance of innovators will be greater at first but will diminish monotonically with time.

    (c) We shall refer to p as the coefficient of innovation and q as the coefficient of imitation.

Since $f(T) =$

$$[p + q\,F(T)][1 - F(T)] = p + (q - p)\,F(T) - q[F(T)]^2,$$

in order to find $F(T)$ we must solve this nonlinear differential equation:

$$dT = dF/(p + (q - p)F - qF^2).$$

The solution is:

$$F = (q - pe^{-(T+C)(p+q)})/q(1 + e^{-(T+C)(p+q)}).$$

Since $F(0) = 0$, the integration constant may be evaluated:

$$-C = (1/(p+q))\mathrm{Ln}(q/p) \quad \text{and} \quad F(T) = (1 - e^{-(p+q)T})/(q/pe^{-(p+q)T} + 1).$$

Then,

$$f(T) = ((p+q)^2/p)[e^{-(p+q)T}/(q/pe^{-(p+q)T} + 1)^2],$$

and

$$S(T) = (m(p + q)^2/p[e^{-(p+q)T}/(q/pe^{-(p+q)T} + 1)^2].$$

To find the time at which the sales rate reaches its peak, we differentiate S,

$$S' = (m/p(p+q)^3 e^{-(p+q)T}(q/pe^{-(p+q)T} - 1))/(q/pe^{-(p+q)T} + 1)^3.$$

Thus,

$$T^* = -1/(p + q) \, Ln(p/q) = 1/(p + q) \, Ln(q/p)$$

and if an interior maximum exists, $q > p$. We note that

$$S(T^*) = (m(p+q)^2/4q \text{ and } Y(T^*) = \int_o^{T^*} S(t)dt = m(q-p)/2q.$$

Since for successful new products the coefficient of imitation will ordinarily be much larger than the coefficient of innovation, sales will attain its maximum value at about the time that cumulative sales is approximately one-half m. We note also that the expected time to purchase, $E(T)$, is $1/q \, Ln((p+q)/p)$.

Subject: New Product

Title: The Analysis of Uncertainty Resolution in Capital Budgeting for New Products

Author: James C. Van Horne

Source: Management Science, Vol. 15, No. 8, April 1969, B-376-386

Summary: A method is developed for analyzing the resolution of uncertainty over time for the individual new product and for the firm's overall product mix. Probability concepts are employed, and it is shown that new products can be evaluated according to their marginal impact upon the resolution of the uncertainty pattern for the firm's total product mix. The analysis is undertaken within a capital-budgeting framework, allowing a GO or NO decision to be reached for the new product under consideration.

Model:

UNCERTAINTY RESOLUTION FOR THE NEW PRODUCT

For most products all expected cash flows are discounted by a risk-adjusted rate to obtain the net-present value of the product, which is

$$NPV = \sum_{t=0}^{n} \frac{A_t}{(1 + k)^t},$$ (1)

where

A_t = expected cash flow in period t

k = risk-adjusted discount rate.

For many new products, a large portion of uncertainty tends to be resolved in the introductory and early growth phases of their lives. The question is how should the expected resolution of uncertainty be measured so that it is useful to management in new-product decisions.

MEASURING RISK FOR THE SINGLE NEW PRODUCT

Assume the expected value of net-present value at time 0 is

$$\overline{NPV} = \sum_{t=0}^{n} \frac{\overline{A}_t}{(1 + i)^t}, \tag{2}$$

where $\overline{A}_t$ is the expected value of net cash flow in period t; and i is the risk-free rate.
The standard deviation at time 0 can be determined by

$$S_0 = [\sum_x NPV_x^2 P_x - (\overline{NPV})^2]^{\frac{1}{2}}, \tag{3}$$

where NPV_x is the net-present value for series x of net cash flows, covering all periods, and P_x is the probability of occurrence of that series.

The statistic used to approximate relative uncertainty at a moment in time is the ratio

$$CV_t = S_t/\overline{NPV}, \tag{4}$$

where S_t represents the "average" standard deviation of the various branches of the probability tree at the end of period t.

COMBINATION OF PRODUCTS

Because overall business risk is important in the valuation of the firm, our concern is with the marginal impact of a new product on the resolution of uncertainty for the firm's entire product mix.

The incremental profitability of a new product is the expected value of net-present value of the combination of existing products plus the new product less the expected value of net-present value for existing products alone.

The standard deviation of the probability distribution of possible total node values for a combination of m products at the end of period t is

$$\sigma_t = [\sum_{j=1}^{m} \sum_{k=1}^{m} \sigma_{jkt}]^{\frac{1}{2}}, \tag{5}$$

where σ_{jkt} is the covariance at time t between possible total node value of products j and k. The covariance is

$$\sigma_{jkt} = r_{jkt} S_{jt} S_{kt}, \tag{6}$$

where r_{jkt} is the expected correlation between possible total node values for products j and k. When $j = k$ in equation (6), σ_{jkt} becomes S_{jt}^2.

Subject: New Product

Title: Forecasting the Demand for New Convenience Products

Author: William F. Massy

Source: D.B. Montgomery and G.L. Urban (eds.), Applications
           of Management Science in Marketing, Englewood Cliffs,
           New Jersey: Prentice-Hall, Inc., 1970, 442-55

Summary: The author's stochastic evolutionary adoption model
           (STEAM) is described. Methods for estimating its
           parameters from panel data covering the first part of the
           introductory period are outlined. A method by which
           the future purchase history of each panel household
           can be simulated and the results projected into a total
           market forecast is reported. Results obtained by applying
           the model and simulation procedure to live data for a
           new product are shown, and compared with the product's
           actual sales during the years after its introduction.

Model:

The Stochastic Model

The stochastic evolutionary adoption model (STEAM)
is designed to predict the probability that a household
will make its next purchase at or before a particular time.
This probability is assumed to depend upon:

1. the household's current depth of trial class
    (i.e., how many purchases it has previously
    made);

2. the time at which the last purchase was made
    (i.e., when the household entered its current
    depth of trial class); and

3. $t - \tau_k$ the time since the household made its
    last purchase (i.e., the amount of time that
    has elapsed since the current depth of trial
    class was entered).

The distribution is assumed to be of a compound Wiebull
form:

Pr[k + 1st purchase is at time $\leq$ t|kth purchase was at τ]

$$= F(t - \tau | \tau) = M \left\{ 1 - \left[\frac{\beta\lambda}{(t - \tau_k)^\lambda + \beta\lambda - 1} \right]^{\alpha\tau\lambda} \right\},$$

$$\tau \geq 1, \ t - \tau \geq 1, \tag{1}$$

where all the parameters are estimated separately for each depth of trial class. The effect of time of the last purchase is determined by the parameter γ and that of the time since the last purchase by λ. The parameters α and β provide estimates of the mean initial purchase probability (i.e., the probability that would exist in the absence of the time effects just discussed): This quantity is given by α/β. They also provide information on how much heterogeneity exists in the population: The variance of the distribution of initial probabilities is α/β^2.

The parameter M is the proportion of households in the given depth of trial class that will eventually make another purchase. That is, a proportion 1 - M of the entrants to the kth depth of trial class will never make the transition to the k + 1st class. Thus, the upper asymptote of the probability distribution given in equation (1) is M rather than unity.

The parameters of STEAM are estimated from consumer panel data by the method of maximum likelihood. Let:

$n_{t\tau}^{(k)}$ = the number of households in the panel who entered the kth DOT class at time τ and make their k + 1st purchase between time t and t + h.

$\bar{n}_\tau^{(k)}$ = the number of households in the panel who entered the kth DOT class at time τ and do not make another purchase during the observation period.

The logarithm of the likelihood function for the data dealing with depth of trial class k is given by :

$$\log L_k = \sum_\tau \{\sum_t n_{t\tau}^{(k)} \log [F(t+h-\tau|\tau) - F(t-\tau|\tau)]$$

$$+ \bar{n}_\tau^{(k)} \log [1 - F(T-\tau|\tau)]\}. \qquad (2)$$

This function must be maximized with respect to the parameters α, β, λ, γ, and M, for each depth of trial class. The maximization can be performed by numerical methods.

The STEAM Projection Simulation

Estimating the parameters of a model from empirical data is only the first step in the forecasting process. It is still necessary to project the model forward in time. Given that we have a nonstationary model whose parameters change with each purchase event, and the time of each event, the technique of Monte Carlo simulation can be applied.

The STEAM projection simulation is of the micro-analytic stochastic type. That is, we simulate individual purchase events for each household in the panel separately rather than trying to predict average rates of activity directly from the formulas of the model.

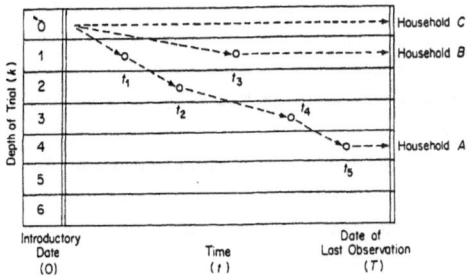

FIGURE 1: Schematic Drawing of Hypothetical Trial and Repeat Purchase Data

Figure 1 depicts the trials and repeat purchases of three hypothetical households. All three households begin their history in depth of trial class (DOT) 0: that is, none of them could have tried the product before its introduction. Household A made its first purchase at time 1, its second at time 2, its third at time 4, and its fourth at time 5. It is, therefore, in DOT class 4 at the end of the observation period. Household B makes only one purchase, at time 3; it is in class 1 at the end of the data. Household C fails to make a purchase; it remains in DOT class 0 throughout the observation period. Suppose that somehow we could know that Household B will make its second purchase at time T + 3, and we want to make predictions about the time of its third purchase. To do this we set the household's τ equal to T + 3 (the time of its most recent purchase) and consult equation (1), using the parameters previously estimated for depth of trial class two. This gives us the probabilities that the third purchase will occur at times $\tau + 1$, $\tau + 2$, etc. If we could know that the third purchase will occur at time T + 10, we would set τ equal to T + 10 and use the parameters estimated for DOT 3 to obtain the probability distribution for the time of the fourth purchase.

Monte Carlo simulation provides a method of generating simulated purchase events to replace the ones cited above, which are, of course, unknown at the time of the simulation. Furthermore, these events can be made to occur in accordance with the laws of market behavior found by fitting STEAM to the available data.

The simulation program begins by calculating the probability distribution for time to next purchase for the first household in the panel, commencing at time T. This can be obtained from the formula:

$$F\left[t - T \mid \tau \quad \text{and} \quad \tilde{P} \text{ in } (\tau, T)\right] = \frac{F(t - T \mid \tau)}{1 - F(T - \tau \mid \tau)}. \qquad (3)$$

This is read as "probability that the next purchase will
occur at or before time t (which is greater than T,
the end of the observation period), given τ (the time of
the last purchase, during the observation period) and
the fact that no purchase occurred between τ and T
(i.e., the household was still in the kth depth of trial
class when the observation period ended)." The F-functions
are obtained from equation (1), using the parameters
corresponding to the household's current depth of trial
class.

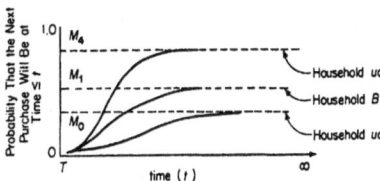

FIGURE 2:   Hypothetical Probability Distributions
for Time to Next Purchase

Equation (3) will produce probability distributions
like those depicted in Figure 2. The sample curves assume
that households in higher depth of trial classes are:
(1) more likely to purchase again--that is, $M_4 > M_1 > M_0$;
and (2) more likely to purchase sooner, given that they
will purchase at all--that is, the curve for Household A
(DOT 4) rises more steeply toward its asymptote than
that for Household B (DOT 1), etc.

The time of the first simulated purchase for any
household can be selected at random from the probability
distribution given by equation (3).

Next the household's τ value is set equal to the time of the simulated purchase, its depth of trial class is incremented, and the time of a second simulated purchase is generated. The process continues until the household falls into the "never purchases again group" or until the simulated time exceeds some preset boundary (e.g., two or three years). Then the program goes on to the next household.

In addition to providing the starting point (τ) for simulating another purchase, each purchase event is entered into a summary table. This table will provide the output for the simulation.

Reference: #110.

Subject: New Product

Title: Dynamic Forecasts of New Product Demand Using a Depth of Repeat Model

Author: Gerald J. Eskin

Source: Journal of Marketing Research, Vol. 10, May 1973, 184-90

Summary: A depth of repeat model that can forecast the demand for new consumer products is developed, an "in-between" model between the STEAM model [#109] and the Fourt-Woodlock model [#97] with some of the desirable attributes of each.

Model:

Definitions

Let

S_t = cumulative sales of a product up to time point t

$R_t(J)$ = cumulative number of consumers repeating at least J times by period t

$R(1)$ = repeat function (consumers made at least 2 purchases)

$R(0)$ = trial function ($J = 0$ means one purchase)

$U_t(J)$ = average units purchased on the Jth repurchase occasion.

Then

$$S_t \equiv \sum_{J=0}^{\infty} R_t(J) U_t(J). \tag{1}$$

Consider $RI_{it}(J)$ as the cumulative fraction that repeat a Jth time by period t given that the $J - 1$ purchase was made during period i. RI differs from R in that it is a fraction rather than an integer and in that it is conditional on the time of entry into the previous state. The algebraic relation between RI and R is:

$$R_t(J) \equiv \sum_{i=1}^{t} RI_{it}(J) \; \Delta R_i(J - 1) \tag{2}$$

$$J = 1, 2, 3, \ldots,$$

where Δ means the first difference operator in time. The equation has no meaning for $J = 0$, thus only positive integers are included in the domain.

Behavioral Assumptions

Behavioral Assumption 1: Geometric-Stretch Function

$$RI_{it}(J) = \alpha_{iJ}(1 - \gamma_{iJ}^{T}) + \delta_{iJ}T \qquad (3)$$

for $T \equiv (t-i) \geqq 0$.

For $T < 0$, RI is defined as 0. For large T the RI function may exceed 1. In such cases RI is defined as the minimum of (3) or the number 1.

(3) is equivalent to that proposed by Fourt-Woodlock as a model of trial. Here we apply it to trial and all repeat levels. It is graphed in Figure 3, where it can be seen that the curve is always below and approaches the ceiling line: $\alpha + \delta T$, the number of consumers entering the repeat class each period being equal to a constant δ plus a fixed proportion of the distance between the current value of the function and the limit line, the proportionality constant being $(1 - \gamma)$. Because γ measures how quickly the curve approaches the limit line, it shall be referred to as the slope factor. The effect of δ is to stretch out the function; hence it will be called the stretch factor.

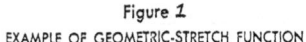

Figure 1
EXAMPLE OF GEOMETRIC-STRETCH FUNCTION

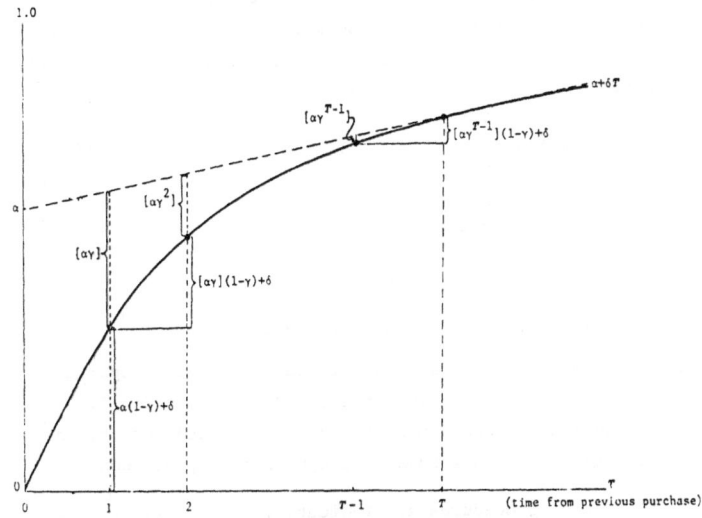

Behavioral Assumption 2: Parameter Stability

    (a) γ's are the same for all repeat levels (except
       trial), i.e., $\gamma_i = \gamma_{iJ}$ for all $J \geq 1$.

    (b) δ's are the same for all repeat levels and time
       of entry cells, i.e., $\delta = \delta_{iJ}$ all $J \geq 1$ and
       all i.

Behavioral Assumption 3: Conversion Proportions

$$RI_{.52}(J) = RI_{.52}(\infty)(1 - \lambda_1^J) \tag{4}$$

for $J = 2,3, \ldots$

Behavioral Assumption 4: Units Per Transaction

$$U_t(J) = \min \begin{bmatrix} \mu_0 + \mu_1 J \\ \overline{U} \end{bmatrix} \tag{5}$$

for $J = 1,2,\ldots$

Characteristics of the Model

Ignoring the stretch factor (i.e., assume $\delta = 0$) and the possibility that everyone will not eventually repeat and differences in time of entry parameters, it can be shown that:

$$\Delta R_t(J) = \begin{pmatrix} t - 1 \\ t - J \end{pmatrix} \gamma^{t-J} (1 - \gamma)^J \tag{6}$$

$$= \begin{pmatrix} T + J - 1 \\ T \end{pmatrix} \gamma^{T}(1 - \gamma)^J$$

for all $\overset{*}{T} \equiv t - J \geq 0$; and all $J \geq 1$,

where the large brackets denote the combination operator; thus (6) is the Negative Binomial Distribution in $\overset{*}{T}$. It can also be shown that for nonsmall values of t, the distribution of R across J is approximately Poisson.

Estimation procedures are indicated.

Reference: #97 and 109.

Subject: New Product

Title: Advertising and Promotion Effects on Consumer Response to New Products

Author: Masao Nakanishi

Source: Journal of Marketing Research, Vol. 10, August 1973, 242-49

Summary: A stochastic model of consumer response to new products is developed which incorporates the effects of overtime variations in advertising and promotion. The model's primary usefulness lies in its ability to generate conditional forecasts of product sales to evaluate alternative marketing programs for new product introduction.

Model:

The model is summarized by the following transition matrix:

TRANSITION MATRIX FOR CONSUMER RESPONSE
TO NEW PRODUCTS

| | States* for n+1st purchase | | | | Conditional purchase propensities |
	PT	T	A	R	
PT	0	θ_0	0	$1 - \theta_0$	$\alpha_0(t)$
States for n^{th} purchase T	0	0	θ_1	$1 - \theta_1$	$\alpha_1(t)$
A	0	0	θ_2	$1 - \theta_2$	$\alpha_2(t)$
R	0	0	0	1	--

*State Description: PT = Pretrial State; T = Trial State (i.e., the state in which the consumer is after the trial (= first) purchase of the new product, but before he decides to accept or reject the product); A = Acceptance State (i.e., the state in which the consumer is after he accepted the product); R = Rejection State (i.e., the state in which the consumer is after he rejected or discontinued the product).

The parameters θ_0 and θ_1 are the probabilities that the consumer does not reject the new product before and after the trial purchase, and θ_2 is the probability that the consumer does not discontinue the product after having accepted it once. The waiting time (or interpurchase time) for transitions between states is a random variable. The probability density function for interpurchase time selected for the model is the nonhomogeneous Poisson process of the form:

$$f_k(t;t') = \alpha_k(t) \exp [- A_k(t;t')],$$

where:

$f_k(t;t')$ = probability density for a purchase at t, given that the consumer is in state k and his last purchase was at t' $(t \geq t')$,

$k = 0,1,2$ (corresponding to state PT, T, and A, respectively),

$\alpha_k(t)$ = conditional purchase propensity function at t, given that the consumer is in state k ($\alpha_k(t) \geq 0$ for all t),

$A_k(t;t') = \int_{t'}^{t} \alpha_k(u) du$ = mean value function at t.

That $f_k(t;t')$ is a valid probability density function can be seen from the fact that $f_k(t;t') \geq 0$ for all $t \geq t'$, and

$$\int_{t'}^{\infty} f_k(u;t') du = \int_{t'}^{\infty} \alpha_k(u) \exp [- A_k(u;t')] du$$

$$= \int_{t'}^{\infty} \exp [- A_k(u;t')] d\{A_k(u;t')\} \, du$$

$$= - \exp [- A_k(u;t')] \Big|_{t'}^{\infty} = 1.$$

The last equality holds under the condition that

$$\lim_{t \to \infty} \exp [- A_k(t;t')] = 0$$

for all k and t. Simply stated, this condition requires
that a consumer who does not reject or discontinue the product
after a purchase must eventually make another purchase, which
is true by definition.   Clearly the most important case for
the marketer of a new product is when $\alpha_k(t)$ is a function of
the level of company marketing activities at .t. For
simplicity's sake, it is assumed that $\alpha_1(t) = \alpha_2(t)$ in the
following analysis. This assumption enables one to apply
this model when the data do not distinguish the first repeat
purchase from other repeat purchases.   For predictive as
well as descriptive purposes, it is necessary to derive
the instantaneous rates of trial and repeat purchases by a
consumer.

Let:

  $q(t)$ = the probability that a consumer has made the
         first purchase (trial) by t (t = 0 for the
         time of new product introduction),

  $\dot{q}(t)$ = $dq(t)/dt$ = the first purchase rate (equal to
         the probability density that a consumer makes
         the transition from state PT to T) at t,

  $r(t;t_1)$ = the cumulative number of repeat purchase incidents
         (not the number of units purchased) by a consumer
         at $t(t \geqq t_1)$, given that his first purchase was
         at t_1,

  $\dot{r}(t;t_1)$ = $dr(t;t_1)/dt$ = the repeat purchase rate (equal to
         the renewal density for repeat purchase incidents)
         at t, given the first purchase at t_1.

It can be shown that :

$$\dot{q}(t) = \theta_0 f_0(t;0) = \theta_0 \alpha_0(t) \exp[-A_0(t;0)]$$
$$\dot{r}(t;t_1) = \theta_1 \alpha_1(t) \exp[-(1-\theta_2) A_1(t;t_1)].$$

(1)

Assuming homogeneity and stability (in terms of the size) of
the consumer population, the expected sales rate at t, s(t),
is given by:

$$s(t) = N[v_1 q(t) + v_2 \int_0^t \dot{q}(t_1) \dot{r}(t;t_1) \, dt_1] \qquad (2)$$

where:

N = the market size, and

v_1, v_2 = average purchase volume per purchase incident for the first and repeat purchases, respectively.

The objective function which is relevant to the analysis of market performance by a new product is given by:

$$PV = \int_0^T [m(t)s(t) - FC(t) - C(x(t))]e^{-\delta t} \, dt$$

where:

PV = the present value of the new product introduction,

m(t) = gross margin per unit,

s(t) = projected sales rate,

FC(t) = fixed cost,

x(t) = vector of marketing activities variables,

C(x(t)) = marketing cost function,

δ = discount rate,

T = appropriate planning horizon.

Subject: Advertising/New Product

Title: An Econometric Analysis of Advertising, Retail Availability,
          and Sales of a New Brand

Author: Leonard J. Parsons

Source: Management Science, Vol. 20, No. 6, February 1974,
          938-47

Summary: Does consumer advertising increase retail availability
          of a new product? Nerlove's procedure is used to
          estimate a dynamic model. Cross-section and time series
          data for one product class are pooled. The managerial
          implications of the model are discussed.

Model: A two-equation model is proposed to describe the market

          mechanism associated with new product introductions.

          The first equation describes the influence of current

          advertising and previous retail availability on current

          retail availability.   Advertising is treated as an

          exogenous variable.   The equation for the retailer

          sector is

$$D_{it} = g_1 D_{it-1} + g_2 A_{it} + u_{it} \quad (i = 1,\ldots,N, \ t = 1,\ldots,T) \quad (1)$$

where D = arc sin of $(\text{distribution})^{1/2}$ expressed in degrees,
distribution is the fraction of retail stores (weighted by
sales volume) handling a product, A = logarithm of
advertising expenditures, u = logarithm of unit sales,
N = number of brands, T = number of time periods. Zero
values for the sales and advertising variables will be
arbitrarily set equal to one before transformation.

The second equation describes the influence of advertising
and retail availability on sales:

$$S_{it} = b_1 D_{it} + g_3 A_{it} + z_{it}. \quad (2)$$

In order to use Nerlove's two-round estimation procedure
["Further Evidence on the Estimation of Dynamic Economic
Relations from a Time Series Data," Econometrica, Vol. 39,
March, 1971 , 359-82], a two-component model is proposed
which includes a brand effect and the remaining effects:

$$u_{it} = \mu_i + v_{it}. \tag{3}$$

A third component representing the time specific effects
is assumed to be absent. The brands were introduced at
different times. The following conditions on the expectations of the two components are assumed:

$$E\mu_i = Ev_{it} = 0, \quad \text{all } i \text{ and } t, \tag{4}$$

$$E\mu_i v_{i't} = 0, \quad \text{all } i, i', \text{ and } t, \tag{5}$$

$$E\mu_i \mu_{i'} = \begin{cases} \sigma_\mu^2, & i = i', \\ 0, & i \neq i', \end{cases} \tag{6}$$

$$Ev_{it} v_{i't'} = \begin{cases} \sigma_v^2, & i = i', \ t = t' \\ 0, & \text{otherwise}. \end{cases} \tag{7}$$

When the disturbances u_{it} are arranged in vector form, first
by brand, then according to period,

$$u = (u_{11}, \ldots, u_{16}, u_{21}, \ldots u_{26}, \ldots, u_{221}, \ldots, u_{226}), \tag{8}$$

the variance-covariance matrix is

$$Euu' = \sigma^2 \begin{bmatrix} A & 0 & \cdots & 0 \\ 0 & A & \cdots & 0 \\ \cdot & \cdot & & \cdot \\ \cdot & \cdot & & \cdot \\ \cdot & \cdot & & \cdot \\ 0 & 0 & \cdots & A \end{bmatrix} = \sigma^2 V = \Omega \tag{9}$$

where

$$\sigma^2 = \sigma_\mu^2 + \sigma_v^2, \quad A = \begin{bmatrix} 1 & \rho & \cdots & \rho \\ \rho & 1 & \cdots & \rho \\ \cdot & \cdot & & \cdot \\ \cdot & \cdot & & \cdot \\ \cdot & \cdot & & \cdot \\ \rho & \rho & \cdots & 1 \end{bmatrix} \quad \text{and} \quad \rho = \sigma_\mu^2/\sigma^2.$$

The parameter ρ is the intra class (brand) correlation coefficient.

Nerlove's two-round estimator involves generalized least squares not using the true value of ρ but the estimator r_c obtained from a least squares regression with individual constant terms. The generalized least squares estimates of the parameter vector are given by

$$\hat{\beta} = (X'V^{-1}X)^{-1} X'V^{-1}y \tag{10}$$

where $[y\ X]$ is the observation matrix and $\sigma^2 V$ is the covariance matrix. An estimate of ρ (for the retail sector equation) is obtained from the least squares regression:

$$D = g_1 D_{-1} + g_2 A + \sum_{i=1}^{22} \mu_i d(i) + v \tag{11}$$

where the μ_i are 22 constants, $d(i)$ is an 132 x 1 vector consisting entirely of zeros except for the ith brand and then of ones. An estimate of σ^2 is obtained from the estimate of the residual variance σ_v^2 and the estimates of the parameters μ_i by

$$\hat{\sigma}^2 = \frac{\sum_i \{\hat{\mu}_i - \sum_i \hat{\mu}_i/22\}^2}{22} + \hat{\sigma}_v^2. \tag{12}$$

As estimate of ρ is obtained as

$$r_c = \frac{\sum_i \{\hat{\mu}_i - \sum_i \hat{\mu}_i/22\}^2}{22}/\hat{\sigma}^2. \tag{13}$$

Profit for a single period t can be expressed as

$$\pi_t(a_1, a_2, \ldots, a_t) = m s_t(a_1, a_2, \ldots, a_t) - a_t \tag{14}$$

where m = the gross margin per unit exclusive of advertising costs, s_t = unit sales in period t, and a_t = advertising expenditures in period t.

Subject: New Product

Title: Ansätze für eine ökonomische Analyse des Konsums neuartiger
          Konsumgüter

Author: Martin J. Beckmann

Source: B. Biervert, K.-H. Schaffartzik, G. Schmölders (eds.),
          Konsum und Qualität des Lebens, Westdeutscher Verlag, 1974,
          237-42

Summary: The demand for new consumer goods is described using
          the example of forecasting the market share of new trans-
          portation modes.

Model: Nach dem Vorgang von Lancaster [#34] werden Güter als
          Bündel von Attributen aufgefasst. Ein Gut konkurriert mit
          bereits bestehenden Alternativen. Es wird nur dann durch
          ein neuartiges Gut ganz verdrängt, wenn es pro Geldeinheit
          von jedem (positiven) Attribut mehr enthält als das
          Vergleichsgut. In allen anderen Fällen muss die Nachfrage
          als Funktion der Attribute gemessen werden. Attribute
          von Verkehrsmitteln sind z.B. die Reisezeiten und die
          Fahrkosten. Sei a_k^i die Menge des Attributes k beim
          Gut i. Wir nehmen die Grenznutzen u_k der einzelnen
          Attribute als konstant an. Der Konsument zieht Gut 1
          dem Gut 2 vor, dann und nur dann wenn

$$u_o + \sum_k a_k^1 u_k > u_o + \sum_k a_k^2 u_k \quad \text{oder} \tag{1}$$

$$\sum_k (a_k^2 - a_k^1) u_k < 0. \tag{2}$$

Die Wahl ist dann völlig deterministisch. Jetzt werde
unterstellt, daß noch weitere Attribute vorhanden sind,
die aber nicht gemessen werden. Dann wird der Nutzen
des Gutes i um die Zufallsvariable i vermehrt, die
den Effekt der nichtgemessenen Attribute wiedergibt. Das
Gut 1 wird dem Gut 2 nicht mit Gewißheit vorgezogen
(oder umgekehrt), sondern mit einer gewissen
Wahrscheinlichkeit. Die Wahrscheinlichkeit einer Präferenz
für Gut 1 gegenüber Gut 2 ist gleich der Wahrscheinlichkeit,
daß

$$\varepsilon_2 - \varepsilon_1 < \sum_k (a_k^1 - a_k^2) u_k, \qquad \text{also}$$

$$pr(1 > 2) = pr(\varepsilon_2 - \varepsilon_1 < \sum_k (a_k^1 - a_k^2) u_k). \qquad (3)$$

Andererseits ist die Wahrscheinlichkeit der Präferenz für Gut 1 gleich dem Marktanteil des Gutes 1

$$pr(1 > 2) = m_1. \qquad (4)$$

Nun werde angenommen, daß die ε_i normal verteilt sind mit Mittelwerten $\mu(\varepsilon_1) = 0$

$$\mu(\varepsilon_2) = \mu_0. \qquad (5)$$

Sei σ die Varianz von $\varepsilon_1 - \varepsilon_2$ und N die kumulierte standardisierte Normalverteilung. Dann ist

$$m_1 = N\left(\frac{\sum_k (a_k^1 - a_k^2) u_k - \mu_0}{\sigma}\right).$$

Approximiert man die Normalverteilung mit der Logistik

$$N_{(x)} = \frac{1}{1 + e^{-x}}, \text{ dann ist} \qquad (6)$$

$$m_1 = \frac{1}{1 + \exp[\mu_0 + \sum_k (a_k^2 - a_k^1) u_k]}$$

oder

$$\log\left(\frac{1}{m_1} - 1\right) = \frac{\mu_0}{\sigma} + \sum_k (a_k^2 - a_k^1) \frac{\mu_k}{\sigma}. \qquad (7)$$

Die Koeffizienten $\frac{\mu_0}{\sigma}$ und $\frac{\mu_k}{\sigma}$ lassen sich jetzt durch lineare Regression schätzen für Güterpaare, zwischen denen eine binäre Wahlmöglichkeit besteht.

Der Konsumzweck kann berücksichtigt werden, indem man den konstanten Teil u_0 vom Zweck p abhängig macht:

$$\mu_o = \mu_o^p.$$

Man kann nun auch interpersonelle Unterschiede in den Präferenzen berücksichtigen, indem man die Grenznutzen u_i als Zufallsvariable, die über die Bevölkerung verteilt sind, auffaßt.

Sei

$$Eu_i = \mu_i$$

$$u_i = \mu_i + \eta_i .$$

Die Zufallsvariablen η_i haben dann Erwartungswert 0. Nun ist

$$pr\ (1 > 2) = pr\ (\varepsilon_2 - \varepsilon_1 + \sum_k (a_k^2 - a_k^1)\ \eta_k$$

$$< \sum_k (a_k^1 - a_k^2)\ \mu_k. \tag{8}$$

Die Zufallsvariablen η_i mögen nun ebenfalls normal verteilt sein, wobei sie möglicherweise miteinander korreliert sind. Dann ist die Zufallsvariable in Gleichung (8) wieder normal verteilt. Ihr Mittelwert ist wieder μ_o, ihre Varianz sei σ.

Dann erhalten wir (7). Der einzige Unterschied liegt in der Interpretation der Varianz.

Reference: #34.

Subject: New Product

Title: PERCEPTOR: A Model for Product Positioning

Author: Glen L. Urban

Source: Management Science, Vol. 21, No. 8, April 1975, 858-71

Summary: The model is structured as a trial and repeat process that produces an estimate of long-run share for a new frequently purchased consumer product. Physical and psychological product attributes are linked to the trial and repeat probabilities through multidimensional scaling procedures. Perceptual maps of existing brands with ideal brand positionings are derived. The product design determines the new brand's position in the perceptual space, and the distance from the new brand to the ideal brand specifies its probability of purchases. Measurement and estimation procedures are discussed.

Model: Let

$$m = ts, \qquad (1)$$

where

    m = long-run market share,

    t = fraction of the target group who ever try the new brand, $0 \leqq t \leqq 1.0$,

    s = share of purchases of new brand among those who have ever tried the brand, $0 \leqq s \leqq 1.0$.

The ultimate trial is defined as:

$$t = qwv, \qquad (2)$$

where

    q = ultimate probability of trial given awareness and availability,

    w = long-run aided awareness of new brand,

    v = long-run per cent of all commodity availability of new brand (i.e., per cent of stores carrying brand weighted by sales volume of store).

The market share of those who have used the brand is modeled as the equilibrium of a two-state Markov process:

$$s = p_{21}/(1 + p_{21} - p_{11}) \qquad (3)$$

where for those who have tried the new brand:

P_{ij} = Probability of purchase of brand j at next purchase opportunity when brand i was purchased last, $0 \leq P_{ij} \leq 1.0$,

i or j = 1 refers to new brand,

i or j = 2 refers to all other brands.

Let

x_{by} = coordinates of brand b on dimension y for the perceptual map of those who have not tried our brand, but who are aware of the concept (b = 1,2, ..., B, where B = new brand; y = 1,2, ..., Y).

I_y = coordinates of average ideal point of dimension y for map of those who have not tried our brand but are aware of the concept.

The coordinates of each brand could be derived by either nonmetric scaling of similarity data or factor or discriminant analysis of brand rating data. For maps developed by factor analysis, the following relationship will result:

$$x_{by} = \sum_{a=1}^{A} f_{ya} r_{ba}, \qquad (4)$$

where

f_{ya} = factor score coefficient for dimension y and attribute scale a (a = 1,2, ..., A),

r_{ba} = standardized average rating of brand b on scale a.

The "distances" on the map for those people who are aware of the new concept but have not yet tried our product are defined as:

$$d_B^2 = \sum_{z=1}^{Z} h_z (x'_{bz} - I'_z)^2, \qquad (5)$$

x'_{bz} = rotated coordinates of brand b on dimension z (z = 1,2, ..., Z), where:

$$x'_{bz} = \sum_{y=1}^{Y} x_{by} T_{yz}, \qquad (5a)$$

T_{yz} = cosine rotational transform of dimension y to z,

x_{by} = coordinates of brand b on dimension y on map for people aware of new concept,

I'_z = rotated coordinates of ideal point on dimension z where:

$$I'_z = \sum_{y=1}^{Y} I_y \, T_{yz}. \tag{5b}$$

I_y = coordinates of ideal point on dimension y on map for people aware of new concept,

h_z = importance weights for dimension z (h_z may be negative).

Equation (5) is the distance of PREFMAP phase I. (J.D. Carrol, "Individual Differences and Multidimensional Scaling," in R.N. Shephard, A.K. Romney, and S. Nerlove (eds.), Multi-dimensional Scaling: Theory and Application in the Behavioral Sciences, New York: Academic Press, 1972, 105-75.) In PREFMAP phase II, no idiosyncratic rotation is done and in phase III axes are equally weighted, so in this case the distance equation is simplified to the Euclidean form. The repeat purchase probability is

$$p_{11} = \tilde{\alpha}_0 + \tilde{\alpha}_1 \tilde{d}^2_B, \tag{6}$$

where

p_{11} = probability of repeat purchase if new brand was purchased last. p_{11} is constrained to the range $0 \leq p_{11} \leq 1.0$,

$\tilde{\alpha}_0, \tilde{\alpha}_1$ = coefficients empirically determined,

$\tilde{d}^2_B$ = distance squared from ideal point to new brand after use (for definition see (5) with coordinates of brand on perceptual map of those who have used the brand).

The new product share is obtained from brands proportional to their appearance in consumer evoked sets and inversely proportional to their distance from the new brand.

$$k_b = m((e_b/D_{dB}^2/ \sum_{B=1}^{B-1} (e_b/D_{dB}^2)), \qquad (7)$$

k_b = loss in market share of existing brand b,

m = market share of new brand,

e_b = fraction of people who have brand b in their evoked set,

D_{dB}^2 = distance squared from brand b to new brand B in users map.

The measurement methodology and estimation procedures are described.

Reference: #117.

Subject: Product Mix

Title: A Mathematical Modeling Approach to Product Line Decisions

Author: Glen L. Urban

Source: Journal of Marketing Research, Vol. 6, February 1969, 40-47

Summary: A model of the interaction between products for normative strategy recommendations is described, an a priori product line model for finding the best marketing mix for each product in a line. The model includes aggregate product group marketing mix, product interdependency, and competitive brand effects.

Model:

Submodels

I. Aggregate Product Class Marketing Mix Effects

$$X_{jI} = a P_{jI}^{EPI} \; A_{jI}^{EAI} \; D_{jI}^{EDI}, \qquad (1)$$

where

X_{jI} = industry sales of Product j,

a = scale constant,

P_{jI} = average price level of all brands in product group j,

A_{jI} = total advertising of all brands in product group j,

D_{jI} = total distribution level for all brands in product group j,

EPI = industry price elasticity for Product j,

EAI = industry advertising elasticity for Product j,

EDI = industry distribution elasticity for Product j.

This function captures marketing mix effects and allows nonlinearity in response to marketing variables. It is estimable by linear logarithmic regression.

II. Product Class Interdependencies

In considering group effects the variables relate to other groups:

$$b = P_{IM}^{CP_j M} \; A_{IM}^{CA_j M} \; D_{IM}^{CD_j M}, \qquad (2)$$

where

b = scale constant,

P_{IM} = average price of product group M,

A_{IM} = total advertising level for product group M,

D_{IM} = total distribution level for product group M,

CP_jM = cross price elasticity for Products j and M,

CA_jM = cross advertising elasticity for Products j and M,

CD_jM = cross distribution elasticity for Products j and M.

The group marketing mix and intergroup product inter-dependencies can be combined to specify the total sales of one product class as:

$$X_{j1} = kP_{jI}^{EPI} A_{jI}^{EAI} D_{jI}^{EDI} (\prod_{M} P_{IM}^{CP_jM} A_{IM}^{CA_jM} D_{IM}^{CD_jM}), \quad (3)$$

where $\prod_{M}$ is product sum over M, M $\neq$ j and k is scale constant.

III. <u>Intra-Group Competitive Brand Effects</u>

Market share expression using (1) and including relative mix effects:

$$\text{market share for Product } j \text{ in Firm } 1 = \frac{P_{1j}^{SP1} A_{1j}^{SA1} D_{1j}^{SD1}}{\sum_i P_{ij}^{SPi} A_{ij}^{SAi} D_{ij}^{SDi}} \quad (4)$$

where

P_{ij} = price of Product j by Firm i

A_{ij} = advertising level for Product j by Firm i

D_{ij} = distribution level for Product j by Firm i

SPi = competitive price sensitivity for Firm i
    and Product j

SAi = competitive advertising sensitivity for Firm i
    and Product j

SDi = competitive distribution sensitivity for Firm i
    and Product j.

Demand, Cost and Profit Models for a Firm

The three submodels can be combined into one equation to describe the sales of one brand of a product class. The sales of Firm l's brand in Product j class is:

$$x_j = kP_{jI}^{EPI} A_{jI}^{EAI} D_{jI}^{EDI} [\prod_M P_{IM}^{CP_jM} A_{IM}^{CA_jM} D_{IM}^{CD_jM}] \left[\frac{P_{1j}^{SP1} A_{1j}^{SA1} D_{1j}^{SD1}}{\sum_i P_{ij}^{SPi} A_{ij}^{SAi} D_{ij}^{SDi}} \right], \quad (5)$$

where $\prod\limits_M$ is product sum over M, $M \neq j$, and other notations as previously defined. Given constant direct and cross elasticities and sensitivities, this equation represents the demand for one product of a firm's product line. This formulation could be extended to include more than three marketing variables by specifying the appropriate direct and cross elasticities and sensitivities.

As to costs, if there are production interdependencies, a linear programming model designed to minimize the cost of producing specified quantities of the products could be used. Successive runs of this model or cost records could provide the data for estimating an interdependent cost function such as:

$$TVC_j = AVC_j(x_j) \prod_M (x_M)^{CC_jM}, \quad (6)$$

where

$\quad TVC_j$ = total variable cost of producing the firms' brand of Product j

$\quad AVC_j$ = average variable cost function for the firm's brand of Product j, if produced independently of other products

$\quad x_j$ = quantity of brand of Product j produced

$\quad x_M$ = quantity of brand of Product M produced, $(M \neq j)$

$\quad CC_jM$ = cross cost elasticity of firm's brands Products j and M, $(M \neq j)$.

Subtracting the variable cost and fixed production, advertising and distribution costs from the total revenue will yield total profit.

Subject: Sales Forecasting

Title: Forecasting Sales by Exponentially Weighted Moving Averages

Author: Peter R. Winters

Source: Management Science, Vol. 6, 1960, 324-42

Summary: Models of the exponential forecasting system are pre-
          sented, along with applications to cooking utensils,
          paint and cellar excarvation.

Model: 1. The Simplest Exponential Model

For the prediction of the expected sales for a product which
has no definite seasonal pattern and no long-run trend
the following procedure is proposed: take a weighted average
of all past observations and use this as a forecast of
the present mean of the distribution, as

$$\tilde{S}_t = AS_t + (1-A)\tilde{S}_{t-1} \tag{1}$$

where

S_t = actual sales during the tth period

$\tilde{S}_t$ = forecast of expected sales in the tth period

$0 \leqq A \leqq 1$.

Then

$$\tilde{S}_{t-1} = AS_{t-1} + (1-A)\tilde{S}_{t-2},$$

so that

$$\tilde{S}_t = AS_t + A(1-A)S_{t-1} + (1-A)^2\tilde{S}_{t-2}. \tag{2}$$

Continuing this process, $\tilde{S}_t$ can be expressed explicitly in
terms of all the past observations of sales:

$$\tilde{S}_t = A \sum_{n=0}^{M} (1-A)^n S_{t-n} + (1-A)^{M+1}\tilde{S}_b \tag{3}$$

where $\tilde{S}_b$ is the beginning value of $\tilde{S}$. M is the number of
observations in the series up to and including the current
period, t. Even for relatively small A, if M is large
enough, $(1-A)^{M+1}$ becomes very small, and the last term
can be ignored.

The expected sales in any period are:

$$E(\tilde{S}_t) = E(S)A \sum_{n=0}^{M} (1-A)^n + (1-A)^{M+1} \tilde{S}_b. \qquad (4)$$

For large M, and most A, $(1-A)^{M+1} \tilde{S}_b$ approaches zero. Under these same conditions $A \sum_{n=0}^{M} (1-A)^n$ approaches one. Thus

$$E(\tilde{S}_t) \simeq E(S)$$

with the degree of approximation depending on the values of M and A.

2. Forecasting with Ratio Seasonals

Often the amplitude of the seasonal pattern is proportional to the level of sales. Using the multiplicative, or ratio, seasonal effect, we obtain

$$\tilde{S}_t = A \frac{S_t}{F_{t-L}} + (1-A)\tilde{S}_{t-1}, \qquad 0 \leq A \leq 1 \qquad (5)$$

for the estimate of the expected deseasonalized sales rate in period t, and

$$F_t = B \frac{S_t}{\tilde{S}_t} + (1-B)F_{t-L}, \qquad 0 \leq B \leq 1 \qquad (6)$$

for the current estimate of the seasonal factor for period t, where L is the periodicity of the seasonal effect. $\tilde{S}_t$ is a weighted sum of the current estimate obtained by deseasonalizing the current sales, $\tilde{S}_t$, and last period's estimate, $\tilde{S}_{t-1}$, of the smoothed and seasonally adjusted sales rate for the series. The value of $\tilde{S}_t$ from (5) is then used in forming a new estimate of the seasonal factor in (6). A forecast of expected sales T periods into the future would be

$$S_{t,T} = \tilde{S}_t F_{t-L+T}, \qquad T \leqq L. \tag{7}$$

The weighted averages in (5) and (6) may be written in terms of past data and initial conditions:

$$\tilde{S}_t = A \sum_{n=0}^{M} (1-A)^n \frac{S_{t-n}}{F_{t-L-n}} + (1-A)^{M+1} \tilde{S}_b \tag{8}$$

and

$$F_t = B \sum_{n=0}^{J} (1-B)^n \left(\frac{S_{t-nL}}{\tilde{S}_{t-nL}}\right) + (1-B)^{J+1} F_{bt}, \tag{9}$$

where $\tilde{S}_b$ is the initial value of $\tilde{S}$, and F_{bt} is the initial value of F for the period in question. J is the largest integer less than or equal to M/L.

3. Forecasting with Ratio Seasonals and Linear Trend

First,

$$\tilde{S}_t = A \frac{S_t}{F_{t-L}} + (1-A)(\tilde{S}_{t-1} + R_{t-1}), \tag{10}$$

where R_{t-1} is the most recent estimate of the additive trend factor.
Then,

$$F_t = B \frac{S_t}{\tilde{S}_t} + (1-B) F_{t-L}, \tag{11}$$

$$R_t = C(\tilde{S}_t - \tilde{S}_{t-1}) + (1-C) R_{t-1}, \tag{12}$$

and

$$S_{t,T} = [\tilde{S}_t + T R_t] F_{t-L+T} \qquad T = 1, 2, \ldots, L. \tag{13}$$

Subject: Sales Forecasting

Title: Use of Consumer Panels for Brand-Share Prediction

Authors: J.H. Parfitt and B.J.K. Collins

Source: Journal of Marketing Research, Vol. 5, May 1968, 131-45

Summary: A method of predicting the market share for newly launched brands and the future equilibrium share of established brands after major proportional activity is described. The method is dependent on the continuous purchasing data obtainable from consumer panels and is developed from case histories.

Model:

Defining t as a discrete time variable, measured after the launch of the new product (Brand α) within a defined product (Field A), then:

$$P(t) = \frac{\sum\limits_{r=0}^{t} N(r)}{\sum\limits_{r=0}^{t} F(r)} \times 100.$$

where

> $P(t)$ = penetration of Brand α within Field A at time t in %,
>
> $N(t)$ = number of new buyers of Brand α introduced at t,
>
> $F(t)$ = number of new buyers of Field A at t,
>
> P = ultimate penetration of Brand α within Field A = $\lim\limits_{t \to \infty} P(t)$.

$$R(t,r,s) = \frac{M(t,r,s)}{E(t,r,s)} \times 100,$$

where

> $R(t,r,s)$ = repeat-purchasing rate for Brand α within Field A at t, for s beginning at r after the first purchase of Brand α in %,

$M(t,r,s)$ = amount of Brand α purchased in period s beginning at time r after the first purchase of Brand α, aggregated over all buyers of Brand α, based on data available at t,

$E(t,r,s)$ = amount of Product field A purchased in s beginning at r after the first purchase of Brand α, aggregated over all buyers of Brand α, based on data available at t,

$R(t,s)$ = ultimate repeat-purchasing rate for Brand α within Product Field A at time t (based on a Purchasing interval s)

$= \lim\limits_{t \to \infty} R(t,r,s)$.

$$B(\alpha,A,t) = \frac{W(\alpha,t)}{W(A,t)},$$

where

$B(\alpha,A,t)$ = buying rate factor for Brand α within Field A at t,

$W(\alpha,t)$ = amount of Field A purchased in s beginning at $t-s$, by all buyers of Brand α,

$W(A,t)$ = amount of Field A purchased in s beginning at $t-s$, by all buyers of Product field A.

Then

% brand-share prediction

$$= \frac{P\% \times R(t,s)\% \times B(\alpha,A,t)}{100}.$$

It is advantageous to produce repeat-purchasing rate analyses by segmented buying groups defined by the period of their first purchase of Brand α.

% Brand-share prediction

$$= \sum_{i=1}^{h} \frac{P_i\% \times R_i(t,s)\% \times B_i(\alpha,A,t)}{100},$$

where $P_i\%$, $R_i(t,s)\%$, and $B_i(\alpha,A,t)$ represent the ultimate penetration, repeat-purchasing rate and buying rate factor for the ith buying group, of which there are h.

For a penetration model the modified exponential form proved the best choice.

Let

$$\Delta P(t) = \frac{P(t+1) - P(t-1)}{2},$$

$$\Delta P(t) = a(K - P(t)) + \varepsilon(t),$$

where

K = ultimate penetration,

a = rate of growth parameter,

$\varepsilon(t)$ = random error associated with the measurements at t.

The deterministic part of this stochastic model may be reduced to:

$$P(t) = K(1 - e^{-a \cdot t}),$$

where e is the exponential function.

Reference: #114.

Subject: Sales Forecasting

Title: Zur Problematik von Wendepunkten in Trendfunktionen,
            dargestellt an einem Modell zur Prognose von Marktanteilen

Authors: Werner Kroebel-Riel and Sighard Roloff

Source: Zeitschrift für betriebswirtschaftliche Forschung, Vol. 24,
            1972, 294-300

Summary: A method for predicting the market share of a product is
            described. Sometimes a trend function can be separated into
            two or more time series with no inflection points. After
            computing the trend lines for the time series, the trend
            curve and its point of inflection can be predicted.

Model: <u>Die Zerlegung einer Trendfunktion</u>

        Bei Parfitt und Collins [#116] wird der Marktanteil
eines Produktes auf zwei Bestimmungsgrößen zurückgeführt:
Auf die kumulative Marktdurchdringung y_1 und auf die
Wiederholungskaufrate y_2, die auf Perioden bestimmter Länge
bezogen werden. Im folgenden wird dagegen ein kontinuierlicher
Verlauf unterstellt.

        Für den Verlauf der kumulativen Marktdurchdringung läßt
sich die Funktion

$$y_1(t) = d(1 - e^{-ft}) \qquad d, f > 0 \qquad (1)$$

unterstellen, mit t als Zeitvariable. Dabei legt d den
asymptotischen Verlauf und f die Krümmung, d. h. die
Geschwindigkeit der Marktdurchdringung fest.
Für die Wiederholungskaufrate wird folgender Verlauf
angenommen:

$$y_2(t) = \frac{a}{t+b} + c \qquad a, b > 0; \quad c \geq 0. \qquad (2)$$

Dabei verändert a die Krümmung der Hyperbel. Der Parameter b
bestimmt die Lage der Polstelle von y_2. Und c legt den
asymptotischen Verlauf von y_2 fest, denn es gilt $\lim_{t \to \infty} y_2(t) = c$.
Die Marktanteilskurve berechnet sich als Produkt der
Marktdurchdringung y_1 und der Wiederholungskaufrate y_2:

$$y(t) = \frac{ad}{t+b}(1 - e^{-ft}) + cd(1 - e^{-ft}). \qquad (3)$$

Eine analytische Untersuchung der Funktion y ergibt folgende
zwei Eigenschaften:

(a) y(t) besitzt für t $\geq$ 0 genau ein Maximum t° > 0 und
    sonst keine weiteren Extrema;

(b) y(t) besitzt für t $\geq$ 0 genau einen Wendepunkt t°° > 0.

Wenn die Einzelfunktionen frühzeitig ein stabiles Verhalten
aufweisen, kann man die Prognose für den Verlauf der
Gesamtfunktion schon zu einem Zeitpunkt vornehmen, zu dem
man einerseits noch keine ausreichenden Daten für die
Schätzung der Gesamtfunktion hat und andererseits die
Gesamtfunktion selbst noch kein stabiles Verhalten aufweist.

Die Anwendung der Zerlegung bei unterschiedlichen
Voraussetzungen für die Schätzung der Trendfunktionen

Es gibt generell zwei Möglichkeiten, um zu den für
eine Extrapolation benötigten Funktionen zu kommen:

(a) Man kennt noch keine erprobten Funktionen, auf die man
    sich bei der Extrapolation von Zeitreihen stützen kann.
    Man weiß aber, daß die zu extrapolierenden Reihen
    Wendepunkte haben müssen. Dann ist die Bestimmung einer
    Extrapolationsfunktion für diese Reihen in der Zeit
    vor Eintritt des Wendepunktes nicht möglich. Wenn es in
    solchen Fällen möglich ist, die Zeitreihe für eine
    Größe G in die Funktionen zweier oder mehrerer einzelner
    Zeitreihen zu zerlegen, die den Verlauf der
    Bestimmungsgrößen g für G wiedergeben, dann läßt sich G
    entsprechend dem dargestellten Verfahren durch
    Extrapolation dieser einzelnen Zeitreihen für g voraussagen.
    Für diese Prognoseverfahren hat man anzunehmen, daß die
    einzelnen Zeitreihen keinen Wendepunkt haben und daß sie
    hinreichend genau durch Extrapolationsfunktionen approximiert
    werden können.

(b) Durch Erfahrungen und durch Rückgriff auf empirische
    Gesetzmäßigkeiten ist man in der Lage, von vornherein
    Funktionstypen für die Extrapolation zu unterstellen.
    Dann kommt es darauf an, durch Parameterschätzungen
    möglichst genau die speziellen Prognosefunktionen für
    die einzelnen Zeitreihen zu bestimmen.

Da schon die Schätzungen für die Parameter von Einzelfunktionen
außerordentlich schwierig sind, ist eine Schätzung der

Parameter von y (höhere Zahl der Parameter) noch
schwieriger, selbst wenn man y in folgender Form darstellen
würde (α: = ad + bdc; β: = cd; γ: = b und δ: = f):

$$y(t) = \frac{(\alpha + \beta t)}{(\gamma + t)} \ (1 - e^{-\delta t}), \quad \alpha, \gamma, \delta > 0; \quad \beta \geq 0. \quad (4)$$

Generell läßt sich sagen, daß die Parameter für die durch
Zerlegung gewonnenen Einzelfunktionen erheblich einfacher
geschätzt werden können als die Parameter der zusammengesetzten
Trendfunktion. Aus diesem Grund ist die Zerlegung auch
dann vorteilhaft, wenn für den Trendverlauf von vornherein
ein bestimmter Funktionstyp vorausgesetzt und für die
Extrapolation genutzt werden kann.

Subject: Sales Forecasting

Title: The RAS Method for Two-Dimensional Forecasts

Author: Baruch Lev

Source: Journal of Marketing Research, Vol. 10, May 1973, 153-59

Summary: A prediction method for two-dimensional forecasts similar
          to the RAS method developed for the adjustment of input-
          output coefficients is presented and applied to data of a
          public utility firm, a retail mail order business, steel
          products, and metal products for agriculture.

Model:

It is assumed that the firm's staff predicts only the
marginal totals, e.g., sales predictions by product type
(all areas combined) and sales predictions by geographic
areas (all products combined). Both sets of predictions
add up to the same number (i.e., predicted total sales), so
that the total number of predictions needed is $m + n - 1$.
The objective is to generate a two-dimensional set of
forecasts that corresponds to the given marginal totals.
Since these totals are considered fixed, we may confine
ourselves to the predictions of proportions (fractions). We,
therefore, predict a bivariate array $[q_{ij}]$, where q_{ij} stands
for sales of the ith product in the jth area, measured as a
fraction of total sales of all products in all areas.
The marginal proportions will be indicated by dots as subscripts so that $q_{i.}$ and $q_{.j}$ stand for the total proportions
of sales of the ith product type and sales in the jth area,
respectively. Predictions will be indicated by hats; hence
$\hat{q}_{ij}$ is the prediction of q_{ij}, while $\hat{q}_{i.}$ and $\hat{q}_{.j}$ are the
marginal predictions prepared by the firm. The $\hat{q}_{ij}$'s should
satisfy the following constraints:

$$\sum_{j=1}^{n} \hat{q}_{ij} = \hat{q}_{i.}, \tag{1}$$

$$i = 1, \ldots, m,$$

$$\sum_{i=1}^{m} \hat{q}_{ij} = \hat{q}_{.j}, \qquad (2)$$

$j = 1, \ldots, n.$

Assume that an array $[p_{ij}]$ of realized corresponding fractions is available from an earlier period. The RAS method uses this array to obtain the $\hat{q}_{ij}$'s as follows:[1]

$$\hat{q}_{ij} = x_i p_{ij} y_j \qquad (3)$$

$i = 1, \ldots, m,$

$j = 1, \ldots, n$

where $x_1, \ldots, x_m$ and $y_1, \ldots, y_n$ are adjusted by means of the constraints (1) and (2):

$$x_i = \hat{q}_{i.} / \sum_{j=1}^{n} p_{ij} y_j \qquad (4)$$

$i = 1, \ldots, m,$

$$y_j = \hat{q}_{.j} / \sum_{i=1}^{m} p_{ij} x_i \qquad (5)$$

$j = 1, \ldots, n.$

---

[1] The RAS method was suggested for the adjustment of input coefficients of an earlier period by means of known marginal totals of these coefficients in a later period. Specifically, the adjustment of coefficients is based on the following three sets of data: (a) the input-output matrix for a base year, (b) the total output of commodities in a later year, and (c) the total of intermediate outputs of commodities and intermediate inputs into commodities in the later year. The A in the RAS stands for the past matrix of coefficients, and R and S for the diagonal matrices by which the former is pre- and postmultiplied. They correspond with p_{ij}, x_i, and y_j, in (3), respectively.

There are $m+n$ variables, x_i, y_j, but it follows from (3) that the predictions, $\hat{q}_{ij}$'s, are unaffected when all x's are multiplied by an arbitrary positive number c, and when at the same time all y's are divided by c. Hence, there are only $m + n - 1$ variables to be adjusted; this equals the number of constraints. Equations (4) and (5) are nonlinear in the x's and y's, but the solution can be obtained by iterative procedure.

Subject: Sales Forecasting

Title: Using Laboratory Brand Preference Scales to Predict Consumer Brand Purchases

Authors: Edgar Pessemier, Philip Burger, Richard Teach, and Douglas Tigert

Source: Management Science, Vol. 17, No. 6, February 1971, B-371-386

Summary: Laboratory measures of brand preferences and survey measures of demographic, media exposure, attitudinal, activity and opinion characteristics of individual consumers have been combined to predict brand purchases in the market. Brands in the tooth paste, liquid household cleaner and cake mix product categories were employed in a set of laboratory experiments. Preference scales derived from the experiments are used in three separate models to predict the subject's purchases recorded in seven months of diary data.

Model: The three parameter model uses all the experimental data available to produce the predictions. The data include:

1. The adjusted scale value for subject i and brand k, $ASV_{i,k}$.

2. The observed relative intransitivity for subject i, RD_i.

3. The expected random error RE.

4. The relative frequency, across all subjects, with which each brand was a first choice selection, FFC_k. This vector is an aggregate experimental measure of brand preference analogous to market share data derived from market purchases.

The expected relative frequency of purchase is

$$RFP_{i,k} = (ASV_{i,k})^{\beta} / \sum_{k} (ASV_{i,k})^{\beta}.$$

The error ratio for subject i is:

$$ER_i = (RD_i/RE)^{\gamma}, \quad RD_i \leq RE, \quad \text{otherwise } ER_i = 1.$$

The three parameter model is the following:

$$FNM_k = \alpha(UP) + (1.0 - \alpha)FFC_k$$

    = predicted purchase probabilities assigned to each brand which is not derived from the subject's scale data. A weighted average of the uniform and experimental first choice probabilities for brand k.

$$ARFP_{ik} = (ER_i)(FNM_k) + (1.0 - ER_i)RFP_{i,k}$$

The third parameter, α, is introduced to permit weighting the FFC and UP predictors.

Subject: Facility Location

Title: A Model for Scale of Operations

Authors: Edward H. Bowman and John B. Stewart

Source: The Journal of Marketing, Vol. 20, No. 3, January 1956,
          242-47

Summary: The problem is stated as, "How large a territory should
          be served by a warehouse to result in a minimum total cost
          for warehousing, trucking between plant and warehouses,
          and delivery from warehouse to customers?" A model of
          the problem is built around a measure of effectiveness
          and includes those variables which most appear to influence
          this measure of effectiveness. The coefficients in the
          model are chosen by multiple regression.

Model:

The Warehouse District Model

It has been determined that the cost per dollar's
worth of goods distributed (the warehouse efficiency) is
equal to certain costs which vary inversely with the
volume plus certain costs which vary directly with the
square root of the area plus certain costs which were
affected by neither of these variables.

$$C = a + \frac{b}{V} + c \sqrt{A} \tag{1}$$

C = cost (within the warehouse district) per
    dollar's worth of goods distributed--the measure
    of effectiveness

V = volume of goods in dollars handled by the
    warehouse per unit of time

A = area in square miles served by the warehouse

a = cost per dollar's worth of goods distributed
    independent of either the warehouse's volume handled
    or area served

b = "fixed" costs for the warehouse per unit of time, which divided by the volume will yield the appropriate cost per dollar's worth distributed

c = the cost of the distribution which varies with the square root of the area; that is, costs associated with miles covered within the warehouse district such as gasoline, truck repairs, driver hours, etc.

After determining for each warehouse C, V, and A by least-squares multiple regression it is possible to determine the values of the parameters a, b, and c.

Minimize:

$$\sum_{i=1}^{i=N} [C_i - (a + \frac{b}{V_i} + c \sqrt{A_i})]^2$$

where C_i, V_i, and A_i indicate actual values in a given (the ith) branch warehouse operation and the $\sum$ indicates a sum total of all (N) warehouses. Actually a set of three simultaneous equations are solved for a, b, and c.

$$\frac{\partial \sum_{1}^{N} [C_i - (a + \frac{b}{V_i} + c \sqrt{A_i})]^2}{\partial a} = 0$$

$$\frac{\partial \sum_{1}^{N} [C_i - (a + \frac{b}{V_i} + c \sqrt{A_i})]^2}{\partial b} = 0$$

$$\frac{\partial \sum_{1}^{N} [C_i - (a + \frac{b}{V_i} + c \sqrt{A_i})]^2}{\partial c} = 0$$

The sales density (K), expressed in dollar volume per square mile of area, is

$$K = \frac{V}{A}. \tag{2}$$

Therefore, V = KA, and it is possible to substitute this expression for V in the original model, giving

$$C = a + \frac{b}{KA} + c\sqrt{A} \tag{3}$$

where a, b, and c are now specific figures determined from the multiple regression calculation.

Cost Minimization With Respect to Area

By differentiation

$$A = \left(\frac{2b}{cK}\right)^{2/3} \tag{4}$$

The Plant District Model

Marginal Model

The type of model set up for this problem is a marginal model. The plant warehouse area should be expanded out to the point where the cost of serving the marginal area (the last addition) from the plant is equal to the cost of serving it from an optimally placed branch warehouse.

Plant direct delivery cost per piece = branch delivery cost per piece (that is, plant to branch plus branch to customer):

$$\frac{2T_o P_d + T_j + T_d H_d}{P_h (H_d - 2P_d H_m - F_t)} = \frac{S_1 + B_e + 2S_o D_p + S_f + 2S_d H_m D_p + I_w}{P_s}$$

$$+ \frac{2T_o D_b + T_f + T_d H_d}{P_h (H_d - 2D_b H_m - F_t)}$$

where

T_o = Truck operation cost per mile

P_d = Plant delivery miles

T_f = Truck fixed costs per day (amortization type charge)

T_d = Truck driver costs per hour

F_t = Fixed driver time per day (check in, check out, coffee break, etc.)

D_b = Miles from branch to delivery

S_1 = Semi load and unload costs

B_e = Branch expense per semi

S_o = Semi-operating costs per mile

D_p = Miles from plant to the branch

S_f = Semi fixed costs per day (amortization type charge)

S_d = Semi driver costs per hour

I_w = Inventory costs per semi per week

H_d = Hours per day

H_m = Hours per mile

P_h = Pieces per hour

P_s = Pieces per semi.

Subject: Facility Location

Title: A Warehouse-Location Problem

Authors: William J. Baumol and Philip Wolfe

Source: Operations Research, Vol. 6, No. 2, March–April 1958, 252–63

Summary: A method for determining a more profitable geographic location pattern for the warehouses that are employed by a firm in delivering known quantities of its finished product to its customers, where the number of warehouses is also permitted to vary. This can normally be expected to be a concave minimization problem. A local optimum can be determined by a method involving a sequence of transportation computations that are shown to converge to the solution. An illustrative small scale computation is included.

Model: Let

X_{ijk} = quantity shipped from factory i $(i = 1, 2, ..., m)$ via warehouse j $(j = 1, ..., n)$, to retailer location k $(k = 1, ..., q)$,

$C_{ijk}(X_{ijk})$ = cost of this shipment including the relevant inventory cost,

Q_i = quantity shipped from plant i,

R_j = capacity of warehouse j,

S_k = quantity required at destination k.

The problem is to minimize the total delivery cost, i.e., to minimize

$$\sum_{i,j,k} C_{ijk}(X_{ijk}), \tag{1}$$

subject to

$$\sum_{j,k} X_{ijk} = Q_i \tag{2}$$

(all goods must be shipped out of the factory),

$$\sum_{i,k} A_{ijk}(X_{ijk}) \leqq R_j \tag{3}$$

(no warehouse capacity can be exceeded), and

$$\sum_{i,j} X_{ijk} = S_k \tag{4}$$

(all customer demands must be met), where all $X_{ijk} \geqq 0$.

The expression $A_{ijk}(X_{ijk})$ in (3) represents the amount
of inventory that will be held as a result of the flow
X_{ijk}.

Assume, to begin with, that our objective function (1)
is linear. An optimal (least cost) solution will
involve shipment of all goods that go from factory i* to
destination k* via that (those) warehouse(s) j* for which

$$C_{i*j*k*} = \text{Min}_j \, C_{i*jk*}. \qquad (5)$$

The solution of the linear program is now simple.
For each factory, destination combination, i*, k*, select
a value j* of j for which (4) is satisfied. That can
be done by simple inspection of the C_{ijk} data. This will
eliminate all but one of the n possible warehouse
locations that can be used to service goods en route from
factory i* to destination k*. We can now revise our
notation, writing $X_{i*j*k*} = X'_{i*k*}$, $C_{i*j*k*} = C'_{i*k*}$ (since
in an optimal solution all other X_{i*jk*}'s will be equal
to zero). Substituting this notation in (1), (2),
and (4) leaves us with a standard transportation problem,
the optimum values of whose variables can be found by
the standard methods.

Nonlinearities may appear in the objective function (1),
and in the warehouse capacity constraints (3), increasing
the cost incurred in using a larger number of warehouses to
handle a given sales volume. Further, the use of additional warehouses increases negotiation, bookkeeping, and
administration costs. In addition, there are usually
economies of large scale in the amount of inventory that
should be kept against the flow of shipments through a
warehouse. Thus, the cost (objective) function can be
approximated by

$$\sum_{i,j,k} K_{ijk} X_{ijk} + \sum_j W_j (\sum_{i,k} X_{ijk})^q + \sum_j V_j i_j, \quad (0 < q < i)$$

where K_{ijk} is the cost of transportation and handling
per unit of X_{ijk}, W_j is the cost of storage per case
per period, and $r_j = 1$ if $\sum_{i,k} X_{ijk} > 0$, and $r_j = 0$
otherwise, i.e., $\sum r_j$ is the number of warehouses used,
and V is the administrative cost to the renting firm
per warehouse employed.

This is a "concave" program. No general method
has been developed for computing the optimum values of
the variables of such a problem. However, it is possible
to find a _local_ optimum through a computing procedure
for dealing with some of the nonlinear aspects of the
Warehouse Location Problem.

Mathematical Description of the Computational Procedure

Let c_{ij} be the unit cost of transportation from
factory i to warehouse j, and let d_{jk} be the unit cost
of transportation from warehouse j to retailer k.
Then $c_{ij} + d_{jk}$ is the C_{ijk} above. Let $Z_j = \sum_{i,k} X_{ijk}$ be
the total flow through warehouse j, where X_{ijk}
represents the flow from factory i through warehouse j to
retailer k. Supposing the home office administrative costs
for the jth warehouse (if used) to be V_j, the total
administrative cost for a given transportation schedule will
be the sum of terms $V_j r(Z_j)$, where

$$r(Z_j) = 0 \quad \text{if} \quad Z_j = 0, \quad r(Z_j) = 1 \quad \text{if} \quad Z_j > 0.$$

The total cost of the transportation schedule X_{ijk}
will then be

$$f(X_{ijk}) = \sum_{i,j,k} (c_{ij} + d_{jk}) X_{ijk}$$

$$+ \sum_j W_j (Z_j)^q + \sum_j V_j r(Z_j).$$

The computing procedure can be outlined in a recursive fashion as follows, where the initial calculation (zeroth) and the nth calculation are given.

0th Calculation

For each pair i,k of factory and retailer, find the least cost of shipment, ignoring the warehouse loading charges and administration costs:

$$C_{ik}^0 = \text{Min}_j \; (c_{ij} + d_{jk}) = (c_{ij_{jk}}^0 + d_{j_{ik}k}^0),$$

where j_{ik} denotes the warehouse routing selected by this criterion.

Using these costs C_{ik}^0, solve the transportation problem of shipping from the factories with known availabilities to the retailers with known demands so as to minimize the cost function $\sum_{i,k} C_{ik}^0 X_{ik}$; and denote the solution of this transportation problem by $\{X_{ik}^0\}$.

nth Calculation

For each pair i,k, let X_{ik}^{n-1} represent the solution of the transportation problem solved in the $(n-1)$th calculation, and determine the warehouse loadings involved in that solution:

$$Z_j^{n-1} = \sum \; \{\text{all } i,k \text{ such that } j_{ik}^{n-1} = j\} \; X_{ik}^{n-1},$$

where j_{ik}^{n-1} denotes the warehouse selected for shipment from factory i to retailer k in the $(n-1)$th calculation.

Now define new transportation costs by

$$C_{ik}^n = \text{Min}_j \; [c_{ij} + d_{jk} + W_j q \, (Z_j^{n-1})^{q-1}]$$

letting j_{ik}^n denote the warehouse for which the Min_j above is assumed.

Using the costs C_{ik}^n, find the solution $\{X_{ik}^n\}$ of the transportation problem with these costs.

Subject: Facility Location

Title: A Heuristic Program for Locating Warehouses

Authors: Alfred A. Kuehn and Michael J. Hamburger

Source: Management Science, Vol. 9, July 1963, 643-66

Summary: A heuristic computer program for locating warehouses is outlined and compared with efforts at solving the problem either by means of simulation or as a variant of linear programming. The heuristic approach outlined offers advantages in the solution of this class of problems in that it (1) provides considerable flexibility in the specification of the problem to be solved; (2) can be used to study large-scale problems, that is, complexes with several hundred potential warehouse sites and several thousand shipment destinations; and (3) is economical of computer time.

Model: The Warehouse Location Problem

Let

$X_{h,i,j,k}$ = quantity of good h (h = 1, ..., p) shipped from factory i (i = 1, ..., q) via warehouse j (j = 1, ..., r) to customer k (k = 1, ..., s),

$A_{h,i,j}$ = per unit transportation cost of shipping good h from factory i to warehouse j,

$B_{h,j,k}$ = per unit transportation cost of shipping good h from warehouse j to customer k,

$C_{h,j}(\sum_{i,k} X_{h,i,j,k})$ = total cost of warehouse operation associated with processing good h at warehouse j. Without loss of generality we may express this function as the sum of $S_{h,j}$ and F_j defined below,

$D_{h,k}(T_{h,k})$ = explicit or imputed cost due to a delay of T time units in delivery of good h to customer k. When the customer imposes a maximum delivery time (constraint), D becomes infinite whenever the indicated time limit is reached,

F_j = fixed cost per time period of operating warehouse j. Note that this is a planned fixed cost to be incurred and not a sunk cost,

$S_{h,j}(\sum_{i,k} X_{h,i,j,k})$ = semivariable cost of operating warehouse j per unit of good h processed, including variable handling and administrative costs, storage cost, taxes, interest on investment, pilferage, and so on (the homogeneous portion of the very general function $C_{h,j}$),

$Q_{h,k}$ = quantity of good h demanded by customer k,

W_j = capacity of warehouse j,

$$Y_{h,i} = \text{capacity of factory } i \text{ to produce good } h,$$

$$Z_j = 1 \text{ if } \sum_{h,i,k} X_{h,i,j,k} > 0 \text{ and zero otherwise,}$$

$$\sum_j Z_j = \text{number of warehouses used.}$$

The problem then becomes one of minimizing total distribution costs:

$$f(X) = \sum_{h,i,j,k} (A_{h,i,j} + B_{h,j,k}) X_{h,i,j,k} + \sum_j F_j Z_j$$

$$+ \sum_{h,j} S_{h,j} (\sum_{i,k} X_{h,i,j,k}) + \sum_{h,k} D_{h,k}(T_{h,k})$$

subject to constraints of the following form:

$$\sum_{i,j} X_{h,i,j,k} = Q_{h,k}$$

(customer k's demand for product h must be supplied),

$$\sum_{j,k} X_{h,i,j,k} \leq Y_{h,i}$$

(factory i's capacity limit on good h cannot be exceeded),

$$I_j (\sum_{h,i,k} X_{h,i,j,k}) \leq W_j$$

(the capacity of warehouse j cannot be exceeded), where $I_j (\sum_{h,i,j} X_{h,i,j,k})$ is a function which denotes the maximum inventory level associated with the flow of all goods from all factories to all customers serviced through warehouse j.

Subject: Facility Location

Title: Brand Switching and Mathematical Programming in Market
           Expansion

Authors: Philip H. Hartung and James L. Fisher

Source: Management Science, Vol. 11, No. 10, August 1965,
           B-231-243

Summary: A method for planning expansion if one industry outlet
           has an influence on another in terms of sales is outlined
           using a mathematical programming approach with non-
           linear objective function. A theoretical model, based
           on brand switching, is validated by experimental evidence
           and these results are then employed in the development
           of a long-term planning model.

Model:

BRAND SWITCHING MODEL

$$X_{t+1} = \alpha X_t + \beta (1 - X_t) \tag{1}$$

$$Y_{t+1} = (1-\beta) Y_t + (1-\alpha)(1 - Y_t) \tag{2}$$

where

X_t = Market share of product A in time period t.

Y_t = Market share of all competitors in t.

α = Probability that if a customer purchases product A
    in the current time period, he will purchase
    product A in the succeeding period.

β = Probability that if a customer purchases from a
    competitor in the current time period, he will
    switch to product A in the succeeding period.

$1-\alpha$ = Probability of a customer of product A switching
    to a competitor from one time period to the next.

$1-\beta$ = Probability of a customer of a competitor remaining
    a customer of a competitor from one time period to
    the next.

Steady-state market share:

$$X = \frac{\beta}{(1-\alpha+\beta)}. \tag{3}$$

MARKET SHARE AND NUMBER OF OUTLETS

It is assumed that an outlet either carries product A or carries a competing brand, but not both.

Let P = number of outlets carrying product A,

O = number of outlets carrying competing brands,

$\left. \begin{array}{c} k_1 \\ k_2 \end{array} \right\}$ = undetermined positive constants

$$\alpha = k_1 \ (\frac{P}{O+P})$$

$$\beta = k_2 \ (\frac{P}{O+P}). \tag{4}$$

Substituting (4) into (3):

$$X = \frac{k_2 P}{O + (1 + k_2 - k_1)P}. \tag{5}$$

Let

S = sales of the product A,

T = total sales of all brands.

Then

$$\frac{S}{P} = \frac{k_2 T}{O + (1 + k_2 - k_1)P}. \tag{6}$$

Define

$\overline{S}$ = average sales per outlet of the company

$\overline{O}$ = average sales per outlet of the industry.

Then

$$\frac{\overline{S}}{\overline{O}} = k_2 + (k_1 - k_2)X. \tag{7}$$

OUTLET EXPANSION

The value of a single outlet in an expansion program is composed of two quantities--one, primary sales made at the outlet and two, additional sales it provides as a result of increasing the market share of the area.

Define:

C_{ij} = construction cost of one outlet in area i, year j,

K_j = total capital allocated for building outlets in year j,

L_{ij} = upper limit on the number of outlets that can be built in area i, year j,

R_{ij} = return from the sale of one unit of the product in area i, year j,

O_{ij} = the number of competing outlets in area i, year j,

T_{ij} = the total market in area i, year j,

r = the interest rate to be used for discounting,

P_{io} = the number of existing company outlets in area i.

N_{ij} = the number of new outlets to be built in area i, year j,

P_{ij} = the total number of company outlets in area i, in year j,

S_{ij} = total company sales in area i, year j,

$\bar{S}_{ij}$ = average sales per outlet.

The planning problem then is
maximize

$$\sum_{i=1}^{M} \left\{ \sum_{j=1}^{Y} \left[N_{ij} \sum_{k=1}^{Y} \frac{1}{(1+r)^k R_{ik} \bar{S}_{ik}} \right] + \frac{P_{io} \sum_{j=1}^{Y} 1}{(1+r)^j R_{ij} \bar{S}_{ij}} \right\} \qquad (8)$$

subject to equations (9), (10), and (11).
Capital constraint:

$$\sum_{i=1}^{M} C_{ij} N_{ij} \leq K_j \qquad j = 1 \ldots Y. \tag{9}$$

Outlet construction constraint:

$$N_{ij} \leq L_{ij} \qquad i = 1 \ldots M, \quad j = 1 \ldots Y. \tag{10}$$

Market share equation:

$$\overline{S}_{ij} = \frac{bT_{ij}}{O_{ij} + (1-a)P_{ij}}. \tag{11}$$

Subject: Facility Location

Title: A Market Potential Model and its Application to Planning Regional Shopping Centers

Authors: Alan M. Voorhees and T.R. Lakshmanan

Source: Science, Technology, and Marketing, R.M. Haas (ed.), Proceedings of the 1966 Fall Conference of the American Marketing Association, Chicago, 1966, 831-45

Summary: Development and application of a market potential model which identifies the scale, composition, location and timing of a number of large retail centers consistent with urban growth processes.

Model: The model states that the sales potential of a retail center is directly related to its size, its proximity to consumers and to the number and consumption level of the consumers; it is an inverse function of its proximity to competing shopping facilities. Using the gravity model framework we have

$$S_{ij} = C_i \; \frac{\dfrac{F_j}{d_{ij}^{\;x}}}{\dfrac{F_i}{d_{ii}^{\;x}} + \dfrac{F_j}{d_{ij}^{\;x}} \cdots \dfrac{F_n}{d_{in}^{\;x}}} = C_i \; \frac{\dfrac{F_j}{d_{ij}^{\;x}}}{\sum_j \dfrac{F_j}{d_{ij}^{\;x}}} \qquad (1)$$

where

S_{ij} = consumer retail expenditures of population in zone i, spent at the shopping center in zone j,

C_i = total consumer retail expenditures of population in zone i,

F_j = size of retail activity in zone j,

d_{ij} = distance between shopping center in zone j and consumers at i,

x = an empirically derived exponent reflecting "resistance" to shopping travel.

The above model states that the retail center in zone j attracts consumer dollars (S_{ij}) from zone i

(a) in direct proportion to the consumer expenditures (C_i) and its size (F_j) and

(b) in inverse proportion to distance to consumers ($d_{ij}^{\;x}$) and competition

$$\left(\sum_j \frac{F_j}{d_{ij}^{\;x}} \right).$$

Equation (1) can be written to state the consumer expenditures available in all zones of the urban area that would probably be spent in the retail center in zone j, that is,

$$S_j = \sum_i c_i \frac{\dfrac{F_j}{d_{ij}^{\,x}}}{\sum_j \dfrac{F_j}{d_{ij}^{\,x}}} \tag{2}$$

where S_j = total sales in retail center j.

(2) implies that there is no trade area boundary but a shopping interaction between all zones, though this may fall off sharply with distance.

Subject: Facility Location

Title: Determining Optimum Distribution Points for Economical Warehousing and Transportation

Author: Arthur W. Napolitan

Source: N.E. Marks and R.M. Taylor (eds.), Marketing Logistics: Perspectives and Viewpoint, New York: John Wiley & Sons, Inc. 1967, 76-82

Summary: Minimization of a total cost formula in order to establish the optimum distribution points for warehouse branches.

Model: The warehousing and transportation problem was to determine the optimum number of branches, the optimum location of these branches, and the optimum area to be serviced by each branch. Minimize

$$f(X) = C_1 X_1 + C_2 X_2 + C_3 X_3 + C_4 X_4$$

where

$f(X)$ = total costs,

$C_1 X_1$ = cost attributable to sales lost because of distance from branches (customer service),

$C_2 X_2$ = cost of operating branches,

$C_3 X_3$ = transportation costs from central warehouse(s) to branches,

$C_4 X_4$ = transportation costs from branches to customers.

A list of potential branch sites was developed by application of certain criteria established by the management--minimal requirements with respect to rail, trucking, and postal services, for example.

With the value of the four factors in the formula known for all possible combinations, the problem of the West Coast branch locations was solved by minimization, using combinatorial programming techniques originated by the OR team for the solution of this problem. The same methods were then used to solve the problem in the area east of the Rockies.

Subject: Facility Location

Title: Market Share, Distance and Potential

Author: Martin J. Beckmann

Source: Regional and Urban Economics, Vol. 1, No. 1, 1971, 3-18

Summary: Suppose that in every location the ratio of sales by
          two firms offering heterogeneous products depends only on
          the difference of economic distance from the two plants,
          suitably defined. Then the market share is a simple
          function of distances. Market areas, defined at those
          subregions where market share of some firm dominates
          that of any other firm, are identical with conventional
          market areas for sellers of an homogeneous product.
          Next curves of equal market share--isoshare lines--are
          introduced, discussed and calculated for special cases.
          It is shown that under certain restrictions on the
          function which specifies the distance effect and on
          the measure of economic distance, the present approach
          is consistent with demand analysis based on utility
          functions. Finally, the hypothesis is related to the
          gravity and potential concepts that have been used in
          travel forecasting and in other contexts of regional
          science.

Model: 1. The Distance Difference Effect

          Consider two firms having a single plant each at
different locations and offerings similar but distinct
products. _Ceteris paribus_ relative sales at a given
location will depend on the distances from the two firms.

$$\frac{s_1 - s_2}{s_1 + s_2} = F(r_1, r_2). \tag{1}$$

We now postulate the existence of a measure of economic
distance $d(r)$ such that the effect of distance on sales
can be expressed in terms of the difference of economic
distances

$$\frac{s_1 - s_2}{s_1 + s_2} = f[d(r_2) - d(r_1)]. \tag{2}$$

Since distance affects sales adversely we stipulate that f
is a monotonically increasing function. The distance
measure d, on the other hand, is a monotonically non-
decreasing function of geometric distance. Measures of
economic distance which have empirical relevance are

d(r) = kr proportionality

d(r) a step function ("zonal tariff")

d(r) = log r a logarithmic measure of geometric distance.

(2)   may be rewritten in terms of the sales ratio s_1/s_2 = $(1+f)/(1-f)$:

$$\frac{s_1}{s_2} = \phi(d(r_2) - d(r_1)) \tag{3}$$

where $\phi = (1+f)/(1-f)$ is again a monotonically increasing function. A slightly more general hypothesis is that the effect of economic distance may be modified by some basic difference of attractiveness which could be due to price differences or quality differences

$$\frac{s_1}{s_2} = \phi(d(r_2) - d(r_1) + a_{12}). \tag{4}$$

For reasons of symmetry, we have

$$\frac{s_2}{s_1} = \phi(d_1 - d_2 + a_{21}) \tag{5}$$

where $a_{21} = -a_{12}$.

Comparing (5) and (4) we see that

$$\phi(-x) = \frac{1}{\phi(x)}. \tag{6}$$

2. Market Areas

Let m_i = market share of firm i. In the case of two sellers we have

$$m_1 = \frac{s_1}{s_1 + s_2} = \frac{1}{1 + \frac{s_2}{s_1}} = \frac{1}{1 + \phi(d_1 - d_2 + a_{21})}.$$

We now define market area to mean that area where the market share of one firm is $\frac{1}{2}$ or greater. In the case of two firms, the entire region is thereby partitioned into two market areas.

The boundary line between market areas is then that
locus where

$$\frac{1}{2} = \frac{1}{1 + \phi(d_1 - d_2 + a_{21})}$$

$$\phi(d_1 - d_2 + a_{21}) = 1$$

and so, since ϕ is strictly increasing, and $\phi(0) = 1$

$$d_2 - d_1 + a_{12} = 0.$$

In the absence of differential attractiveness a_{12} it follows
that the market boundaries are where

$$d_2 = d_1,$$

i.e., where the economic distances from the suppliers are
equal. In the case of a strictly monotonic distance measure
it follows that then geometric distances are also equal

$$r_2 = r_1$$

and this is the same criterion as for homogeneous products.

The concept of market area may be extended to more than
two suppliers. We define the market area of a supplier to be
that region where this supplier's market share is dominant,
i.e., larger than that of any other supplier. Now m_i is
defined by

$$m_i = \frac{s_i}{s_1 + \ldots + s_i + \ldots + s_n}$$

or

$$m_i = \frac{1}{\frac{s_1}{s_i} + \ldots + \frac{s_i}{s_i} + \ldots + \frac{s_n}{s_i}}$$

$$= \frac{1}{\phi[d_i - d_1] + \phi[d_i - d_2] \ldots + \phi[d_i - d_i] \ldots + \phi[d_i - d_n]}$$

assuming away any differential attractions a_{ij}. The market area of supplier 1 is then where

$$\phi(d_1 - d_1) + \phi(d_1 - d_2) + \ldots \phi(d_1 - d_n) \leqq$$
$$\phi(d_j - d_1) + \ldots \qquad + \qquad \phi(d_j - d_n) \quad \text{all} \quad j. \tag{7}$$

Consider now the point set where

$$d_1 - d_i \leqq d_j - d_i \quad \text{for all} \quad i. \tag{8}$$

This is the conventional market area of firm 1 as defined for homogeneous commodities. Substitution of (8) in (7) shows that the newly defined market areas (7) include the conventional market area (8). Since the conventional market areas are a partitioning of the entire space, the new and conventional market areas must be identical. Thus also in the case of multiple firms, areas of dominant market share may be identified with conventional market areas of nearest supplier.

3. Iso-Share Lines

Let iso-share lines be loci of constant market shares. Consider the case of two sellers. From

$$m_1 = \frac{1}{1 + \phi(d_1 - d_2 + a_{21})}$$

we see that iso-shares lines are loci of constant difference of economic distance. If economic distance is a linear function of geometric distance this clearly implies that iso-share lines are hyperbola.

Consider next the case where the measure of economic distance is proportional to the logarithm of geometric distance

$$d_i = a \log r_i.$$

In that case

$$d_1 - d_2 = \text{constant} = a \log \mu \quad \text{(say)}$$

implies

$$\frac{r_1}{r_2} = \mu$$

the iso-share lines are loci on which the distance ratio is constant.

The location of the two firms are themselves loci of constant market shares. Depending on the commodity under consideration it may be appropriate to assume that market share is unity at the home location of the firm.

A market share of 1 for $r_1 = 0$ implies

$$1 = \frac{1}{1 + \phi(-\infty)}$$

or

$$\phi(-\infty) = 0, \quad \phi(\infty) = \infty$$

in view of (6).

If on the other hand $\phi(-\infty) = \alpha$, then the maximal market shares at the firm location, is

$$m_{max} = \frac{1}{1+\alpha}.$$

As we vary μ we obtain the entire family of iso-share lines. Using Euclidian distance the iso-share line equation is:

$$\mu[(x+1)^2 + y^2] = (x-1)^2 + y^2$$

from which

$$(x - \frac{1+\mu}{1-\mu})^2 + y^2 = \frac{4\mu}{(1-\mu)^2} \quad 0 \leq \mu \leq \infty \quad \text{(see Figure 1)}.$$

To each circle in the left-half-plane, representing a market share $m_1 > \frac{1}{2}$ there corresponds a symmetric circle in the right-hand plane representing a share of $1-m_1$.

The family of iso-share lines is thus a family of circles symmetric to the normal bisector between the two given plant locations.

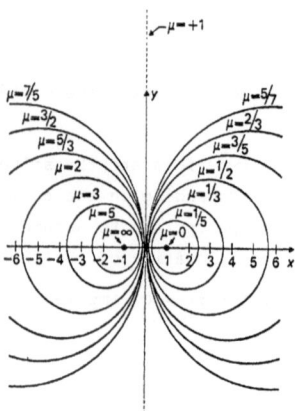

Figure 1

Further it is shown that in special cases the sales ratio
(4) is consistent with conventional demand curves
resulting from utility maximization by consumers.

Finally, the gravity approach yields a relationship
between total sales and potential of population:

$$S_i(\omega) \sim \frac{P(\omega)A_i}{D_i(\omega)}$$

where

ω = location
P = population
A = attractiveness
D = economic distance.

Subject: Facility Location

Title: A System Construct for Evaluating Retail Market Locations

Authors: L.A. White and J.B. Ellis

Source: Journal of Marketing Research, Vol. 8, February 1971, 43-46

Summary: A systems theory framework for predicting supermarkets' yearly sales. The model can be used to predict sales of a hypothetical store or changes in sales resulting from major road or store changes.

Model:

ORIGINS

$$Y_i = 52 \times C_i \times P_i$$

where

Y_i = yearly flow of money from origin area i in dollars

C_i = food cost per week per capita for origin area i

P_i = population of origin area i.

LINKS

$$X_{ij} = R_{ij} Y_{ij}$$

where

X_{ij} = pressure required to cross the link connected from node i to node j

R_{ij} = resistance to flow through link ij

Y_{ij} = yearly net flow of shopping money through link ij.

In most cases, driving time would be the best measure of R_{ij}.

DESTINATIONS

$$Y_j = A_j X_j$$

where

Y_j = yearly sales of supermarket j

A_j = attraction of supermarket j

X_j = propensity to shop at supermarket j.

26 characteristics of each store, such as floor area, number of checkouts, number of products sold, store hours, etc. were recorded. Using a stepwise, linear regression analysis procedure, an equation for A_j in terms of the store characteristics was obtained. Many operations were performed on the data, such as squaring the variables, taking logarithms, etc., before arriving at the final form for A_j. For technical reasons, the regression was established for the variable $1/A_j$ rather than A_j itself:

$$1/A_j = -90.6059 + 136925.0/(\text{floor area})$$
$$+ 3.8049/(\text{number of checkouts})$$
$$+ 84.4889 \ (\text{relative price})$$
$$+ 0.2953/(\text{specials})$$
$$+ 1.1303 \ (\text{location coefficient}).$$

For this regression equation, the multiple correlation coefficient = 0.934; goodness of fit, $F(5,18) = 24.6813$ standard error = 7.049.
The average per cent error was 12.04.

Subject: Facility Location

Title: A Model for Allocating Retail Outlet Building Resources
          across Market Areas

Authors: Gary L. Lilien and Ambar G. Rao

Source: Operations Research, Vol. 24, No. 1, January-February
          1976, 1-14

Summary: A model is developed to help plan retail outlet building.
          An S-shaped, outlet share-market share relation is
          hypothesized and estimated satisfactorily from company
          data. This relation is then one input in a resource
          allocation algorithm that produced optimal or near-
          optimal plans.

Model: Assume

          (a) An S-shaped relation between outlet shares s,
              and market share m;

          (b) in any particular market the share enjoyed by a brand
              depends upon the age of its outlets as compared to
              competition.

          A variable "aggressiveness" is defined as

$$a = \dfrac{\dfrac{\text{No. of recently built company outlets}}{\text{Total company outlets}}}{\dfrac{\text{No. of recently built industry outlets}}{\text{Total industry outlets}}}$$

          The firm's market share m is related to the aggressiveness
          a   and share of outlets by a relation

$$m = g(a,s). \qquad (1)$$

          In general, m, a, s as well as g will be known for a
          particular market. Thus, the following is assumed:

          1: In equation (1) m, a and the function g
          are known with certainty, while s is to be determined
          from the equation.

          The objective of the following algorithm is to maximize
          the total net present value (NPV) of a Y-year building
          programs subject to restrictions on the total number of
          outlets that can be built (a) within a market, (b) across
          all markets in a given year, and (c) during the Y-years,
          where NPV is defined as

$$NPV = \sum_{j=1}^{j=J} \sum_{i=1}^{i=\tau} CF_{ij}/(1+R)^{i-1}, \tag{2}$$

where CF_{ij} = cash flow associated with market area j in
year i, R = discount rate, J = market areas considered
in the plan, and τ = planning horizon ($\tau > Y$). To do this
maximization, the procedure selects the group of outlets
in the market that has the highest average NPV per outlet.
It then selects the next highest NPV group and so on until
all allowable outlets have been allocated.

Assume

2: 'New' outlets, used in the definition of aggressiveness, are defined as those four years old or newer.
In year 3 of the building plan, outlets built in years -1
(last year), 0 (this year), 1, and 2 are included in the
definition of aggressiveness.

The building plan is designed for Y years (where Y
usually equals 5); the planning horizon is set for τ
(generally 20) years.

Further assume

3: After Y years the firm will build enough outlets
to maintain its market share: $m_k = m_Y$, $k = Y + 1$, ...
Thus aggressiveness is assumed equal to 1: $a_k = 1$, $k = Y + 1$...

An allocation algorithm for a single building plan can
now be developed. We will then extend it to Y years.

Let X_i = number of outlets built in market i, n_i =
market building constraint, T = overall building constraint,
V_{ik} = incremental net present value (NPV) of the kth station
in market i, $\sum_{k=i}^{k=j} V_{ik}$ = cumulative NPV of the first j
stations in market $i, j = 1, 2, \ldots, n_i$,

$$W_{ij} = \begin{cases} \sum_{k=1}^{k=j} V_{ik}/j = \text{average NPV of the first } j \text{ stations} \\ \qquad\qquad \text{in market } i, j = 1, \ldots, n_i, \\ \\ -B \qquad\qquad = j > n_i, \text{ B is large positive number,} \end{cases}$$

M = Number of markets, and $N = \max_i [n_i]$.

The single-year problem is

$$\max Z = \sum_{i=1}^{i=M} \sum_{k=1}^{k=X_i} V_{ik},$$

$$0 \le X_i \le n_i, \qquad X_i \text{ integer, } i = 1, \ldots, M, \quad (3)$$

$$\sum_{i=1}^{i=M} X_i \le T.$$

The solution algorithm is the following:

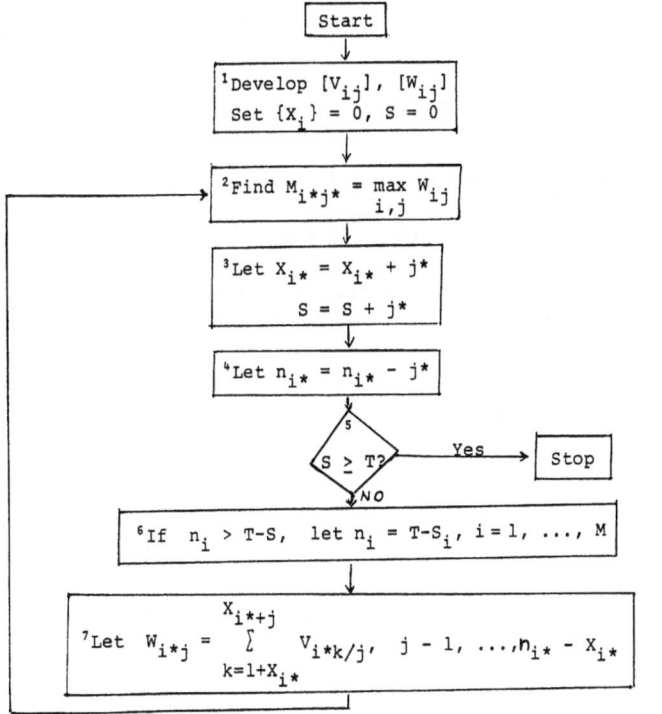

The rationale behind the algorithm is the following theorem. If, in every market, NPV is a concave function of the number of outlets built, then a simple allocation according to incremental NPV yields an optimal building plan.

Now consider a building program that can span several years ($Y > 1$). Define X_{it} = number of outlets built in market i in year t, V_{ijt} = incremental NPV of the jth outlet built in market i, given it is built in year t, and T_t = cumulative number of outlets that can be built up through year t; $t = 1, \ldots, Y$.

All other quantities are altered by adding a subscript t to the price symbol. The problem becomes

$$\max Z = \sum_{i=1}^{i=M} \sum_{t=1}^{t=Y} \sum_{j=1}^{j=X_{it}} V_{ijt},$$

$$0 \leq X_{it} \leq n_{it}, \quad X_{it} \text{ integer, } i = 1, \ldots, M, \, t = 1, \ldots, Y, \quad (4)$$

$$\sum_{k=1}^{k=t} \sum_{i=1}^{i=M} X_{ik} \leq T_t, \quad t = 1, \ldots, Y.$$

Assume 4: V_{ijt} is independent of the time at which outlets $j-1$ were built. 5: $V_{ijt} > V_{ij(t+1)}$--the earlier an outlet is built, the greater its NPV. Then the algorithm for the multi-year case is very similar to that for the one-year case.

Subject: Sales Force

Title: Spatial Allocation of Selling Expense

Author: J.A. Nordin

Source: Journal of Marketing, Vol. 7, No. 3, January 1943, 210-19

Summary: A method for the allocation of selling effort among sales districts is described.

Model: Assumptions

1. There is one product, sold in two districts.

2. The planner is to plan for the first period, on the supposition that the sales of every subsequent period are independent of the sales within the period selected for analysis.

3. The facts of the period just concluded are expected to continue unchanged.

4. The object of making adjustments is to maximize the total sales of the two districts taken together, while maintaining the constant total of selling expense.

5. The only form of selling expense is the salaries of salesmen. Salary per man is constant and personal efficiency is assumed the same.

6. The price is known, and is the same for both the districts.

7. The problem of deciding the optimum selling expense for the given period will not be considered.

Define $y_1 = f(x_1) = ax_1^a$ and $y_2 = g(x_2) = bx_2^a$ where the y's denote marginal selling expense, the x's denote unit sales, and the subscripts 1 and 2 distinguish District 2. The constants a and b will be determined by the model and a is a constant set by judgment.

If x_1 is the total unit sales in District 1, then the total cost of selling x_1 units is given by

$$\int_0^{x_1} f(x_1)dx_1.$$

Similarly for District 2, the total cost is given by

$$\int_0^{x_2} g(x_2)dx_2.$$

The task is that of maximizing $x_1 + x_2$, while preserving the condition

$$\int_0^{x_1} f(x_1) dx_1 + \int_0^{x_2} g(x_2) dx_2 = K,$$

where K is the total selling expense available for allocation between the two districts.

Let P_1 and P_2 be the dollars of promotion spent in Districts 1 and 2.

$$P_1 = \int_0^{x_1} f(x_1) dx_1 = \int_0^{x_1} ax_1^{\alpha} dx_1.$$

We determine P_2 by a similar procedure, and obtain:

$$P_1 = \frac{ax_1^{\alpha+1}}{\alpha+1} \quad \text{and} \quad P_2 = \frac{bx_2^{\alpha+1}}{\alpha+1} \tag{1}$$

The ratio of P_1 to x_1 is:

$$\frac{P_1}{x_1} = \frac{ax_1^{\alpha}}{\alpha+1}. \quad \text{Similarly,} \quad \frac{P_2}{x_2} = \frac{bx_2^{\alpha}}{\alpha+1}. \tag{2}$$

In equilibrium,

$$ax_1^{\alpha} = bx_2^{\alpha}. \tag{3}$$

Since P_2 is in terms of x_2,

$$P_2 = \frac{K}{(\frac{b}{a})^{1/\alpha} + 1}. \quad \text{Similarly,} \quad P_1 = \frac{K}{(\frac{a}{b})^{1/\alpha} + 1}. \tag{4}$$

This allocation approximates the optimum.

Subject: Sales Force

Title: A Study of Sales Operations

Authors: Arthur A. Brown, Frank T. Hulswit, and John D. Kettelle

Source: Operations Research, Vol. 4, No. 3, June 1956, 296-308

Summary: In a case study an experiment is executed to determine customer responses to various levels of effort. The experiment shows how much effort should be assigned to get customers, and how much to hold customers. The effort required, and the probabilities of success, determine the proper allocation of accounts to salesmen. By referring to the graph of customer size distribution, sizes of sales forces resulting from various allowed percentages for sales expense are derived. In addition, one method is presented for evaluating the expected profits from improved customer service.

Model: Let $C(x)$ be the steady-state probability that a customer subjected to x hours per month of sales effort will respond with a substantial increase in purchases, and $H(x)$ be the holding effort or the probability of maintaining business. The problems are: first, what is the appropriate rate of effort per customer in each group and second, how many customers in each group should be assigned to an individual salesman. Experimental data established a 20 per cent chance of conversion at the ten-hour-per-month level of conversion effort and a holding rate of two hours per month.

To determine the size of the sales force, let

T_C = expected time (months) required for conversion,

T_H = expected holding time,

x = dollars per hour for cost of salesman,

E = average sales expense per month per customer,

then

$$E = x(10T_C + 2T_H)/(T_C + T_H). \qquad (1)$$

If, in addition,

R = average revenue per month from a customer,

P = potential total monthly revenue from a customer,

a = fraction of potential revenue expected from a customer during the holding period,

then

$$R = aPT_H/(T_C + T_H).\tag{2}$$

Suppose that the profits are large enough so that a fraction f of additional revenue may be spent on sales effort to procure that revenue. Then the potential of the smallest customer who is worth promoting may be determined by combining (1) and (2) with

$$E = fR\tag{3}$$

to obtain

$$P = x(10T_C + 2T_H)/afT_H.\tag{4}$$

Subject: Sales Force

Title: On a Linear Programming, Combinatorial Approach to
          the Travelling-Salesman Problem

Authors: G.B. Dantzig, D.R. Fulkerson and S.M. Johnson

Source: Operations Research, Vol. 7, No. 1, January-February 1959 ,
          58-66.

Summary: Linear programming used to reduce the combinatorial
          magnitude of travelling-salesman-problems. To illustrate
          the method, a step-by-step solution of Barachet's
          ten-city example is presented.

Model: In Barachet's ten-city example (OR, Vol. 5, 1957, 841-5), let
          the variables x_{ij} (i < j) be nonnegative and

$$\sum_{i<k} x_{ik} + \sum_{i>k} x_{ki} = 2, \quad (k = 0,1, \ldots, 9) \quad (1)$$

$$x_{ij} \leq 1. \tag{2}$$

To get a starting set of conditions with respect to
which this tour is a basic solution, we impose, in addition
to (1), an upper bound on x_{12}, and add the basic variable x_{68}
(with value 0). Alternatively, we may think of all upper
bound relations (2) as being present in the problem, and
view x_{12} as a 'nonbasic variable' at its upper bound value
of unity (instead of zero).

The first step is to compute 'potentials' π_i (simplex
multipliers or prices) satisfying

$$\pi_i + \pi_j = d_{ij} \tag{3}$$

corresponding to basic variables x_{ij}. Next compute

$$\delta_{ij} = \begin{cases} \pi_i + \pi_j - d_{ij}, & \text{(for zero nonbasic variables)} \\ -(\pi_i + \pi_j - d_{ij}). & \text{(for positive nonbasic variables)} \end{cases} \tag{4}$$

If $\delta_{ij} \leq 0$ for all nonbasic variables, the tour is established
as optimal. Otherwise, we select some nonbasic variable

corresponding to $\delta_{ij} > 0$ for entry into the basic set. Throughout this problem we take the largest δ_{ij} and attempt to increase the corresponding nonbasic variable x_{ij} if it has value zero, or decrease it if its value is unity, thereby obtaining a new basic set of variables.

For Barachet's ten-city example, the tour (1,2,3,4,5,0,9,8,6,7) turns out to be uniquely optimal.

Subject: Sales Force

Title: An Optimal Plan for Salesmen's Compensation

Author: John U. Farley

Source: Journal of Marketing Research, Vol. 1, May 1964, 39-43

Summary: Under a sales compensation plan paying equal commission rates on the gross margins of the products in a salesman's line a salesman trying to maximize his commissions simultaneously maximizes his contribution to company profits. The plan also handles problems such as loss leaders and capacity constraints with minor modifications.

Model:

Let

π = company gross profit in dollars

S = salesman's commission income in dollars

B_i = commission rate, in per cent, on product i

t_i = time devoted to selling product i

Q_i = quantity sold of i

P_i = selling price of i

K_i = variable nonselling cost per unit of i

$M_i = P_i - K_i$ = gross margin

C = total time the salesman devotes to selling, in appropriate units, such as month.

P_i and K_i are assumed constant for all Q_i. Also assumed is the relation $Q_i = f_i(t_i)$, where $\dfrac{df_i(t_i)}{dt_i} > 0$ over a wide range of t.

The salesman allocates his time (C) among the n products he sells:

$$C = t_1 + t_2 + \ldots + t_n, \quad \text{or} \quad C = \sum_{i=1}^{n} t_i.$$

Maximum Profits for the Firm

The firm's gross profit on the sales of a salesman is:

$$\pi = \sum_{i=1}^{n} [P_i Q_i - Q_i K_i - B_i (P_i - K_i) Q_i]$$

$$= \sum_{i=1}^{n} Q_i [(P_i - K_i)(1 - B_i)].$$

Maximize:

$$\pi = \sum_{i=1}^{n} Q_i [(P_i - K_i)(1 - B_i)]$$

subject to $C = \sum_{i=1}^{n} t_i$, the salesman's time constraint.

Result:

$$\frac{f_j'(t_j)}{f_i'(t_i)} = \frac{M_i (1 - B_i)}{M_j (1 - B_j)} \qquad \text{for all } i,j.$$

Maximum Commission for the Salesman

If the company pays commissions as a percentage of gross margins on sales, the salesman's commission is

$$S = \sum_{i=1}^{n} M_i Q_i B_i.$$

Maximizing this equation subject to

$$C = \sum_{i=1}^{n} t_i$$

yields

$$\frac{f_j'(t_j)}{f_i'(t_i)} = \frac{M_i B_i}{M_j B_j} \qquad \text{for all } i,j.$$

Optimum Commission Rates

Both the company's gross profit and the salesman's commission income are maximized when

$$\frac{B_i M_i}{B_j M_j} = \frac{M_i (1 - B_i)}{M_j (1 - B_j)} .$$

This implies

$B_i = B_j$, that is the commission plan is optimal when equal commission rates are paid on gross margins of each product.

Subject: Sales Force

Title: Surveillance of Multi-Component Systems: A Stochastic Traveling Salesman's Problem

Authors: Cyrus Derman and Morton Klein

Source: Naval Research Logistics Quarterly, Vol. 13, 1966, 103-12

Summary: The linear programming formulation of Markovian decision processes are used to develop optimal visit sequences.

Model: Suppose that the system consists of L+1 components labeled $0, 1, \ldots, L$, and that the cost to go from component i to j is equal to w_{ij}. If the component being inspected during period t is i, then the next component for inspection is j with probability D_{ij} $(i, j = 0, 1, \ldots, L)$. Thus, we allow use of the class of all randomized decision rules, denoted as C', characterized by a matrix $D = \{D_{ij}\}$, where $D_{ij} \geq 0$ and $\sum_j D_{ij} = 1$, $i = 0, \ldots, L$.

Given such a rule, let $\{Y_t : t = 0, 1, \ldots\}$ represent the sequence of components inspected over the infinite time horizon. Then, under D, $\{Y_t : t = 0, 1, \ldots\}$ is a Markov chain, with state space $\{0, 1, \ldots, L\}$ and with stationary transition probabilities $P_{ij} = D_{ij}$. We assume that the chain starts in state 0.

Assuming that the rule used is such that the chain has only one ergodic class, then $Q(D)$, the expected value of the average cost per unit time, is

$$Q(D) = \sum_i \sum_j \pi_i D_{ij} w_{ij}, \tag{1}$$

where the π_j's uniquely satisfy the system of equations

$$
\begin{cases}
\pi_j \geq 0; \\
\sum_i \pi_i D_{ij} = \pi_j; \quad j = 0, 1, \ldots, L. \tag{2} \\
\sum_j \pi_j = 1.
\end{cases}
$$

The problem is to choose a rule $D \in C'$ such that $Q(D)$ is minimized, subject to the constraints that $\{Y_t: t = 0,1, \ldots\}$ has only one ergodic class, that the equations (2) are satisfied, and that

$$\pi_j = b_j, \quad j = 0,1, \ldots, L, \tag{3}$$

where the b_j's are the specified average frequencies of inspections of the indicated components. We assume that

$$h_j > 0, \quad j = 1, \ldots, L, \tag{4}$$

$$\sum_j b_j = 1. \tag{5}$$

Since the requirement for one ergodic class is not expressible in the form of linear constraints, as an approximation we substitute the restrictions $D_{oj} \geq \varepsilon_{oj}$ and $D_{jo} \geq \varepsilon_{jo}$, for all $j = 0,1, \ldots, L$; where the ε's are given positive numbers.

The approximate problem then is to find D so as to minimize

$$Q(D) = \sum_i \sum_j b_i D_{ij} w_{ij}, \quad D \in C'; \tag{6}$$

subject to

$$\begin{cases} D_{oj} \geq \varepsilon_{oj}, & j = 0, \ldots, L; \\ D_{jo} \geq \varepsilon_{jo}, & j = 0, \ldots, L; \end{cases} \tag{7}$$

and

$$\begin{cases} D_{ij} \geq 0, & i,j = 1, \ldots, L; \\ \sum_j D_{ij} = 1, & i = 0, \ldots, L; \\ \sum_i b_i D_{ij} = b_j, & j = 0, \ldots, L. \end{cases} \tag{8}$$

This problem is linear in the D_{ij}'s and always has a feasible solution provided that

$$\sum_{j=1}^{L} b_j < 1.$$

If the b_j's have a common value, then the above model can be thought of as a stochastic version of the "traveling salesman's" problem. If we also allow the ε's to be zero and if the irreducibility constraint is dropped, the linear programming problem becomes an "assignment" problem. A methodology for controlling the variation between inspections of the same component is provided.

Subject: Sales Force

Title: Sales Planning and Control Using Absorbing Markov Chains

Authors: William W. Thompson, Jr. and James U. McNeal

Source: Journal of Marketing Research, Vol. 4, February 1967, 62-65

Summary: A stochastic model that generates data for sales planning and control is described. An example is presented that shows how these data are used to plan short-run sales activities and train effective salesmen. Changes in customer propensities to buy are treated as Markov processes.

Model: ## Assumptions

1. The salesman sells only one product, the price of which is constant.

2. Sales are made on an ad hoc basis. There is no continuing customer-salesman relationship.

3. The expected cost of calling on a customer is equal for all customers.

4. The expected time consumed in a sales call is equal for all customers.

5. Customers may be classified on the basis of their relative propensities to buy as exhibited during the most recent previous call.

The system described here may be treated as a finite absorbing Markov process. It includes two absorbing states-- sale completed and sale lost--and n-2 nonabsorbing or transient states. These are ordered so as to establish a range of discrete classes, based on perceived differences in customer propensities to buy. We define one nonabsorbing state for the initial classification of new customers. The set of n states is exhaustive and mutually exclusive. The transition of customers from one state S_i to another S_j occurs according to the transition probability

$$\sum_{j=1}^{n} p_{ij} = 1 \quad (\text{for } i = 1, 2, \ldots, n). \tag{1}$$

For a given salesman, and for a given point in time, probability values are determined as follows:

$$p_{ij} = \frac{u_{ij}}{\sum_{k=1}^{m} u_{ik}}, \tag{2}$$

where u_{ij} is the number of times a transition from S_i to S_j is expected to occur in m trials. Consider a case where $n = 6$ and assume the following probability values:

$$P = \begin{bmatrix} 1 & 0 & 0 & 0 & 0 & 0 \\ 0 & 1 & 0 & 0 & 0 & 0 \\ .10 & .30 & 0 & .25 & .20 & .15 \\ .05 & .45 & 0 & .20 & .20 & .10 \\ .15 & .10 & 0 & .15 & .25 & .35 \\ .20 & .05 & 0 & .15 & .30 & .30 \end{bmatrix}. \tag{3}$$

Transforming (3) into canonical form:

$$P = \begin{array}{c} \\ S_1 \\ S_2 \\ S_3 \\ S_4 \\ S_5 \\ S_6 \end{array} \begin{bmatrix} S_1 & S_2 & S_3 & S_4 & S_5 & S_6 \\ 1 & 0 & 0 & 0 & 0 & 0 \\ 0 & 1 & 0 & 0 & 0 & 0 \\ \hline .10 & .30 & 0 & .25 & .20 & .15 \\ .05 & .45 & 0 & .20 & .20 & .10 \\ .15 & .10 & 0 & .15 & .25 & .35 \\ .20 & .05 & 0 & .15 & .30 & .30 \end{bmatrix}. \tag{4}$$

From (4)

$$N = (I-Q)^{-1} = \begin{bmatrix} 1 & .555 & .667 & .626 \\ 0 & 1.455 & .590 & .504 \\ 0 & .547 & 1.890 & 1.024 \\ 0 & .547 & .939 & 1.976 \end{bmatrix}. \tag{5}$$

The expected total number of calls that must be completed before a customer's absorption is,

$$\tau = N\xi, \tag{6}$$

where ξ is a column vector with all unity elements. For the example,

$$\tau = \begin{bmatrix} 2.848 \\ 2.549 \\ 3.461 \\ 3.462 \end{bmatrix}. \tag{7}$$

Thus (7) shows that a customer who is classified in S_3 will require an additional 2.848 calls before a sale is made or lost. The matrix,

$$B = NR, \tag{8}$$

gives probabilities b_{ij} that a process presently in the non-absorbing state S_i will be absorbed in the absorbing state S_j. From (5) and (4),

$$B = \begin{bmatrix} .352 & .648 \\ .261 & .739 \\ .515 & .485 \\ .562 & .438 \end{bmatrix}. \tag{9}$$

If the cost of making a call is assumed constant for all calls, the expected total costs of moving customers from each of the nonabsorbing states to absorbing states are given by the vector,

$$C = cr, \tag{10}$$

where c is a constant and r is the vector of (7). Assume that the cost of a call is \$15. Then,

$$C = \begin{bmatrix} 42.72 \\ 38.24 \\ 51.92 \\ 51.93 \end{bmatrix}. \tag{11}$$

Expected revenues for customers in each of the nonabsorbing states are given by

$$R = BU, \tag{12}$$

where B is the matrix of (8), and U is a column vector having revenue elements corresponding to the two absorbing states.

If, for example, a sale provides a constant \$190 in revenues, and a lost sale yields no revenue, then

$$U = \begin{bmatrix} 190 \\ 0 \end{bmatrix}, \tag{13}$$

and

$$R = \begin{bmatrix} 66.88 \\ 49.59 \\ 97.85 \\ 106.78 \end{bmatrix}. \tag{14}$$

Expected values of customers by states are given by the vector

$$V = R - C. \tag{15}$$

For the example,

$$V = \begin{bmatrix} 24.17 \\ 11.35 \\ 45.93 \\ 54.85 \end{bmatrix}. \tag{16}$$

This vector may be interpreted as follows: A customer now classified in S_4 has an expected value, in net dollar return, of only \$11.35. Similarly, a customer in S_6 is worth \$54.85. The priority ordering S_6, S_5, S_3, S_4 provides an economic rule for allocating the salesman's time.

Subject: Sales Force

Title: The Traveling Salesman Problem: A Survey

Authors: M. Bellmore and G.L. Nemhauser

Source: Operations Research, Vol. 16, No. 3, May-June, 1968,
           538-58

Summary: A survey and synthesis of research on the traveling
           salesman problem is given.   The problem is defined
           and several theorems are presented. This is followed
           by a general classification of the solution techniques
           and a description of some of the proven methods.
           A summary of computational results is given.

Model: In the traveling salesman problem we are given a non-

           negative integer n and n-dimensional square matrix

           $C = \{c_{ij}\}$.   Any sequence of $p+1$ integers taken from

           $(1,2, \ldots, n)$, in which each of the n integers

           appears at least once and the first and last integers

           are identical is called a tour. A tour may be written as

$$t = (i_1, i_2, i_3, \ldots, i_{p-1}, i_p, i_1).$$

           By a feasible solution to the traveling salesman
           problem, we mean a tour. An optimal solution is a tour
           such that

$$z(t) = \sum_{(i,j)\in t'} c_{ij} \text{ is minimized,}$$

           where $t' = [(i_1,i_2), (i_2,i_3), \ldots, (i_{p-1},i_p), (i_p,i_1)]$

           is the ordered pair representation of t.

           The n integers correspond to cities or nodes,
           the ordered pairs (i,j) are links or arcs joining the nodes,
           and c_{ij} is the 'distance' from node i to node j, or
           the length of arc (i,j). The tour t is a closed path
           passing through each node at least once. The length of
           the tour, denoted by $z(t)$, is the sum of the arc lengths
           over the arcs included in the tour.

A subtour s,

$$s = (i_1, i_2, i_3, \ldots, i_k, i_1)$$

is a closed path that does not pass through all of the n
nodes [$(i_1, i_2, \ldots, i_k)$ are distinct and $k < n$]. The
length of every subtour must be nonnegative, i.e.,

$$z(s) = \sum_{(i,j) \epsilon s'} c_{ij} \geqq 0,$$

where s' is the ordered pair representation of s.
The main theorems are stated without proof. They are:

Theorem 1

If C satisfies the triangle inequality, there is an
optimal tour in which each node is visited once and only
once.

Theorem 2

There exists an optimal tour that does not cross
itself when C satisfies the Euclidean distance measure.

Theorem 3

Let G be the convex hull of the points in two-dimensional
Euclidean space. There exists an optimal tour in which
the relative order of the points on the boundary of G is
preserved.

Theorem 4

If C is symmetric and t_1 and t_2 are two tours in which
the nodes are visited in reverse order, that is

$$t_1 = (i_1, i_2, \ldots, i_n, i_1), \quad t_2 = (i_1, i_n, \ldots, i_2, i_1)$$

then $z(t_1) = z(t_2)$.

In general, there are $(n-1)$ tours, but for symmetric
problems only $(n-1)!/2$ need to be considered.

Theorem 5

Let t be any tour in which each node is visited exactly once and $x_{ij} = 0$ if (i,j) is not in the tour and $x_{ij} = 1$ if (i,j) is in the tour. Then $\{x_{ij}\}$ is a feasible solution to the assignment problem with

$$z(t) = \sum_{i=1}^{i=n} \sum_{j=1}^{j=n} c_{ij}x_{ij} = w.$$

The "assignment problem" may be stated as

$$\min w = \sum_{i=1}^{i=n} \sum_{j=1}^{j=n} c_{ij}x_{ij},$$

subject to:

$$\sum_i x_{ij} = 1, \qquad (j = 1, \ldots, n)$$

$$\sum_j x_{ij} = 1, \qquad (i = 1, \ldots, n)$$

$$x_{ij} = 0,1. \qquad (\text{all } i \text{ and } j)$$

The converse is not true; feasible solutions to the assignment problem may not be tours.

Theorem 6

Let k_p and k_q be real numbers associated with a fixed pair of nodes p and q such that:

$$c'_{pj} = c_{pj} - k_p, \qquad (j = 1, \ldots, n; \; j \neq q)$$

$$c'_{iq} = c_{iq} - k_q, \qquad (i = 1, \ldots, n; \; i \neq p)$$

$$c'_{pq} = c_{pq} - k_p - k_q,$$

$$c'_{ij} = c_{ij}, \quad \text{otherwise,}$$

and $z'(t)$ be the length of tour t under C'. Then $z'(t) = z(t) - k_p - k_q$.

Theorem 7

If tour t does not contain arc (p,q) $z(t) \geq h_p + h_q$ where $h_p = \min_{j \neq p,q} c_{pj}$ and $h_q = \min_{i \neq p,q} c_{iq}$.

Theorem 8

Let p_1 and p_2 be two different permutations of the integers $(2,3, \ldots, k+1)$;

$$p_1 = (i_1, i_2, \ldots, i_k), \qquad p_2 = (j_1, j_2, \ldots, j_k)$$

and let

$$z(p_m) = \sum_{(i,j) \epsilon p_m'} c_{ij}, \qquad (m = 1,2)$$

where p_m' is the ordered pair representation of p_m. Then, if

$$c_{1i_1} + z(p_1) + c_{i_k s} < c_{1j_1} + z(p_2) + c_{j_k s},$$

then $(1, p_2, s)$ cannot be a segment of an optimal tour.

Theorem 9

Let S, $\bar{S}$ be a partition of the integers $i = 1, \ldots, n$: i.e., $S \cap \bar{S} = \phi$ and $S \cup \bar{S} = \{1, 2, \ldots, n\}$. For symmetric distances let $x_{ij} = 0$ if undirected arc (i,j) is not in a tour and $x_{ij} = 1$ if undirected arc (i,j) is in a tour. An optimal tour can be found by solving the integer program

$$\min z = \sum_{j=2}^{j=n} \sum_{i=1}^{j-1} c_{ij} x_{ij},$$

subject to:

$$x_{ij} = 0,1, \quad (i = 1, \ldots, j-1; \ j = 2, \ldots, n)$$

and the loop constraints

$$\sum_{i \epsilon S} \sum_{j \epsilon \bar{S}} x_{ij} \geq 2,$$

for all nonempty partitions $(S, \bar{S})$ such that if $(S, \bar{S})$ is considered $(\bar{S}, S)$ is not.

There are $2^{n-1} - 1$ of constraints in an n-city problem. Asymmetric problems require twice as many variables and loop constraints. Specifically, the loop constraints are

$$\sum_{i \varepsilon S} \sum_{j \varepsilon \overline{S}} x_{ij} \geq 1,$$

for all nonempty partitions $(S,\overline{S})$, where $x_{ij} = 1$, (0) if directed arc (i,j) is in (not in) a tour.

Theorem 10

Let $x_{ij} = 1$, (0) if directed arc (i,j) is in (not in) a tour. An optimal tour can be found by solving

$$\min z = \sum_{i=1}^{i=n} \sum_{j=1}^{j=n} c_{ij} x_{ij},$$

$$\sum_{i=1}^{i=n} x_{ij} = 1, \qquad \text{all } j \text{ except } j = i_o \ (i_o \text{ arbitrary}),$$

$$\sum_{j=1}^{j=n} x_{ij} = 1, \qquad \text{all } i \text{ except } i = i_o,$$

$$u_i - u_j + nx_{ij} \leq n-1, \qquad \text{all } i \text{ and } j \text{ except } i = j = i_o.$$

Theorem 11

Given the nodes $(i = 1, \ldots, n)$, arcs (i,j) and distance matrix C construct a new network containing the nodes and arcs from the original network plus one additional node, denoted by α, and an additional arc (j,α) for each j such that $(j,1)$ is an arc in the original network. The distances d_{ij} in the new network are:

$d_{ii} = 0, \quad$ for all i,

$d_{j1} = -\infty, \quad$ for all $j \neq 1$,

$d_{j\alpha} = k - c_{j1}, \quad$ for all $j \neq \alpha$,

$d_{ij} = k - c_{ij}, \quad$ otherwise,

where k is any finite number > sum of n largest c_{ij}.
A longest path from 1 to α in the new network contains
every intermediate node $(2, \ldots, n)$ and if $(1, i_1, \ldots,$
$i_{n-1}, \alpha)$ is such a longest path $(1, i_1, \ldots, i_{n-1}, 1)$
is an optimal tour.

Subject: Sales Force

Title: A Multiple-Product Sales Force Allocation Model

Authors: David B. Montgomery, Alvin J.,Silk, and Carlos E. Zaragoza

Source: Management Science, Vol. 18, No. 4, Part 2, December
           1971, P-3-24

Summary: When several products are marketed by the same sales force,
           it frequently becomes impossible or impractical for salesmen
           to promote all items in the product line extensively in
           each and every time period. Management's problem is to
           decide how the available selling effort should be allocated
           across products and over time. The opportunity costs
           associated with using limited selling resources to promote
           certain products but not others must be evaluated. This
           paper describes a decision calculus-type modeling system
           for dealing with this question.

           The problem is analyzed by a two-step procedure. First,
           a response function is defined which relates selling effort
           to sales and profit results in a manner which represents some
           behavioral phenomena considered to be important. An inter-
           active conversational program elicits judgmental data from
           managers which are used to parameterize the response model.
           A separate response function is specified for each product
           in the firm's line by this method. The set of response
           functions so obtained becomes the input for the second
           component of the system, an allocation heuristic. An
           incremental search procedure is employed to find an allocation
           of the sales force's time to the various products and over
           several time periods which is "best" in terms of total
           contribution to company profits. The model is presented in
           the context of an ethical drug manufacturer's multiple-product
           sales force allocation problem.

Model: Assumptions

           1. A firm sells a set of products in a competitive market.
              It is assumed that cross elasticities of demand with
              respect to sales effort for the firm's products are zero.
              Personal selling is a primary tool for promoting the
              product line.

           2. Past and potential buyers are numerous and heterogeneous.
              The duration of a sales call is ordinarily shorter than
              the seller would like.

           3. For the planning horizon under consideration (e.g., a
              year), the size of the sales force is essentially fixed.
              Each salesman operates in an exclusive territory. The
              major portion of his compensation consists of a fixed
              salary.

The area of investigation is the field of ethical drugs.
The company has come to believe that no more than three
products can be effectively presented in a single sales
visit. Thus, each salesman contacts N doctors and promotes
three products in each call.

Let the number of products receiving "complete," "half,"
(drug is promoted in 50% of the calls) and "quarter" coverage
be represented by $D(C)$, $D(H)$, and $D(Q)$, respectively. The
values of these latter quantities must satisfy the following
relation:

$$3N = D(C)N + D(H)N/2 + D(Q)N/4. \tag{1}$$

Defining the "relative exposure value" of various detailing
policies, let

$$REV(X_t) = f(X_t), \tag{2}$$

where

$REV(X_t)$ = Relative Exposure Value of a particular detailing
policy,

X_t = a discrete detailing policy alternative used in
time period t, i.e., X_t = complete coverage,
half-coverage, quarter coverage, or no detailing.

To account for accumulation and decay of REV, we employ an
exponential forgetting function:

$$ATE(t) = f[REV(X_t)] + (1-\lambda)ATE(t-1), \quad 0 < \lambda < 1, \tag{3}$$

where

$ATE(t)$ = level of REV in period t that has accumulated
as a result of current and past detailing,

$f[REV(X_t)]$ = effect on $ATE(t)$ of detailing policy carried out
in period t,

λ = forgetting parameter.

Sales in period t are given by

$$S(t) = SP(t)*SI(t), \qquad (4)$$

where

    $S(t)$ = unit sales in period t,

    $SP(t)$ = sales potential in period t,

    $SI(t)$ = sales index in period t.

The sales index at t is given by:

$$SI(t) = \alpha + \beta[ATE(t)]^2 - \gamma[ATE(t)]^3 \text{ if } \alpha + \beta[ATE(y)]^2$$
$$- \gamma[ATE(t)]^3 \leqq 100 \qquad (5)$$

or

$$= 100 \quad \text{if} \quad \alpha + \beta[ATE(t)]^2 - \gamma[ATE(t)]^3 > 100,$$

where α, β, and γ are all nonnegative constants. Since $0 \leqq ATE(t)$, we have $0 \leqq SI(t) \leqq 100$. The upper bound of 100 is set to be consistent with (4) above. That is, no more than 100% of sales potential may be realized.

The detailing policy in each period must satisfy the equality constraint:

$$\sum_{i=1}^{I} X(i,t) = 12. \qquad (6)$$

The objective function is total product line contribution over the planning horizon

$$TGP = \sum_{i=1}^{I} GM(i) \sum_{t=1}^{T} SP(i,t)*SI(i,t) \qquad (7)$$

where

    TGP = total gross product line profits over the planning horizon,

    $GM(i)$ = % gross margin on product i, and $SP(i,t)$ and $SI(i,t)$ are as defined in (4) and (5).

The allocation is carried out by a heuristic algorithm. Reference: #67.

Subject: Sales Force

Title: CALLPLAN: An Interactive Salesman's Call Planning System

Author: Leonard M. Lodish

Source: Management Science, Vol. 18, No. 4, Part 2, December
          1971, P-25-40

Summary: CALLPLAN is an interactive computer system designed to
          aid salesmen or sales management in allocating sales
          call time more efficiently. CALLPLAN uses as input
          the salesman's own best estimates of expected contribution
          of all possible call policies for each account and
          prospect. The computer can help the estimating procedure
          by fitting curves through estimated points on a response
          function or by obtaining expected values from probability
          estimates. The system solves a mathematical program
          which determines the best time allocation to maximize
          contribution according to these estimates. Factors
          considered by the system include travel time and costs
          to get to geographical areas within the territory, amount
          of time required per call on an account within an area,
          account profitability, and minimum and maximum account
          call frequency limitations. An incremental analysis
          routine is discussed as a solution procedure for the
          mathematical program.

Model: Let

x_i = the number of calls to be made during an effort
period on account i, constrained to be within
minima and maxima input by the salesman.

Min_i = the minimum number of calls which must be made
on account i during an effort period regardless
of sales or contribution resulting from the calls.

Max_i = the maximum number of calls which can be made
on account i during an effort period.

$r_i(x_i)$ = the expected sales to account i during the response
period if x_i calls are made during an average effort
period.

Our fitting procedure uses pieces of two four-parameter curves
for different areas of the response curve. The four-parameter
curve is of the form:

$$r_i(x_i) = ZER + (SAT - ZER)\ \frac{x_i^{\sigma}}{\gamma + x_i^{\sigma}} \qquad (1)$$

where

ZER = expected sales during the response period with 0
calls during an average effort period,

SAT = expected sales with saturation sales call effort.

The parameters ZER, SAT, σ and γ are uniquely determined by four input data points. This curve is used twice to obtain the complete response function.

Let

    a = account specific adjustment factor,

    j = geographic area,

    t = time unit,

    u = average time unit for each trip to a geographical area,

    c = out of pocket expenses,

    NT = number of trips during average effort period,

    e = effort periods in each response period,

    g_i = the geographic area in which account i is located.

In terms of mathematical programming, the objective is to find x_i for $i = 1, \ldots, I$ to maximize z, the total adjusted expected sales from all accounts and prospects minus travel costs over the response period, where

$$z = \sum_{i=1}^{I} a_i r_i(x_i) - e \sum_{j=1}^{J} NT_j c_j. \tag{2}$$

The amount of time spent on selling and traveling must be less than T, the amount of time available during an average effort period.

$$\sum_{i=1}^{I} t_i x_i + \sum_{j=1}^{J} NT_j u_j \leq T. \tag{3}$$

The number of trips to an area is a function of the number of calls made to each account in the area.

$$NT_j = \text{Max} \{x_i \text{ such that } g_i = j\} \quad \text{for} \quad j = 1, \ldots, J. \tag{4}$$

Also the number of calls per effort period must lie within the stated bounds

$$\text{Min}_i \leq x_i \leq \text{Max}_i \quad \text{for} \quad i = 1, \ldots, I. \tag{5}$$

The problem is to find x_i for $i = 1, \ldots, I$ to maximize z subject to constraints (3), (4), and (5). An incremental analysis routine which solves a "loose" version of the problem is discussed.

Subject: Marketing Mix

Title: Some Correlates of Coffee and Cleanser Brand Shares

Author: Seymour Banks

Source: Journal of Advertising Research, Vol. 1, No. 4, June 1961, 22-28

Summary: A model of market demand for brands of convenience goods (coffee and cleanser) is described and tested.

Model: Market share is taken to mean a brand's share of the total volume of sales of the given product class in a certain geographical area.

The general demand model may be written as:

$$P_i = f_c(A_1, A_2, \dots) + f_r(B_1, B_2, \dots) + f_w(D_1, D_2, \dots)$$
$$C_i \qquad\qquad R_i \qquad\qquad W_i$$
$$+ f_m(E_1, E_2, \dots)$$
$$M_i$$

where

        P = a brand's share of the market,

        C = consumer evaluation of the intrinsic attributes of a brand,

        A = criteria by which consumers evaluate brand qualities (flavor, bouquet, etc.),

        R = selling effort by retailers,

        B = retailer's performance of activity (special displays, repairs, etc.),

        W = selling effort by wholesalers,

        D = wholesaler's performance of activity (demonstrations, credit concessions, etc.),

        M = selling effort by manufacturers,

        E = manufacturer's performance of activity (new consumer advertising campaigns, special prices, etc.).

The model is tested on data for coffee and scouring cleanser using regression analysis.

Subject: Marketing Mix

Title: Mathematical Model for a Duopolistic Market

Authors: K.S. Krishnan and S.K. Gupta

Source: Management Science, Vol. 13, No. 7, March 1967, 568-83

Summary: A model for a marketing situation with two competitors when each competitor has two control variables, price and promotional effort is presented. Mills [#84] assumed that each competitor has only one control variable, viz., promotional effort and the profit margins of the two competitors are known. In this paper the profit margins are also control variables. The paper derives conditions under which nonboundary equilibrium solutions exist and the sensitivity of the model for small deviations in the decision variables from their equilibrium values is tested. Mills' results are found to be valid only under certain conditions.

Model:

There are two competitors X_1 and X_2 with c_1 and c_2 as unit manufacturing costs. Consider the two-person game with the following payoff functions denoted by R_i $(i = 1,2)$

$$R_i = A[\alpha_i S_i/(\alpha_1 S_1 + \alpha_2 S_2) + k(p_1 + p_2 - 2p_i)](p_i - c_i) - S_i$$

$$(1)$$

where

  A = total market potential (constant)
  α = effectiveness of promotional effort per dollar
  S = promotional expenditure
  p = selling price
  k = positive constant.

Each competitor tries to maximize his payoff. The problem is to find equilibrium values of p_i and S_i ($i = 1,2$) for the game defined by (1) in the sense that if any competitor deviates from the equilibrium value, his payoff decreases.

Non-Boundary Solutions

When both S_1 and S_2 are positive, the necessary and sufficient conditions for the maximum profit of competitor X_i are

$$\partial R_i/\partial p_i = \partial R_i/\partial S_i = 0 \qquad \text{(for } i = 1,2) \qquad (2)$$

$$\partial^2 R_i/\partial p_i^2 < 0, \ \partial^2 R_i/\partial S_i^2 < 0 \qquad\qquad (3)$$

and $(\partial^2 R_i/\partial p_i^2)(\partial^2 R_i/\partial S_i^2) > [\partial^2 R_i/\partial p_i \partial S_i]^2$ (for $i = 1,2$).

For competitor X_1

$$\partial R_1/\partial p_1 = Af_1 - Ak(p_1 - c_1) \qquad \text{where}$$

$$f_1 = \alpha_1 S_1/(\alpha_1 S_1 + \alpha_2 S_2) + k(p_2 - p_1)$$

$$\partial R_1/\partial S_1 = A\alpha_1 \alpha_2 S_2 (p_1 - c_1)/(\alpha_1 S_1 + \alpha_2 S_2)^2 - 1$$

$$\partial^2 R_1/\partial p_1^2 = -2Ak$$

$$\partial^2 R_1/\partial S_1 \partial p_1 = A\alpha_1 \alpha_2 S_2/(\alpha_1 S_1 + \alpha_2 S_2)^2$$

$$\partial^2 R_1/\partial S_1^2 = -2A\alpha_1^2 \alpha_2 S_2 (p_1 - c_1)/(\alpha_1 S_1 + \alpha_2 S_2)^3.$$

Hence, from (2) and (3) optimum p_1 and S_1 satisfy the following relations:

$$p_1 = c_1 + f_1/k; \quad 4k(p_1 - c_1) > \alpha_2 S_2/(\alpha_1 S_1 + \alpha_2 S_2) \quad (4)$$

and

$$(p_1 - c_1)S_2 = (\alpha_1 S_1 + \alpha_2 S_2)^2/A\alpha_1 \alpha_2 \quad \text{where } S_2 > 0.$$

Similarly, the optimum decision of competitor X_2 satisfies the following relations:

$$p_2 = c_2 + f_2/k; \quad 4k(p_2 - c_2) > \alpha_1 S_1/(\alpha_1 S_1 + \alpha_2 S_2) \quad (5)$$

and

$$(p_2 - c_2)S_1 = (\alpha_1 S_1 + \alpha_2 S_2)^2/A\alpha_1\alpha_2 \quad \text{where} \quad S_1 > 0.$$

When both S_1 and S_2 are strictly positive, then p_1, p_2, S_1 and S_2 can be obtained from equations (4) and (5).

Let

$$k(c_2 - c_1) = \lambda$$

$$\alpha_1/\alpha_2 = \alpha$$

and

$$S_1/S_2 = y.$$

Then[1]

$$p_1{}^* - c_1 = (1 + \lambda + \alpha y^*/(1 + \alpha y^*))/(3k)$$

$$p_2{}^* - c_2 = (1 - \lambda + 1/(1 + \alpha y^*))/(3k)$$

$$(p_1{}^* - c_1)S_2{}^* = (p_2{}^* - c_2)S_1{}^*.$$

The above three relations by eliminating $p_1{}^*$ and $p_2{}^*$ can be simplified as

$$y^2(1 - \lambda)\alpha + y[2 - \lambda - (2 + \lambda)\alpha] - (\lambda + 1) = 0$$

$$S_1{}^* = (A\alpha y^*/(1 + \alpha y^*)^2)(1 + \lambda + \alpha y^*/(1 + \alpha y^*))/(3k)$$

$$S_2{}^* = (A\alpha y^*/(1 + \alpha y^*)^2)(1 - \lambda + 1/(1 + \alpha y^*))/(3k).$$

Also, the market share must be less than unity. This leads to

$$\lambda \leqq 1 + 1/(1 + \alpha y^*).$$

The revenues are given by

---

[1]Asterisks denote nonboundary values.

$$R_1^* = Ak(p_1^* - c_1)^2 - S_1^*$$

$$R_2^* = Ak(p_2^* - c_2)^2 - S_2^*.$$

Boundary Solutions

(i) When $S_1 = 0$ and $S_2 > 0$, the payoff functions are

$$R_1 = Ak(p_2 - p_1)(p_1 - c_1)$$

$$R_2 = A[1 + k(p_1 - p_2)](p_2 - c_2) - S_2.$$

By differentiation, one can verify that R_1 and R_2 are maxima when[2]

$$p_1^{**} - c_1 = (1 + \lambda)/(3k)$$

$$p_2^{**} - c_2 = (2 - \lambda)/(3k).$$

The value of S_2 should be such that

$$\partial R_1/\partial S_1 \leqq 0 \qquad \text{for all } S_1.$$

Then

$$(A\alpha/3k)(1 + \lambda)/(S_2^{**})) - 1 \leqq 0 \quad \text{or} \quad S_2^{**} \geqq A\alpha(1 + \lambda)/3k.$$

Hence, competitor X_2 must spend $A\alpha(1 + \lambda)/(3k)$ on promotional effort. The revenues are given by

$$R_1^{**}(S_1 = 0) = A(1 + \lambda)^2/(9k)$$

$$R_2^{**}(S_1 = 0) = A(2 - \lambda)^2/(9k) - A\alpha(1 + \lambda)/(3k).$$

(ii) When $S_1 > 0$ and $S_2 = 0$, the payoff functions are

$$R_1 = A[1 + k(p_2 - p_1)](p_1 - c_1) - S_1$$

$$R_2 = Ak(p_1 - p_2)(p_2 - c_2). \quad .$$

---

[2] Double asterisks denote boundary values when either or both $S_i = 0$, $i = 1, 2$.

The solutions are given by

$$p_1^{**} - c_1 = (2 + \lambda)/(3k); \quad S_1^{**} = A(1 - \lambda)/(3k\alpha)$$

$$p_2^{**} - c_2 = (1 - \lambda)/(3k); \quad S_2^{**} = 0$$

$$R_1^{**}(S_2 = 0) = A(2 + \lambda)^2/(9k) - A(1 - \lambda)/(3\alpha k)$$

$$R_2^{**}(S_2 = 0) = A(1 - \lambda)^2/(9k).$$

(iii) When $S_1 = S_2 = 0$, the payoff functions are

$$R_1 = 0$$

$$R_2 = 0.$$

The equilibrium solutions are found by comparing the non-boundary solutions with the boundary solutions.
The result of Mills can be derived by putting $\alpha = 1$, when $\lambda \leqq \frac{1}{9}$.

References: #84 and 141.

Subject: Marketing Mix

Title: Mathematical Models in Marketing

Author: Shiv K. Gupta and K.S. Krishnan

Source: Operations Research, Vol. 15, No. 6, November-December 1967, 1040-50

Summary: Mathematical models are developed expressing the relation between net revenue and the variables that affect it. These models are extensions of the ones discussed by Mills [#84] and Krishnan and Gupta [#140]. Mills assumed each competitor has only one controllable variable, viz., promotional effort. Krishnan and Gupta obtained equilibrium solutions for two competitors, each having two controllable variables. This paper deals with multi-competitors and discusses four models when the market potential is independent of both price and promotional effort and when it is dependent on either or both of the controllable variables.

Model: ASSUMPTIONS

1. There are n competitors X_1, X_2, ..., X_n and each competitor has two controllable variables, viz., price and promotional effort.

2. Manufacturing cost of each competitor is directly proportional to the production quantity and each competitor knows manufacturing costs of all other competitors.

3. The effectiveness of promotional effort per dollar is different for different competitors.

4. Each competitor's share of market depends on his relative effective promotional effort and the difference between his price and average price of all other competitors.

5. Collusion is not permitted and each competitor tries to maximize his net revenue.

PROBLEM

The Payoff (net-revenue) for X_j is given by

Payoff = (total volume of sales) x (margin of profit) - (promotional expenditure),

$$R_j = \begin{cases} Af_j m_j - s_j & \text{if } \sum_{i=1}^{i=n} s_i > 0, \\ 0 & \text{if } s_1 = s_2 = s_3 \ldots = s_n = 0, \end{cases}$$

where, for X_j,

R_j = net revenue,

A = market potential which is a function of price and promotional efforts of all competitors,

$f_j = v_j + ux_j + k/(n-1)(\sum_{i=1}^{i=n} p_i - np_j)$,

v_j = contribution of factors like quality of product, distribution channels, that influence sales,

$x_j = (\alpha_j s_j)^{e_j} / \sum_{i=1}^{i=n} (\alpha_i s_i)^{e_i}$,

α_j = effectiveness of promotional effort per \$ of X_j,

p_j = price,

s_j = expenditure on promotional effort,

$m_j = p_j - c_j$ = margin of profit,

c_j = unit cost of production,

e_i, α_i, v_i, u, and k are all constants and $\sum_{i=1}^{i=n} v_i + u = 1$.

It is assumed that production is equal to sales of any competitor. The problem is to find equilibrium points (p^*, s^*) for each competitor in the sense that if any competitor deviates from the equilibrium values, his payoff goes down. The equilibrium values of p_j and s_j are obtained by maximizing payoff for X_j subject to the constraints that $s_j \geq 0$ and $m_j \geq 0$.

SOLUTIONS

Model I: Market Potential is Independent of Price and Promotional Effort

Applying Kuhn-Tucker conditions, the maximum value of R_j subject to $s_j \geq 0$ is obtained from

$$\partial R_j / \partial p_j = 0,$$
$$\partial R_j / \partial s_j \leq 0, \qquad\qquad (1)$$
$$s_j \partial R_j / \partial s_j = 0.$$

If $s_j > 0$, then $\partial R_j/\partial s_j = 0$. (1) is solved for:

     (i) $s_j > 0$;

     (ii) $s_j = 0$; $\sum_{i=1}^{i=n} s_i > 0$.

$\partial R_j/\partial p_j = 0$ gives

$$f_j = km_j = v_j + ux_j + [k/(n-1)] \left(\sum_{i=1}^{i=n} p_i - np_j \right). \tag{2}$$

$\partial R_j/\partial s_j = 0$ gives

$$s_j = Am_j ue_j x_j (1 - x_j). \tag{3}$$

Solving (2) and (3) for $j = 1, 2, \ldots, n$ the solutions corresponding to $s_j > 0$ and $m_j > 0$ are given by (asterisks denote these values):

$$\sum_{i=1}^{i=n} p_i^* = \sum_{i=1}^{i=n} c_i + 1/k,$$

$$m_j^* = [(n-1)(v_j + ux_j^*) + 1 + \lambda_j]/k(2n-1)$$

where

$$\lambda_j = k \left(\sum_{i=1}^{i=n} c_i - nc_j \right),$$

$$s_j^* = Aue_j m_j^* x_j^* (1 - x_j^*),$$

$$R_j^* = Akm_j^{2^*} - s_j^*,$$

when $s_j = 0$ and $\sum_{i=1}^{i=n} s_i > 0$, then

$$R_j = A[v_j + k/n-1) \left(\sum_{i=1}^{i=n} p_i - np_j \right) m_j,$$

$$R_i = A[v_i + ux_i + k/(n-1) \left(\sum_{w=1}^{w=n} p_w - np_i \right)] m_i - s_i. \quad (i \neq j)$$

By differentiating R_j and R_i with respect to the respective controllable variables and simplifying, the solutions corresponding to $s_j = 0$ are given by (** denote these solutions):

$$m_j^{**} = [(n-1)v_j + 1 + \lambda_j]/k(2n-1),$$

$$m_i^{**} = [(n-1)(v_i + ux_i^{**}) + 1 + \lambda_i]/k(wn-1),$$

$$s_i^{**} = Aue_i m_i^{**} x_i^{**}(1 - x_i^{**}),$$

$$R_j^{**} = Ak \cdot m_j^{2**},$$

$$R_i^{**} = Akm_j^{2**} - s_i^{**}.$$

If $s_1 = s_2 = \ldots = s_n = 0$, then the payoff for each competitor is zero.

Particular Case with Two Competitors

Let $n=2$, $e_1 = e_2 = 1$, $v_1 = v_2 = 0$ and $u = 1$. Then the payoff for X_j is given by

$$R_j = A[(\alpha_j s_j/\alpha_1 s_1 + \alpha_2 s_2) + k(p_i - p_j)]m_j - s_j, \quad (i \neq j) \quad (4)$$

$$\text{if } s_1 + s_2 > 0$$
$$= 0 \quad \text{if } s_1 = s_2 = 0.$$

This model has been discussed in detail by Krishnan and Gupta and is an extension of Mills' model. Mills assumed that each competitor has only one controllable variable, whereas here each competitor has two controllable variables, viz., price and promotional effort.

Model II: Total Market Potential is a Function of Promotional Effort Only

$$\partial A/\partial p_j = 0,$$

$$\partial A/\partial s_j = \alpha_j \, dA/dz.$$

If $s_j > 0$ and $m_j > 0$, the maximum value of R_j is given by

$$\partial R_j/\partial p_j = 0, \quad (5)$$

$$\partial R_j/\partial s_j = 0.$$

If $s_j = 0, \sum_{i=1}^{i=n} s_i > 0$ and $m_j > 0$, then the maximum R_j is given by

$$\partial R_j / \partial p_j = 0,$$

$$s_j = 0.$$

If $s_1 = s_2 = \ldots = s_n = 0$, the net revenue $R_j = 0$.

Let $A = a[1 - \exp(-\gamma z)]$ where a and γ are positive constants.

If $s_j > 0, m_j > 0$

$$m_j^* = [(n-1)(v_j + ux_j) + 1 + \lambda_j]/k(2n-1),$$

$$s_j^* = Aue_j m_j^* x_j^* (1 - x_j^*) + km_j^{2*} \alpha_j \gamma (a-A);$$

if $s_j = 0, \sum_{i=1}^{i=n} s_i > 0$ and $m_j > 0$,

$$m_j^{**} = [(n-1)v_j + 1 + \lambda_j]/k(2n-1).$$

By comparing profits in the two situations, the equilibrium solutions can be found.

Model III: Total Market Potential is a Function of Price Only

$$\partial A / \partial s_j = 0 \quad \text{for} \quad j = 1, 2, \ldots, n.$$

If $s_j > 0$ and $m_j > 0$, the maximum value of R_j is obtained when

$$\partial R_j / \partial p_j = 0,$$

and

$$\partial R_j / \partial s_j = 0. \tag{6}$$

If $s_j = 0, \sum_{i=1}^{i=n} s_i > 0$ and $m_j > 0$, then maximum value of R_j is given by

$$\partial R_j / \partial p_j = 0. \tag{7}$$

When $A = a[1 - \sum_{i=1}^{i=n} w_i P_i]$, (6) and (7) simplify as follows:

If $s_j > 0$ and $m_j > 0$,

$$m_j^* = (A - aw_j m_j^*)[v_j + ux_j^* + k/(n-1)(\sum_{i=1}^{i=n} p_i^* - n_j^*)]/Ak. \quad (8)$$

$$s_j^* = Aue_j m_j^* x_j^* (1 - x_j^*). \quad (9)$$

If $s_j = 0, \sum_{i=1}^{i=n} s_i > 0$ and $m_j > 0$ then

$$m_j^{**} = (A - aw_j m_j^{**})[v_j + k/(n-1)(\sum_{i=1}^{i=n} p_i^{**} - np_j^{**})]/Ak. \quad (10)$$

$$(j = 1,2, \ldots, n).$$

The equilibrium solutions can be obtained by comparing the profits in the two situations.

Model IV: Total Market Potential is a Function of both Price and Promotional Effort

The equilibrium solutions of X_j are obtained, as before, by comparing maximum payoff when $s_j > 0$ with that when $s_j = 0$.

Let $A = a[1 - \sum_{i=1}^{i=n} w_i p_i][1 - \exp(-\gamma z)]$,

where

p_i = price charged by X_i,

$z = \sum_{i=1}^{i=n} (\alpha_i s_i)$,

and a, w_i, γ are constants. If $s_j > 0$ and $m_j > 0$, the maximum value of R_j is obtained when

$$(1 - w_j m_j^* - \sum_{i=1}^{i=n} w_i p_i^*)[v_j + ux_j^* + k/(n-1)(\sum_{i=1}^{i=n} p_i^* - np_j^*)]$$

$$= km_j^*(1 - \sum_{i=1}^{i=n} w_i p_i^*),$$

$$s_j^* = a(1 - \sum_{i=1}^{i=n} w_i p_i^*)m_j^*[ue_j x_j^*(1 - x_j^*)$$

$$+ \gamma \alpha_j s_j^* \exp(-\gamma z^*) f_j^*]. \quad (j = 1,2, \ldots, n) \quad (11)$$

If $s_j = 0$, $\sum_{i=1}^{i=n} s_i > 0$ and $m_j > 0$, then the maximum value
of R_j is obtained when

$$(A - aw_j m_j^{**}) f_i^{**} = Akm_j^{**}. \qquad (j = 1, 2, \ldots, n) \quad (12)$$

By comparing the profits, the equilibrium solution can be
found.

References: #84 and 140.

Subject: Marketing Mix

Title: Determinants of Market Share

Author: Doyle L. Weiss

Source: Journal of Marketing Research, Vol. 5, August 1968,
          290-95

Summary: Analysis of market share movements for a low-cost,
          frequently purchased consumer product. An examination
          of the influence of price and advertising on sales
          is presented.

Model:

Model 5

$$S_{B,t} = a_o + a_1 P_{B,t} + a_2 A_{B,t} + a_3 Q_1 + a_4 Q_2,$$

where

$S_{B,t}$ = market share for Brand B at time t

$P_{B,t}$ = price ($\$$ per oz) for Brand B at t

$A_{B,t}$ = advertising expenditures (thousands of $)
      for Brand B at t.

Q_1 = 1 when B = 2 and 0 otherwise

Q_2 = 1 when B = 1 and 0 otherwise.

The dummy variables represent proxies for quality and
distribution measures.

Model 6

$$S_{B,t} = a_o + a_1 (P_{B,t} - \overline{P}_t)$$
$$+ a_2 (A_{B,t} - \overline{A}_t) + a_3 Q_1 + a_4 Q_2,$$

where

$\overline{P}$ = average price (weighted by volume) for all
      three brands for period t,

$\overline{A}_t = \sum_B A_{B,t}/3$.

Model 7

$$S_{B,t} = a_o + a_1 (P_{B,t}/\overline{P}_t)$$

$$+ a_2 (A_{B,t}/\overline{A}_t) + a_3 Q_1 + a_4 Q_2.$$

Model 8

$$S_{B,t} = a_o \{(P_{B,t}/\overline{P}_t)^{a_1}\}$$

$$\{(A_{B,t}/\overline{A}_t)^{a_2}\}\{e^{(a_3 Q_1 + a_4 Q_2)}\}.$$

When a log transformation (Base e) is made
Model 8 becomes:

$$\log (S_{B,t}) = \log (a_o) + a_1 \log (P_{B,t}/\overline{P}_t)$$

$$+ a_2 \log (A_{B,t}/\overline{A}_t) + a_3 Q_1 + a_4 Q_2.$$

Models 6, 7, and especially 8 produced the best fits.

Reference: #148.

Subject: Marketing Mix

Title: A Theory of Market Segmentation

Authors: Henry J. Claycamp and William F. Massy

Source: Journal of Marketing Research, Vol. 5, November, 1968, 388-94

Summary: A normative theory of market segmentation is presented as a multistage mathematical model of the full range of segmentation possibilities from the perfectly discriminating monopolist to the mass marketer. The theory's major implications for the philosophy and application of the market segmentation strategy are discussed.

Model: The model uses only price and promotional variables but could be revised to include any of the marketing mix elements.

We assume a market with firms sufficiently decoupled such that strategies can be planned without direct reference to problems of possible competitive retaliation. The analysis considers profit maximization strategies for a single product. The model will be developed in five stages.

Stage 1: Segmentation by Perfect Discrimination among Customers

Suppose that a firm attempts to market its product to N customers, each with the demand function

$$d_i = f_i(p_i, x_i), \quad i = 1, \ldots N,$$

where p_i is the price and x_i a vector of m nonprice promotional variable offered to the ith customer. If the unit cost of distribution (not including promotion) to the ith customer is c_i, the firm's gross revenue equation can be written as:

$$R = \sum_{i=1}^{N} (p_i - c_i) d_i = \sum_{i=1}^{N} (p_i - c_i) f_i(p_i, x_i).$$

The firm's cost equation includes the costs of supplying and promoting the product:

$$C = g \{ \sum_{i=1}^{N} d_i \} + \sum_{i=1}^{N} q_i,$$

where d_i is the product demand and q_i is the total cost of implementing the promotional package, denoted by x_i, both for the ith customer. The function $g\{\cdot\}$ does not include distribution or promotion costs. The cost equation is rewritten as an explicit function of p_i and x_i. Let v_i^1 be the vector of per unit costs of promotion to the ith customer, so that $q_i = v_i^1 x_i$.

We can write the following profit equation for the firm:

$$\Pi = R - C = \sum_i (p_i - c_i) f_i (p_i, x_i) - g\{\sum_i f_i (p_i, x_i)\} - \sum_i v_i^1 x_i. \quad (1)$$

The firm's optimal marketing mix will be obtained when equation (1) is maximized with respect to p_i and the elements of x_i (for $i = 1, \ldots N$).

By differentiating partially and setting the derivatives equal to zero, we get the decision rules:

$$(p_i - c_i - MC) \frac{\partial f_i}{\partial p_i} = -f_i, \quad i = 1, \ldots N,$$

$$(p_i - c_i - MC) \frac{\partial f_i}{\partial x_{ij}} = v_{ij}, \quad \begin{array}{l} i = 1, \ldots N \\ j = 1, \ldots m; \end{array} \quad (2)$$

where MC is the cost function derivative with respect to total demand. The $(m+1)N$ equations must be solved to determine the optimal price-promotional mix for each market (here, an individual customer) supplied by the firm.

Stage 2: Customer Segmentation with Institutional Constraints

Suppose that the firm faces a fixed set of promotional vehicles (through which it must exercise its nonprice marketing efforts) denoted by the vector $y_1, y_2, \ldots y_n$, which we shall call "media." The elements of the nonprice promotional vector for the ith customer can be related to the media by

$$x_{ij} = \Psi_i(y) \quad j = 1, \ldots m; \quad i = 1, \ldots N.$$

We shall assume that the Ψ-functions are all linear and write

$$x_{ij} = \sum_{k=1}^{n} b_{ijk} y_k,$$

where the "media characteristic parameters" b_{ijk} represent the contribution of the kth kind of promotional input for the ith customer. In matrix form:

$$x_i = \underline{B}_i y, \quad i = 1, \ldots N, \tag{3}$$

where $\underline{B}_i$ is the mxn matrix of media characteristic parameters. Suppose $p_i = p$ for all i and $c_i = c$ in all cases. Let the vector $w' = w_1, w_2, \ldots w_n$ be the per unit costs of using the media. Then,

$$\Pi = R - C = (p-c) \sum_i f_i(p, B_i y) - w'y - g\{\sum_i f_i(p, B_i y)\}. \tag{4}$$

Differentiation with respect to price leads to

$$(p - c - MC) \sum_i \frac{\partial f_i}{\partial p} = -\sum_i f_i.$$

The derivative with respect to a given medium variable y_k is:

$$\frac{\partial \Pi}{\partial y_k} = 0 = (p-c) \sum_i \sum_j \frac{\partial f_i}{\partial x_{ij}} \frac{\partial x_{ij}}{\partial y_k} - w_k$$

$$- MC \sum_i \sum_j \frac{\partial f_i}{\partial x_{ij}} \frac{\partial x_{ij}}{\partial y_k}.$$

Transposing and recognizing that $\partial x_{ij}/\partial y_k = b_{ijk}$, we have

$$(p - c - MC) \sum_i \sum_j b_{ijk} \frac{\partial f_i}{\partial x_{ij}} = w_k, \quad k = 1, \ldots n. \tag{5}$$

This result differs from the one for Stage 1 because weighted averages of the response derivatives $\partial f_i / \partial x_{ij}$ are used in aggregated equations, instead of individual terms in individual equations.

Stage 3: Microsegmentation

Suppose that media circulation is known only for a total of M mutually exclusive and exhaustive consumer classes. These classes will be called media descriptor classes or microsegments. The media characteristic coefficient matrices now refer to the descriptor classes rather than to individual customers--we have $\underline{\underline{B}}_\ell$, $\ell = 1, \ldots M$, where, for example, a given matrix might refer to "high-income, high-educated persons over 65." In principle, these matrices can be determined from audience survey information. Introducing descriptor classes leads to the following modification of (5):

$$(p - c - MC) \sum_j \sum_\ell b_{ijk} \sum_{i \in \ell} \frac{\partial f_i}{\partial x_{ij}} = w_k; \; k = 1, \ldots n, \quad (6)$$

where the notation $i \in \ell$ means all persons within the ℓth descriptor cell.

Stage 4: Macrosegmentation

The macrosegment h consists of the customers in media descriptive cells $\ell \in h$. The media characteristics for macrosegments h can be found by simple aggregation.

Modifying (6) to accommodate the higher level of aggregation,

$$(p - c - MC) \sum_j \sum_h \{ \sum_{\ell \in h} b_{\ell jk} \} \{ \sum_{\ell \in h} \sum_{i \in \ell} \frac{\partial f_i}{\partial x_{ij}} \} = w_k, \; k = 1, \ldots n. \quad (7)$$

Stage 5: The "Mass Market" Concept

In the case in which no segmentation strategy is

practiced at all, profit maximization leads to the
following decision rule for promotion:

$$(p - c - MC) \sum_j \{ \sum_{\ell=1}^{M} b_{\ell jk} \} \{ \sum_{\ell=1}^{M} \sum_{i \in \ell} \frac{\partial f_i}{\partial x_{ij}} \} = w_k, \quad k = 1, \ldots, n, \quad (8)$$

where the first term in the brackets represents the total
impact of medium k in terms of promotion type j for
all numbers of the population, and the second term is the
derivative of the total market demand function.

Subject: Marketing Mix

Title: Market Measurement and Planning With a Simultaneous-Equation Model

Author: Randall L. Schultz

Source: Journal of Marketing Research, Vol. 8, May 1971, 153-64

Summary: Demand and market share response functions are estimated from empirical data for airlines in one two-city market. These structures are then utilized in a normative model of marketing decision making. The results show profit-maximizing levels for number of flights and dollars of advertising.

Model:

I. Demand Model

A. With No Lags

(1) Demand (T) =

$$\beta_0 + \beta_1 \text{ price (T)} + \beta_2 \text{ advertising (T)} + \beta_3 \text{ population (T)}$$
$$+ \beta_4 \text{ business income (T)} + \beta_5 \text{ personal income (T)}$$
$$+ \beta_6 \text{ discounts (T)} + \beta_7 \text{ time (T)} + \beta_8 \text{ seasonality (T)}$$
$$+ \beta_9 \text{ GNP (T)} + \varepsilon_t.$$

B. With Lags

(2) Demand (T) =

$$\beta_0 + \beta_1 \text{ price (T)} + \beta_2 \text{ advertising (T)}$$
$$+ \beta_3 \text{ advertising (T-1)} + \beta_4 \text{ population (T)}$$
$$+ \beta_5 \text{ business income (T)}$$
$$+ \beta_6 \text{ business income (T-1)} + \beta_7 \text{ personal income (T)}$$
$$+ \beta_8 \text{ personal income (T-1)} + \beta_9 \text{ discounts (T)}$$
$$+ \beta_{10} \text{ time (T)} + \beta_{11} \text{ seasonality (T)} + \beta_{12} \text{ GNP (T)}$$
$$+ \beta_{13} \text{ GNP (T-1)} + \varepsilon_t.$$

II. <u>Marketing System Model</u>

   B. <u>Lags in First Equation</u>

      <u>Structural Equations (Normalized Form)</u>

$$M_t = Ar_t \gamma_1 + Ar_{t-1} \beta_{23} + As_t \gamma_2 + As_t \beta_{29} + F_t \gamma_3 \tag{1}$$

$$+ Cr_t \beta_1 + Cs_t \beta_2 + S_t \beta_3 + \beta_{19} + u_{1t}.$$

$$Ar_t = F_t \gamma_4 + D_{t-1} \beta_4 + M_{t-1} \beta_5 + Ar_{t-1} \beta_6 + R_{t-1} \beta_7 \tag{2}$$

$$+ P_{t-1} \beta_8 + \beta_{20} + u_{2t}.$$

$$As_t = F_t \gamma_5 + D_{t-1} \beta_9 + M_{t-1} \beta_{10} + As_{t-1} \beta_{11} + R_{t-1} \beta_{12} \tag{3}$$

$$+ P_{t-1} \beta_{13} + \beta_{21} + u_{3t}.$$

$$F_t = D_{t-1} \beta_{14} + M_{t-1} \beta_{15} + F_{t-1} \beta_{16} + P_{t-1} \beta_{17} + E_t \beta_{18} \tag{4}$$

$$+ \beta_{22} + u_{4t}.$$

where:

M_t = market share

Ar_t = advertising share--City R

As_t = advertising share--City S

F_t = frequency share

Cr_t = population share--City R

Cs_t = population share--City S

S_t = service

D_t = demand

R_t = revenue

P_t = profit

E_t = equipment.

The nonlinear objective function is:

$$\sum_{t=1}^{4} \pi_t = PD_I \; \left(\frac{X_1}{X_1 + K}\right)^{\alpha} \; \left(\frac{Y_0}{Y_0 + L}\right)^{\beta} \; - \; cX_1 \; - \; Y_1$$

$$+ \; PD_2 \; \left(\frac{X_2}{X_2 + K}\right)^{\alpha} \; \left(\frac{Y_1}{Y_1 + L}\right)^{\beta} \; - \; cX_2 \; - \; Y_2$$

$$+ \; PD_3 \; \left(\frac{X_3}{X_3 + K}\right)^{\alpha} \; \left(\frac{Y_2}{Y_2 + L}\right)^{\beta} \; - \; cX_3 \; - \; Y_3$$

$$+ \; PD_4 \; \left(\frac{X_4}{X_4 + K}\right)^{\alpha} \; \left(\frac{Y_3}{Y_3 + L}\right)^{\beta} \; - \; cX_4 \; - \; Y_4$$

where:

    X = the company's number of daily flights

    K = the competitors' number of daily flights (a constant)

    Y = the company's level of quarterly advertising in City R
       in dollars

    L = the competitors' level of quarterly advertising
       in City R in dollars (a constant).

The first order optimality conditions for the nonlinear model are:

$$\alpha PD_j Y_i^{\beta} X_j^{\alpha-1} K \; - \; c(Y_i + L)^{\beta}(X_j + K)^{\alpha+1} = 0,$$

$$i = 0, \; 1, \; 2, \; 3$$

$$j = 1, \; 2, \; 3, \; 4, \quad \text{and}$$

$$\beta D_j X_j^{\alpha} Y_i^{\beta-1} L \; - \; (X_j + K)^{\alpha}(Y_i + L)^{\beta+1} = 0,$$

$$i = 1, \; 2, \; 3$$

$$j = 2, \; 3, \; 4, \quad \text{with}$$

$$Y_4^{*} = 0.$$

Subject: Marketing Mix

Title: A Computer On-Line Marketing Mix Model

Author: Jean-Jacques Lambin

Source: Journal of Marketing Research, Vol. 9, May 1972, 119-26

Summary: This approach to developing and implementing a dynamic,
          competitive marketing mix model for a major oil company
          combines econometric methods, simulation techniques, and
          subjective judgments. Regression coefficients provide
          estimates of the response functions of the different
          inputs.

Model: Assume that a brand's market share is determined by
          the relationship of its marketing expenditures to the total
          for the industry. Without considering dynamic factors,
          the "relative marketing pressure" concept can be expressed as:

$$MS_{i,t} = \frac{DS_{i,t}^{\varepsilon_{1,i}} \cdot DO_{i,t}^{\varepsilon_{2,i}} \cdot ST_{i,t}^{\varepsilon_{3,i}}}{\sum_i DS_{i,t}^{\varepsilon_{1,i}} \cdot DO_{i,t}^{\varepsilon_{2,i}} \cdot ST_{i,t}^{\varepsilon_{3,i}}} \tag{1}$$

where

MS = market share,

DS = number of service stations,

DO = number of other outlets,

ST = total advertising expenditures,

ε = sensitivity coefficient.

This model is not linear and cannot be estimated by standard
econometric methods. A trial and error or heuristic approach
based on the manager's best judgment and the computational
power of the computer was applied and did not prove useful.

An alternative form is:

$$MS_{i,t} = \left(\frac{DS_{i,t}}{\sum DS_{i,t}}\right)^{\eta_{1,i}} \cdot \left(\frac{DO_{i,t}}{\sum DO_{i,t}}\right)^{\eta_{2,i}} \cdot \left(\frac{ST_{i,t}}{\sum ST_{i,t}}\right)^{\eta_{3,i}} \tag{2}$$

in which all the variables are expressed in terms of market
share. A dynamic linear version of (2) is:

$$MS_t = k + b_1 DS_t^* + b_2 DO_t^* + c_1 ST_t^* + c_2 ST_{t-1}^* + \lambda MS_{t-1} + u_t \quad (3)$$

derived by Koyck transformations:

$$MS_t = k + c_1 ST_t^* + c_2 ST_{t-1}^* + c_2 \sum_{i=1}^{\infty} \lambda^i ST_{t-i-1}^* + v_t. \quad (4)$$

Model (3) differs from (2) in that the lagged dependent variable appears in the right-hand side of the equation (the asterisks denote share variables).

It is assumed in (3) that the variables follow the same exponential decay pattern, so the estimated value of λ includes not only the effects of past advertising, but also the assets accumulated through the brand's distributive network as well as a combination of these two factors. Normally, λ can be defined as:

$$\lambda = \lambda_1 \delta + \lambda_2 \theta + \lambda_3 (\delta\theta) \quad (5)$$

where

   δ = relative weight of the goodwill capital created by the distributive network (in per cent),

   θ = relative weight of the goodwill capital created by advertising (in per cent),

   $\delta\theta$ = relative weight of the goodwill capital created by the interaction of advertising and distribution (in per cent).

Neglecting the product term, (5) simplifies to:

$$\lambda = \lambda_1 \delta + \lambda_2 \theta \quad (6)$$

where $\theta + \delta = 1$.

Thus λ_1 and λ_2 are parameters to be estimated. If both advertising and distribution have lagged effects, the exponentially distributed lag model is:

$$MS_t = k + b_1 \sum_{i=0}^{\infty} \lambda_1^i DS_{t-1}^* + b_2 DO_t^* + c_1 ST_t^*$$

$$+ c_2 ST_{t-1}^* + c_2 \sum_{i=1}^{\infty} \lambda_2^i ST_{t-i-1}^* + u_t. \quad (7)$$

Reference: #148.

Subject: Marketing Mix

Title: Solving the "Marketing Mix" Problem Using Geometric
           Programming

Authors: V. Balachandran and D.H. Gensch

Source: Management Science, Vol. 21, No. 2, October 1974,
           160-71

Summary: Optimal allocation of the marketing budget within the
           marketing-mix decision variables so that sales (or profit)
           is maximized in a planning horizon. Since the influence
           of marketing mix variables upon sales are, in reality,
           nonlinear and interactive, a geometric programming
           algorithm is used. A procedure to estimate a functional
           of sales on the marketing mix and environmental variables
           utilizing the judgments of the firm's executives and
           the raw data is provided. The derived functional is
           later optimized by the Geometric Programming algorithm
           under a constraint set consisting of budget and
           strategy restrictions imposed by a firm's marketing
           environment, and conditions under which the optimal solu-
           tion is either local or global are identified.

Model:

   Let

        A = relative advertising expenditure

        I = relative in-store promotion

        P = relative price (retail price of firm's product
              divided by average price charged by competing
              firms)

        C = firm's relative price differential from previous
              time period

        T = discounts to wholesalers and retailers

        S = salesman's effort (salary and commission)

        D = distribution (availability of product)

        W = relative customer service

        B = relative packing appeal

        Q = relative quality

       Ac = age composition

       In = personal disposable income.

   Problem: Find a posynomal (a polynomial with positive
   coefficients) which is a reasonable estimator of sales
   and then optimize the functional using geometric programming.
   Analyzing the raw data, the following two-state regression
   procedure was developed:

Step One--Log Regression

From the data, log sales were regressed on the logs of all singles, pairs, and triples of the predictor variables. Also, terms were included representing combinations of variables that the marketing department felt a priori to be useful. The log regression coefficients gave the elasticities.

Step Two

Screening of the terms before proceeding: Those terms which yielded unreasonable elasticities were modified or dropped. Similarly, those terms that the executives felt were completely unreasonable were also dropped. Changes of all of the signs of the terms from positive to negative.

Step Three

The useful terms are to be combined into a linear model. For that, one needs independence of the selected terms. Thus, look at matrix of correlation among the terms from step two.

Step Four

Stepwise regression procedure of sales on terms of step two. Two conditions: First, when additional terms are included in the derived functional, the adjusted $\bar{R}^2$ should not decrease. Second, estimate a functional with at most one positive regression coefficient of a term (constant term excluded).

In the case of beer advertising, A_t is separated into three variables:

$A1_t$ = advertising emphasizing price

$A2_t$ = image advertising

$A3_t$ = quality advertising.

The following functional form is obtained:

$$\text{Sales} = b_o + b_1 [(A1)_{t-1} (PTC)_t]$$

$$+ b_2 [(A2)_{t-1} B_t] + b_3 (AcQ)_t + b_4 [I_t S_{t-1}] + b_5 (PD)_t \quad (1)$$

where $b_o, \ldots, b_5$ are the regression coefficients.

Solution procedure for solving the marketing mix decision variables by optimizing the functional, subject to a set of constraints, is provided.

Reference: #41.

Subject: Marketing Mix

Title: A Market Share Theorem

Authors: David E. Bell, Ralph L. Keeney, and John D.C. Little

Source: Journal of Marketing Research, Vol. 12, No. 2,
May 1975 , 136-41

Summary: Many marketing models use variants of the relationship:
Market share equals marketing effort divided by total
marketing effort. Replacing marketing effort with its
resulting "attraction," the relationship is derived from
the assumptions: (1) attraction is nonnegative,
(2) equal attractions imply equal shares, and (3) a
seller's share is affected the same if the attraction of
any other seller increases a fixed amount.

Model: Given a finite set $S = \{s_1, \ldots, s_n\}$ of sellers which
includes all sellers from whom a given customer group makes
its purchases, suppose that for each seller $s_i \ \varepsilon \ S$ an
"attraction" value $a(s_i)$ is calculated. Let the competitive
situation be completely determined by the vector of
attractions:

$$a = (a(s_1), a(s_2), \ldots a(s_n)) = (a_1, a_2, \ldots, a_n).$$

That is, the market share $m(s_i)$ of a seller is fully determined by a.

Attraction may be a function of the seller's advertising
expenditure and effectiveness, the price of his product, the
reputation of the company, the service given during and
after purchase, location of retail stores, and much more.
Indeed, the attraction of an individual seller can, if we
wish, be a function of these qualities for all the other
sellers, or

$$a(s_i) = \phi_i(q_1, \ldots q_n; p_1, \ldots, p_n; \ldots),$$

where q_j may be quality of service of seller j, p_j might
indicate seller j's price, and so on.

Since, by definition, attraction completely determines
market share,

$$m(s_i) = f_i(a). \qquad i = 1, \ldots, n,$$

for some function f_i where $m(s_i)$ is the market share of seller i. Clearly,

$$\sum_{i=1}^{n} m(s_i) = 1$$

and

$$0 \leq m(s_i) \leq 1, \qquad i = 1, \ldots, n,$$

but otherwise the functions f_i are as yet arbitrary. The aim here is to give conditions on the relationship between attraction and market share which force the simple linear normalization model

$$f_i(a) = \frac{a_i}{\sum_{j=1}^{n} a_j}.$$

The assumptions of the theorem are:

1: The attraction vector is nonnegative and nonzero,

$$a \geq 0 \quad \text{and} \quad \sum_{i=1}^{n} a_i > 0.$$

2: A seller with zero attraction has no market share,

$$a_i = 0 \rightarrow m(s_i) = 0.$$

3: Two sellers with equal attractions have equal market share,

$$a_i = a_j \rightarrow m(s_i) = m(s_j).$$

4: The market share of a given seller will be affected in the same manner if the attraction of any other seller is increased by a fixed amount Δ, i.e.,

$$f_i(a + \Delta e_j) - f_i(a), \quad \text{for} \quad j \neq i,$$

is independent of j, where e_j is the jth unit vector.

Theorem

If a market share is assigned to each seller based only on the attraction vector and in such a way that assumptions 1-4 are satisfied, then market share is given by:

$$m(s_i) = \frac{a(s_i)}{\sum\limits_{j=1}^{n} a(s_j)}, \quad \text{for} \quad i = 1, 2, \ldots, n.$$

The theorem is proved.

Subject: Marketing Mix

Title: BRANDAID: A Marketing-Mix Model, Part 1: Structure

Author: John D.C. Little

Source: Operations Research, Vol. 23, No. 4, July-August 1975,
            628-55

Summary: BRANDAID, an expansion of ABUDG [#67], is a flexible, on-
            line model for assembling marketing decision elements to
            describe the market and evaluate strategies. The structure
            is modular so that individual decision areas can be added
            or deleted at will. The model is of the aggregate response
            type, in which decision variables relate closely to
            specific sales performance measures. The major submodels
            are advertising, promotion, price, salesmen, and retail
            distribution. The advertising submodel employs a long-
            run sales response to advertising function and a linear
            log process. Promotional effects are built up from a
            characteristic time pattern for the type of promotion
            and a response curve. Salesmen affect sales through a
            response process structurally similar to that for
            advertising. Retail distribution variables are
            intermediaries that the company affects and that in turn
            affect customer response. Submodel outputs combine
            multiplicatively. Competition enters in a modular,
            symmetric way through a matrix of competitive coefficients
            that determine the source of sales for each brand as
            it seeks to increase its market position.

Model: Let

$s_b(t)$ = sales rate of brand b in time period t (sales
units/customer/yr),

$S(t)$ = sales rate of the product class in period t (sales
units/customer/yr),

$m_b(t)$ = market share of brand b in t (fraction). Then

$$S(t) = \sum_b s_b(t), \tag{1}$$

$$m_b(t) = s_b(t)/S(t). \tag{2}$$

Let $p_b(t)$ = profit rate of brand b in t ($\$/customer/yr$),

$g_b(t)$ = gross contribution of brand b in t ($\$/sales unit$),

and $c_b(t)$ = cost rate in t for brand b resulting from
the ith marketing activity ($\$/customer/yr$). Then

$$p_b(t) = g_b(t)s_b(t) - \sum_i c_b(i,t). \tag{3}$$

Let s_o = reference brand sales rate ($/customer/yr.),
$e(i,t)$ = effect on brand sales of ith sales influence
(index), and I = the set of influences on brand sales.
Then

$$s(t) = s_o \prod_{i \in I} e(i,t). \qquad (4)$$

Let m_o = reference brand market share (fraction),
and S_o = reference product class sales rate ($/customer/yr.).
Then

$$s_o = m_o S_o. \qquad (5)$$

Advertising

Let $a(t)$ = advertising rate at t (index), $r(a)$ = long-
run sales response to advertising (index), $e(t)$ = effect of
advertising on sales at t (index), and $\alpha(a)$ = carry-over
rate for advertising effect on sales (fraction/period).
Then

$$e(t) = \alpha e(t-1) + (1-\alpha) r[a(t)] \qquad (6)$$

where $e(i,t)$ is shortened to $e(t)$.

Let $h(t)$ = media efficiency in time period t (exposures,$),
$k(t)$ = copy effectiveness in t (dimensionless), $x(t)$ =
advertising spending rate in t ($/customer/yr.). Using the
subscript 0 to denote the reference value of these quantities,
we model advertising rate by

$$a(t) = h(t)k(t)x(t)/h_o k_o x_o. \qquad (7)$$

Let $\hat{a}(t)$ = effective advertising at t (index), and β =
memory constant for advertising (fraction/period). Then

$$\hat{a}(t) = \beta \hat{a}(t-1) + (1-\beta)a(t). \qquad (8)$$

$\hat{a}$ could substitute for a in (3).
Let $w(j)$ = weight for jth type of advertising (dimensionless).
Media efficiency, copy effectiveness, spending rate, and
reference conditions now vary with advertising type.

Equation (7) generalizes to

$$a(t) = \sum_j h(j,t)k(j,t)w(j,t)x(j,t) / \sum_j h_o(j)k_o(j)w_o(j)x_o(j). \quad (7a)$$

Promotion

The term promotion covers a wide variety of sales stimulating devices, including temporary price reductions, premiums, coupons, and sampling. Let $q(\tau)$ = time pattern: the sales index for a reference promotion in the τth period after the start (index), $a(t)$ = promotional intensity of a promotion starting in t; $a = 1$ for a reference promotion, and $r(a)$ = sales response to promotional intensity.

If we suppose that the sales of the product line with no promotion is S_{np}, then the effect at t of a reference promotion at t_p is a net sales gain of $S_{np}[q(t-t_p)-1]$. If the promotional intensity is a, the net gain is $S_{np}[q(t-t_p)-1]r[a(t_p)]$. Let ℓ = portion of the line promoted (fraction), and b = fraction of sales gain in promoted portion cannibalized from rest of line (fraction). Then the sales gain at t of the promoted portion is $S_{np}\ell r[a(t_p)][q(t-t_p)-1]$, of which only a fraction $(1-b)$ is a gain for the whole line. The total sales of the line are therefore $S_{np}\{1 + \ell r[a(t_p)][q(t-t_p)-1](1-b)\}$.

Let e_o = effect on reference sales if all promotions are deleted (index), and $e(t)$ = effect of promotion on sales at t (index). Then, for a promotion run at t_p,

$$e(t) = e_o\{1+\ell r[a(t_p)][q(t-t_p) - 1](1-b)\}.$$

Index the schedule of promotions by p. The promotional sub-model is

$$e(t) = e_o\{1 + \sum_p \ell_p r_p[a_p(t_p)][q_p(t-t_p) - 1](1-b_p)\}. \quad (9)$$

Let $x(t)$ = promotional offer at t($/sales unit), $h(t)$ = coverage efficiency at t (fraction of customers), and $k(t)$ = consumer effectiveness at t (dimensionless). Using the subscript zero to denote the reference promotional offer, we have for the promotional intensity

$$a(t) = h(t)k(t)x(t)/h_o k_o x_o. \quad (10)$$

Let $c(t)$ = cost of promotion at t ($/customer/yr.), and
$c_{fp}(t)$ = fixed cost of promotion p at t ($/customer/yr.).
Consider the case where the variable cost is incurred on all
the normal sales in the portion of the line being promoted
and on all incremental sales. Normal sales in the absence
of promotion are $s(t)[e_o/e(t)]$. Let $[\tau_1, \tau_2]$ = interval
during which promotional allowance is paid on sales, and
$I(t) = \{p|(t-t_p)\varepsilon[\tau_1, \tau_2]\}$. Then

$$c(t) = \sum_p c_{fp}(t) + \sum_{p\varepsilon I(t)}$$

$$\cdot \{x_p(t_p)\ell_p s(t)[e_o/e(t)][1+[q_p(t-t_p)-1]r_p[a_p(t_p)]]\}. \quad (11)$$

Price

The price under consideration is the basic wholesale price
charged by the manufacturer. Let $x(t)$ = manufacturer's brand
price ($/unit), x_o = reference brand price ($/unit), and $a(t)$ =
normalized brand price (index), i.e.,

$$a(t) = x(t)/x_o. \quad (12)$$

Let $r(a)$ = share response to brand price (index), $\Psi(x)$ = additional effect of retail price-ending (index) and $e(t)$ = effect
of brand price on share at t (index). Then

$$e(t) = \Psi[x(t)]r[a(t)]. \quad (13)$$

Salesmen

Let $x(t)$ = salesmen effort rate ($/customer/yr.), $h(t)$ =
coverage efficiency (calls/$), $k(t)$ = effectiveness on store
(effectiveness/call), and $a(t)$ = normalized salesman effort
rate (index). Again letting the subscript 0 denote
maintenance or reference effort, we take

$$a(t) = h(t)k(t)x(t)/h_o k_o x_o. \quad (14)$$

Let $\hat{a}(t)$ = effective effort at t, including remembered
effort (index), and β = carryover constant for remembered effort
(fraction/period). We have

$$\hat{a}(t) = \beta\hat{a}(t-1) + (1-\beta)a(t). \quad (15)$$

Let $e(t)$ = effect of salesman effort on sales (index), α = carry-
over constant for product loyalty (fraction/period), and $r(\hat{a})$ =
long-run sales response to salesman effort (index). Then

$$e(t) = \alpha e(t-1) + (1-\alpha)r(\hat{a}). \quad (16)$$

Other Influences on Sales

Two other sales influences are seasonality and trend. Seasonality enters as a direct index affecting product class sales and, for a few products, share. Product class sales may have a trend, which can be treated by either a direct index or a growth rate. In the latter case, let e_o = starting point for trend (index), and $r(t)$ = growth rate in t (fraction/period). Then

$$e(t) = e_o \prod_{\tau=1}^{\tau=t} [1.0 + r(\tau)]. \tag{17}$$

Competition

Consider a single sales influence, say price or promotion. Brand inputs, as modeled earlier, produce e_b' = unadjusted effect index for brand b (index), and consequently

$$s_b' = s_{ob} e_b', \tag{18}$$

where s_b' = unadjusted sales rate for brand b (sales units/customer/yr.), and s_{ob} = reference sales rate for brand b. The competitive source-of-sales coefficient between brands is denoted by γ_{bc} = fraction of brand c's unadjusted incremental sales that comes from brand b (fraction). Brand c's unadjusted incremental sales relative to reference are $s_c' - s_{oc}$, so that adjusted sales for brand b become

$$s_b = s_b' - \sum_{c \neq b} \gamma_{bc} (s_c' - s_{oc}). \tag{19}$$

Dividing by s_{ob}, we obtain

$$e_b = e_b' - \sum_{c \neq b} (s_{oc}/s_{ob}) \gamma_{bc} (e_c' - 1). \tag{20}$$

Generalizing to an arbitrary sales influence i and making time dependence explicit, we obtain

$$e_b(i,t) = e_b'(i,t) - \textstyle\int_{c\neq b} [s_{oc}/s_{ob}] \gamma_{bc}(i) [e_c'(i,t)-1]. \quad (21)$$

(19) expresses the fundamental model of competitive interaction. (21) puts it in calculated form for use in the general expressions

$$s_b(t) = s_{ob} \, \Pi_i \, e_b(i,t),$$
$$S(t) = \textstyle\int_b s_b(t),$$
$$m_b(t) = s_b(t)/S(t). \quad (2)$$

Define

$$\gamma_{cc} = 1 - \textstyle\int_{b\neq c} \gamma_{bc}, \quad (22)$$

where γ_{cc} is the fraction of unadjusted incremental sales of brand c coming from a product class sales gain.

Often a brand draws its incremental sales from competing brands proportional to their reference sales. Then $\gamma_{bc} = (const)s_{ob}$ or, normalizing,

$$\gamma_{bc} = s_{ob}(1-\gamma_{cc})/\textstyle\int_{c\neq b} s_{oc}. \quad (23)$$

Retail Distribution

By retail distribution we mean a cluster of marketing activities that the retailer conducts and that affect the sales of a brand; examples are retail price, retail advertising, availability, quality of shelf position and facings, number of in store promotional displays. Let $I_M = \{i_1, \ldots, i_M\}$ = set of manufacturer activities, $I_R = \{i_1, \ldots, i_R\}$ = set of retail activities, and $I_R = \{i_1, \ldots, i_E\}$ = set of environmental and other influences. Thus sales for a given brand are

$$s(t) = s_o \ \Pi_{i \epsilon I_M} \ c(i,t) \ \Pi_{i \epsilon I_R} \ e(i,t) \ \Pi_{i \epsilon I_E} \ c(i,t). \quad (24)$$

Let $d(i,t)$ = the ith retail variable, $i \epsilon I_R$, and $f(i,d)$ = response submodel for $d(i,t)$. Then

$$e(i,t) = f[i,d(i,t)], \quad i \epsilon I_R \quad \text{and} \quad (25)$$

$$d(i,t) = d_o(i) \ \Pi_{k \epsilon D(i)} e(k,i,t), \quad (26)$$

where $e(k,i,t)$ = effect of kth manufacturer, environmental, or other influence on retail variable $d(i,t)$ (index), $D(i)$ = set of influences on $d(i,t)$, and $d_o(i)$ = reference value of $d(i,t)$. We build an availability submodel where availability is taken to include such items as the presence or absence of the product, its shelf position, and the number of its facings.

Suppressing subscripts, let $d(t)$ = availability of brand at t (fraction), $e(t)$ = effect on consumer sales rate at t (index), and $r(d)$ = sales response to availability (index). Then

$$e(t) = r[d(t)]. \quad (27)$$

Let d_o = reference availability (fraction), D = set of manufacturer and other activities influencing $d(t)$, and $e(k,t)$ = effect of kth activity on $d(t)$ (index). We have

$$d(t) = d_o \ \Pi_{k \epsilon D} \ e(k,t). \quad (28)$$

Let $v(t)$ = consumer sales rate at reference retailer activity as a fraction of reference sales (index). From (24)

$$v(t) = \Pi_{i \epsilon I_{E \cup I_M}} e(i,t). \quad (29)$$

Suppressing the activity label, let $r(v)$ = long-run response of availability measure to sales rate (index), α = carry-over constant (fraction/period), and $e(t)$ = effect of sales rate on availability measure (index). Then

$$e(t) = \alpha e(t-1) + (1-\alpha)r[v(t)]. \tag{30}$$

References: #67, 80, 142, 145.

Subject: Miscellaneous

Title: Simulation of Market Processes

Authors: Frederick E. Balderston and Austin C. Hoggatt

Source: Institute of Business and Economic Research, University of California, Berkeley, California, 1962, 4-27

Summary: Computer program with manufacturers, wholesalers, and retailers. Besides the usual set of economic variables the model possesses the following features: (1) It represents a multi-stage market; (2) the firms constituting the market face uncertainty and operate with limited information; (3) transactions occur by means of sequences of steps that are reminiscent of the Walrasian "tâtonnements"; and (4) the system is dynamic.

Model:

Assume two alternative systems of communication linkages in a hypothetical market: (a) Direct links between manufacturers and retailers, (b) indirect links (wholesalers).

Further assume:

1. Each link has cost of q/time period, regardless of the amount of information flowing through it, distance, or other factors.

2. Commodity flow from manufacturers to retailers is given and constant.

3. Every manufacturer must be connected with every retailer directly or indirectly.

4. The number of manufacturers, M, and the number of retailers, R, are given and constant.

A system of <u>direct links</u> involves no intermediaries, each manufacturer being connected to every retailer. The total communication cost of such a system is

$$T_1 = q(M \times R) \qquad (1)$$

where M = number of manufacturers and R = number of retailers.
Injecting one wholesaler into the situation, we see that
the total linkage costs become:

$$T_2 = q(M + R).\qquad(2)$$

Now suppose that additional wholesalers can enter the
trade, but that each new one can only replicate the same
network as was possessed by the first. The total linkage
cost of the system can be expressed as a function of the
number of wholesalers, W:

$$T_3 = W[q(M + R)].\qquad(3)$$

The cost relation T_3 and the economic profit relation
$\pi = f(W)$ can now be summed to give us:

$$T_4 = W[q(M+R)] + f(W).\qquad(4)$$

This function represents the overall effective cost of
the system as viewed by its users, that is, inclusive of
economic profits arising from limits on the number of
wholesalers. By simple differentiation, we can obtain a
minimum of this cost function and find the number of
wholesalers associated with this minimum.

Subject: Miscellaneous

Title: Marketing Analysis Training Exercise

Authors: Alfred A. Kuehn and Doyle L. Weiss

Source: Behavioral Science, Vol. 10, No. 1, January 1965, 51-67

Summary: MATE (Marketing Analysis Training Exercise) simulates a
            market for one multi-branded product--packaged detergents--
            in form of a game. It can be used to train marketing
            students and junior executives as well as to screen
            and experiment with marketing strategies prior to trial runs
            in the actual market, thus paralleling the use of a pilot
            plant.

Model: Let three firms market from one to three brands of

            detergent in four geographical regions. The firms may revise

            or alter once a month within each region: price, advertising

            expenditure, sales force, retail allowance. A firm may

            purchase market survey reports containing estimates of:

            Total Retail Sales and Market Shares, Retail Distribution

            and Stockouts, Advertising Expenditures. Each firm owns

            one factory and a factory warehouse and rents warehouse space.

            The total case demand for packaged detergent is calculated

            for each region by:

$$Q_t = s_t \, Q_o \, K^t \, \left(\frac{\overline{P}_t}{P_o}\right)^{\eta_P} \cdot \left(\frac{\overline{E}_t}{q_3 \, s_t \, \overline{E}_o \, K^t}\right)^{\eta_E} \left(\frac{Y_t}{Y_o}\right)^{\eta_Y}$$

where:

   Q = total quantity demanded expressed in cases

   s = seasonal index where $\sum_{t=1}^{12} s_t = 12.0$

   K = growth term

   $\overline{p}$ = average industry price (each brand's price being
        weighted by its market share)

   $\overline{E}$ = total sales-promotion expenditures for industry

   Y = average income per capita

   t = week subscript

   o = subscript implying a base industry value for the
        subscripted variable.

The three exponents are elasticities. The habitual demand

from brand loyal customers is:

$$Q_{B,t}^{(H)} = (s_t/s_{t-1}) \cdot r_1 \cdot X_{B,t-1} \cdot \beta_{t,B} \tag{1}$$

where:

$Q_B^{(H)}$ = habitual or holdover demand for brand B

s = seasonal factor

r_1 = percentage of last period's customers (weighted by volume) attempting to make a repeat purchase through habit

X_B = brand B's sales

t = month subscript

$$\beta_t = 1. - \left| \frac{P_{B,t} - P_{B,t-1}}{e_2 P_{B,t}} \right|^{e_1}$$

with:

$$0 \le \beta \le 1; \quad 0 \le e_1 \le 1 \quad \text{and} \quad e_2 > 0.$$

The relative consumer appeal of a brand in terms of its washing power (C_1) and gentleness (C_2) characteristics is represented by:

$$C_{1,B} = \frac{(1 - a_1) \ \exp. |\hat{C}_1 - C_B^{(1)}|}{\sum\limits_{B=1}^{k} \{(1-a_1) \ \exp. |\hat{C}_1 - C_B^{(1)}|\}}$$

$$C_{2,B} = \frac{(1 - a_2) \ \exp. |\hat{C}_2 - C_B^{(2)}|}{\sum\limits_{B=1}^{k} \{(1-a_2) \ \exp. |\hat{C}_2 - C_B^{(2)}|\}}$$

where

a_1, a_2 = consumer discrimination parameters having a range between zero and one,

$\hat{C}_1, \hat{C}_2$ = arbitrary (limiting) values for each of the characteristics, representing values equal to or greater than the maximum values attainable by the research laboratories,

$C_B^{(1)}, C_B^{(2)}$ = actual values of the gentleness and washing power characteristics for brand B.

Also,

$$\sum\limits_{i=1}^{n} (a_4 z_{1,i} + a_5 z_{2,i}) = 1,$$

where $z_{1,i}$ is the relative demand for the ith degree of sudsiness resulting from the automatic washer and general purpose submarkets and $z_{2,i}$ is the relative demand for the ith degree of sudsiness resulting from the dishwashing submarket.

$$Q_{1,i,t} = Q_t \, a_4 z_{1,i}$$

$$Q_{2,i,t} = Q_t \, a_5 z_{1,i}.$$

Equation (1) representing habitual demand can now be developed for each cell of the sudsiness demand vector:

$$Q_{B,i,t}^{(H)} = (s_t/s_{t-1})^{r_1} \, X_{B,i,t-1} \, \beta_{B,t}. \qquad (2)$$

$$z_{B,i,t} = \frac{(1 - a_6) \, \exp. |C_B^{(3)} - i|}{\sum\limits_{B=1}^{k} (1 - a_6) \, \exp. |C_B^{(3)} - i|}, \quad i = 1 \ldots n$$

with

$C_B^{(3)}$ = a brand's sudsiness value, which must be an integer between 1 and 20 to fit the discrete preference distribution for sudsiness,

a_6 = a consumer discrimination parameter where $0 \leq a_6 < 1$. If a_6 is zero, the consumer is unable to discriminate between products differing only with respect to sudsiness. As $a_6 \to 1$, the consumer's discrimination with respect to differences in sudsiness approaches perfection,

k = number of brands on the market.

The Share of Potential Shifters Demand (S) for a particular brand B can now be defined as:

$$S_{B,i,t} = (1 - b_{pda}) \left[\frac{z_{B,i,t} \, \Omega_{B,t}}{\sum\limits_{B=1}^{k} z_{B,i,t} \, \Omega_{B,t}} \right]$$

$$+ \, b_{pda} \left[\frac{z_{B,i,t} \, \Omega_{B,t} \, \dfrac{A_{B,t}^{\varepsilon_A}}{\sum\limits_{B=1}^{k} A_{B,t}^{\varepsilon_A}}}{\sum\limits_{B=1}^{k} z_{B,i,t} \, \Omega_{B,t} \, \dfrac{A_{B,t}^{\varepsilon_A}}{\sum\limits_{B=1}^{k} A_{B,t}^{\varepsilon_A}}} \right] \qquad (3)$$

where Ω is defined as:

$$\Omega_{B,t} = \left[C_{1,B,t} \quad C_{2,B,t} \quad \frac{D_{B,t}^{\varepsilon_D}}{\sum\limits_{B=1}^{k} D_{B,t}^{\varepsilon_D}} \quad \frac{P_{B,t}^{\varepsilon_p}}{\sum\limits_{B=1}^{k} P_{B,t}^{\varepsilon_p}} \right]$$

where

    D = sales effectiveness of a brand's availability,

  b_{pda} = probability of a potential brand shifter being influenced by advertising,

$\varepsilon_A, \varepsilon_p, \varepsilon_D,$ are interbrand elasticity constants.

A brand's sales (X) within each cell of the sudsing-demand vector is calculated from:

$$X_{B,i,t} = [1 - R_{B,t}] \, Q_{B,i,t}. \tag{4}$$

R is the per cent stockout experienced by a brand and determines the sales lost if a brand is unavailable. Demand for a brand is defined as

$$Q_{B,i,t} = Q_{B,i,t}^{(H)} + Q_{B,i,t}^{(S)}, \tag{5}$$

where

$$Q_{B,i,t}^{(S)} = S_{B,i,t} \, \{ [Q_{i,t} - \sum_{B=1}^{k} Q_{B,i}^{(H)}] + \sum_{B=1}^{k} (R_{B,t-1} Q_{B,i,t-1}) \}. \tag{6}$$

Let A = advertising dollars. Then,

$$\overline{A}_{B,t} = (1 - a_7)(A_t) + a_7 \overline{A}_{B,t-1}.$$

In addition, the effect of a change in a product's characteristics upon customer using the product, the ordering rules of wholesalers and retailers, the introduction of a new brand, as well as new product research are described.

Subject: Miscellaneous

Title: Trade Area Boundaries: Some Issues in Theory and
           Methodology

Author: Louis P. Bucklin

Source: Journal of Marketing Research, Vol. 8, February 1971,
           30-37

Summary: Consumer store choice preferences are linked to retail
           trading areas by a new measure of the forces of
           geographical competition in marketing. Initial examina-
           tion suggests that existing methodology does not model
           the overlap between intra-urban trading areas correctly.
           The new measure may also offer improved opportunities
           for the study of consumer propensity to search.

Model:

The Meaning of Trade Area Overlap

Given any two centers competing for consumer patronage,
a trade area contour profile can be derived from the
patronage probabilities associated with the locus of points
on the straight line between the two facilities. As one
moves from one center to the other along this line, the
probabilities of a shopper visiting the first location
decline regularly. Algebraically, this condition can be
expressed as a function of the likelihood that a shopper
will visit the first center, designated as α:

$$P_{\alpha} = f(D_{\alpha}, T), \quad D_{\alpha} = 0, \ldots T \tag{1}$$

where P_{α} is the unknown probability, D_{α} some specific
distance from α, and T the total distance between α
and a competitor, b.

In the symmetric case, where the probabilities of
visiting either center are .5, the degree of overlap may
be defined mathematically as:

$$0 = (M - \int_{0}^{M} f(D_\alpha) dD_\alpha)/(M/2) \qquad (2)$$

where M is the distance to the midpoint, and 0 is the degree of overlap.

The particular function employed to estimate the different contour profiles is the simple gravity model:

$$P_\alpha = (A_\alpha/D_\alpha{}^\lambda)/(A_\alpha/D_\alpha{}^\lambda + A_b/(T - D_\alpha)^\lambda) \qquad (3)$$

where A_α and A_b represent some basic attraction to consumers of the two centers. The exponent controls the shape of the contour profile entirely. The total distance between the centers plays no role.

Two hypotheses:

1. The degree of overlap between competing trade centers declines as the distance between them increases.
2. The overlap between competing trading centers will be higher where consumers buy products for which the propensity to search is high.

Several consumer shopping surveys were assembled. For each, distance was related to consumer choice for several pairs of competing retail facilities.

To estimate the degree of overlap between each pair, an estimate of the shape of the contour profile was made with discriminant analysis.

The contour profile was estimated by converting discriminant scores for 21 points directly between the competing retail centers into probabilities by Dixon's method according to the following equation:

$$P_\alpha = e^{S_\alpha}/(e^{S_\alpha} + e^{S_b})$$

where S_α and S_b are the two discriminant scores.

To estimate the overlap, the best equation proved to be:

$$P_\alpha = .5 - \frac{1}{\pi} \arctan \Lambda(D_\alpha - M). \qquad (4)$$

The two parameters of (4), Λ and M, were fitted to each of the contour profiles by means of a nonlinear regression algorithm.

The overlap generated from (4) may be measured from its integral. In general, because the patronage probability of .5 will not fall midway between the two centers, the overlap areas will have to be evaluated. M^*, a negative number, is the distance from the origin to center α, M^{**} the distance to center b. The total overlap, O_T, is equal to the weighted sum of the overlaps from both quadrants:

$$\begin{aligned}
O_T = 1 - \frac{2}{\pi T} (&M^{**} \arctan \Lambda M^{**} \\
& - M^* \arctan \Lambda M^*) \\
& + \frac{1}{\Lambda \pi T} [\log (1 + \Lambda^2 M^{**2}) \\
& + \log (1 + \Lambda^2 M^{*2})],
\end{aligned} \qquad (5)$$

where $T = M^{**} - M^*$ and the logarithms are to the base e.

Table 1

Average Trade Area Overlaps for Distance
and Search Categories

Distance (miles)	Search Propensity		
	High (autos)	Medium (household goods)	Low (food)
3.4 and up	73.6	35.2	n.d.
1.4 to 3.3	75.1	54.4	37.7
Below 1.4	⁻100.0	n.d.	59.4

In Table 1 the entries in the cells represent the
overlaps computed from (5) and the actual distances
between the facilities.

These data show that as distance increases, the
extent of overlap is indeed enhanced, holding the type
of search propensity constant.

AUTHOR INDEX

JOURNAL INDEX

SUBJECT INDEX